T0384247

BCC-Algebras

The subjects described in the book are BCC-algebras and an even wider class of weak BCC-algebras. The aim of the book is to summarize the achievements to date in the subject and to present them in the form of a logically created theory. Through appropriate grading and a precise description of the steps of the proofs, this theory is easily assimilated, and it should not take too long for the reader to learn about it.

We begin with the motivation for their creation, many examples and basic results used later in the book. Then we deal with the constructions of BCC-algebras and calculate the numbers of their subalgebras.

The author describes the so-called solid weak BCC-algebras. They have some properties of BCI-algebras, but this requires completely new, often difficult, proofs. The important subclasses of weak BCC-algebras and the relationships between them are presented with many examples.

This book is intended for researchers dealing with abstract algebra and for logicians working on the border between logic and algebra. The monograph is also of interest to students interested in the theory of (weak) BCC-algebras or simply in abstract algebra.

The structure of the book makes it possible to discover topics that require further research, which, depending on the degree of difficulty, may be completed with a thesis or dissertation.

Janus Thomys received his MS in mathematics at the Jan Długosz University in Częstochowa, Poland. He earned his PhD in mathematical sciences at the Wrocław University of Science and Technology.

BCC-Algebras

Janus Thomys

CRC Press is an imprint of the
Taylor & Francis Group, an **informa** business
A CHAPMAN & HALL BOOK

First edition published 2023
by CRC Press
6000 Broken Sound Parkway NW, Suite 300, Boca Raton, FL 33487-2742

and by CRC Press
4 Park Square, Milton Park, Abingdon, Oxon, OX14 4RN

CRC Press is an imprint of Taylor & Francis Group, LLC

Library of Congress Cataloging-in-Publication Data

Names: Thomys, Janus, author.
Title: BCC-algebras / Janus Thomys.
Description: First edition. | Boca Raton : CRC Press, 2023. | Includes bibliographical references and index.
Identifiers: LCCN 2022028634 (print) | LCCN 2022028635 (ebook) | ISBN 9781032363615 (hardback) | ISBN 9781032363653 (paperback) | ISBN 9781003331568 (ebook)
Subjects: LCSH: Algebraic logic.
Classification: LCC QA10 .T56 2023 (print) | LCC QA10 (ebook) | DDC 511.3/24--dc23/eng20220917
LC record available at https://lccn.loc.gov/2022028634
LC ebook record available at https://lccn.loc.gov/2022028635

ISBN: 978-1-032-36361-5 (hbk)
ISBN: 978-1-032-36365-3 (pbk)
ISBN: 978-1-003-33156-8 (ebk)

DOI: 10.1201/9781003331568

Publisher's note: This book has been prepared from camera-ready copy provided by the authors.

Contents

Preface

By an algebra of type (2,0) we mean a nonempty set X together with one binary operation (usually denoted by $\cdot$ or $*$) and a distinguished element 0. Such defined algebra is denoted by $\langle X; \cdot, 0\rangle$ or by $\langle X; *, 0\rangle$.

Many algebras of such type are inspired by some logical systems (cf. [8, 14, 73, 74]). For example, BCK-algebras are inspired by a BCK logic, i.e., an implicational logic based on modus ponens and the following axioms scheme (for details see for example [14]):

$$\begin{array}{ll}
\textsf{Axiom B} & \mathcal{A} \supset \mathcal{B}. \supset .(\mathcal{C} \supset \mathcal{A}) \supset (\mathcal{C} \supset \mathcal{B}), \\
\textsf{Axiom C} & \mathcal{A} \supset (\mathcal{B} \supset \mathcal{C}). \supset .\mathcal{B} \supset (\mathcal{A} \supset \mathcal{C}), \\
\textsf{Axiom K} & \mathcal{A} \supset (\mathcal{B} \supset \mathcal{A}), \\
\textsf{Axiom I} & \mathcal{A} \supset \mathcal{A}.
\end{array}$$

An implicational logic satisfying only B, C and I is called a BCI positive logic.

This inspiration is illustrated by the similarities between the names. We have BCK-algebra and BCK positive logic, BCI-algebra and BCI positive logic, positive implicative BCK-algebra and positive implicative logic, implicative BCK-algebra and implicative (classical) logic, and so on. In many cases, the connection between such algebras and their corresponding logics is much stronger. In this case one can give a translation procedure which translates all well-formed formulas and all theorems of a given logic $\mathcal{L}$ into terms and theorems of the corresponding algebra (cf. [125]). In some cases one can also give an inverse translation procedure which translates all terms and all theorems of this algebra into well-formed formulas and theorems of a given logic $\mathcal{L}$. In this case we say that the logic $\mathcal{L}$ and its corresponding algebra are isomorphic. W.M. Bunder [14] proved, for example, that a BCK-algebra and a BCK positive logic are isomorphic, but a BCI-algebra and a BCI positive logic are not isomorphic.

Nevertheless, the study of algebras motivated by known logics is interesting and very useful for corresponding logics, also in the case when these structures are not isomorphic.

The class of all BCK-algebras does not form a variety. To prove this fact Y. Komori [105] introduced (see also [106]) the new class of algebras called BCC-algebras or BIK$^+$-algebras. Nowadays, many mathematicians, especially from China, Japan and Korea, have been studying various generalizations of BCC-algebras such as, for example, B-algebras, BE-algebras, difference algebras, implication algebras, GB-algebras, WFI-algebras, Hilbert algebras, d-algebras and many others. All these algebras have one distinguished element and satisfy some common identities playing a crucial role in these algebras. Selecting these common identities, we obtain the axiom system for algebras which are now called weak BCC-algebras or BZ-algebras. From the mathematical point of view, the first name is more appropriate; however, the second one is still more popular, especially in China (cf. [144]).

Relations between these algebras are illustrated in the following diagram, where $A \longrightarrow B$ means that A is B but B may not be A:

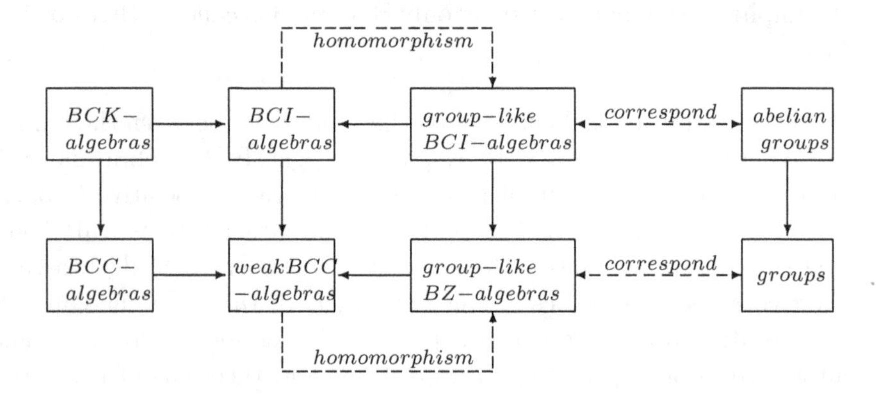

In all these algebras we can introduce a natural partial order which has many interesting properties and gives decomposition into so-called branches saving some properties of these algebras. Branches are mutually disjoint sets formed by one minimal element (minimal with respect to this natural partial order) and all elements comparable to it. The set of all minimal elements and some special subsets called ideals also play an important role in the study of these algebras.

This book, containing four chapters, is devoted to reviewing BCC- and weak BCC-algebras and their various subclasses.

Chapter 1 gives a survey of primary results concerning weak BCC-algebras used in the subsequent chapters. We provide an axiom system of BCK- and BCI-algebras and extend it to the BCC- and weak BCC-algebras as the widest presented classes. We also describe the connections between weak BCC-algebras with BCK-algebras and BCI-algebras. The problem of existence of weak BCC-algebras is solved by showing our extension methods for weak BCC-algebras. Some construction methods of weak BCC-algebras and new weak BCC-algebras from given BCC-algebras or weak BCC-algebras are presented, too. We conclude this section by providing a way to estimate the number of subalgebras of BCC-algebras.

In BCI-algebras, the identity $(x * y) * z = (x * z) * y$ is satisfied by all elements, but, as W.A. Dudek observed, the part of results obtained for BCI-algebras can also be proven in the case when this identity is restricted to branches of weak BCC-algebras, i.e., it is satisfied only for x, y belonging to the same branch and for an arbitrary element z. Such weak BCC-algebras are called *solid* or *solid BZ-algebras* and they are discussed in Chapter 2.

In later sections of Chapter 2 we define and show the properties of the objects needed to define and describe solid weak BCC-algebras. We start with some results describing atoms and the minimal elements called initial elements of weak BCC-algebras. The initial elements are the building blocks of the branches that make up each weak BCC-algebra. We discuss in detail the map $\varphi(x) = 0 * x$ that will also be used later in this book.

We will show why a new look at the ideals of BCC-algebras was needed. We will also deal with congruences due to their mutual connections with ideals and their equivalence class corresponding to branches.

In BCC-algebras, each BCC-ideal is a subalgebra, but in weak BCC-algebras it is not the case. We describe the connections between ideals, subalgebras and sets of initial elements of weak BCC-algebras and investigate the conditions for which a subalgebra is an ideal. The properties of such ideals are strongly connected with some endomorphism initiated by φ. This motivated us to investigate elements with a finite "order". Such elements are called nilpotent. They form a subalgebra. Moreover, elements from the same branch have the same "nilpotency index".

The interesting fact that initial elements create a group-like weak BCC-algebra that many mathematicians have studied leads to the description of this subclass of weak BCC-algebras having strong connections with the groups.

With the knowledge of these objects, we can deal with solid weak BCC-algebras themselves, which we do in the last section of Chapter 2.

Chapter 3 is devoted to subclasses of BCC-algebras. The results reported in Chapter 2 motivated us to study various subclasses of weak BCC-algebras as it is done in the case of BCK-algebras. It turned out that many results similar to those obtained for BCK-algebras can be proven for solid weak BCC-algebras, but the methods of proving are entirely different. Used methods suggest that we need to examine weak BCC-algebras with properties restricted to branch only. BCK-algebra can be embedded into a solid weak BCC-algebra as one of its branches. In this way we obtain a class of branchwise commutative, branchwise weakly commutative and branchwise positive implicative weak BCC-algebras. Moreover, solid φ-implicative weak BCC-algebras play an important role in these classes of weak BCC-algebras.

In the final Chapter 4, we deal with ideals. Contrary to Chapter 2, the objects of our interest will be special kinds of ideals combined with the subclasses of weak BCC-algebras presented in the preceding chapter. We present, for example, a method of studying of ideals that together with a given element also contain the whole branch with this element. Our method is based on the map φ. This method is universal, and it can be used to investigate the connections between different types of ideals in weak BCC-algebra and in other types of algebras inspired by logic.

All cited works (and many more) are listed in the Bibliography.

I am very grateful to my teacher of algebra Professor Wiesław A. Dudek for his suggestions and corrections of the contents of the book.

Janus Thomys
Düsseldorf
September 2022

1

BCC-Algebras - Introduction

In this chapter we define weak BCC-algebras as a class containing the classes of BCI- and BCK-algebras. We explain the correlations between them and provide the basic facts needed to understand the subsequent results in this book.

Section 1.2 contains multiple examples and constructions to illustrate the essence of BCC-algebras. We also give a theorem stating that for every number n, there exists a BCC-algebra with n-elements. Finally, we estimate the number of subalgebras of a given finite BCC-algebra.

1.1 Basic Definitions and Facts

First, let us begin with a few notes on the notation used in this book.

Algebras will be denoted by German capital letters, for example $\mathfrak{X}, \mathfrak{Y}$, and the base sets by the corresponding italic capital letters, for example X, Y.
The binary operation will be denoted by juxtaposition. Dots will be only used to avoid repetitions of brackets. For example, the formula $((xy)(zy))(xz) = 0$ will be abbreviated as $(xy \cdot zy) \cdot xz = 0$.

We begin by defining the key objects considered in this book.

Definition 1.1.1. An algebra $\mathfrak{X} = \langle X; \cdot, 0\rangle$ of type $(2,0)$ is called a *BCK-algebra* if it satisfies the following axioms for any $x, y, z \in X$:

(BCK1) $(xy \cdot xz) \cdot zy = 0$,

(BCK2) $xx = 0$,

(BCK3) $x0 = x$,

(BCK4) $0x = 0$,

(BCK5) $xy = yx = 0 \implies x = y$.

The class of BCI-algebras is broader than the class of BCK-algebras.

Definition 1.1.2. An algebra $\mathfrak{X} = \langle X; \cdot, 0\rangle$ of type $(2,0)$ is called a *BCI-algebra* if it satisfies the following axioms for any $x, y, z \in X$: (BCK1), (BCK2), (BCK3) and (BCK5).

Now we can define the *BCC-algebra* as a generalization of BCK-algebras.

Definition 1.1.3. A *BCC-algebra* is an algebra $\mathfrak{X} = \langle X; \cdot, 0\rangle$ of type $(2,0)$ satisfying the following axioms for $x, y, z \in \mathrm{X}$:

(BCC1) $(xy \cdot zy) \cdot xz = 0$,

(BCC2) $xx = 0$,

(BCC3) $x0 = x$,

(BCC4) $0x = 0$,

(BCC5) $xy = yx = 0 \implies x = y$.

The widest class of algebras presented in this book are the *weak BCC-algebras.*

Definition 1.1.4. A weak BCC-algebra is an algebra $\mathfrak{X} = \langle X; \cdot, 0\rangle$ of type $(2,0)$ satisfying the axioms (BCC1), (BCC2), (BCC3) and (BCC5).

Since there is no equivalent equation to (BCK5), the class of all BCK-algebras is not a variety (cf. [137]). Therefore, the class of all weak BCC-algebras also does not form a variety (cf. [106]). But there are some subclasses of these algebras that form a variety. These subclasses will be presented in Chapter 3.

A BCC-algebra which is not a BCK-algebra is called *proper.* Similarly, a weak BCC-algebra which is not a BCC-algebra is called *proper* if it is not a BCI-algebra.

We show a few simple examples of BCC-algebras.

Example 1.1.5. An algebra $\langle \{0,1,2,3\}; \cdot, 0\rangle$ with the operation $\cdot$ defined by the following Cayley table:

$\cdot$	0	1	2	3
0	0	0	0	0
1	1	0	0	0
2	2	2	0	1
3	3	3	3	0

Table 1.1.1

is a proper BCC-algebra.

Indeed, this is a BCC-algebra. It can be easily seen that axioms (BCC2), (BCC3), (BCC4) and (BCC5) are satisfied. It remains to show that (BCC1) holds in the above algebra. It is the case for $x = 0$ or $y = 0$. Similarly, it is the case if $x = y$ or $x = z$ or $y = z$.
For $z = 0$ (BCC1) takes the form $xy \cdot x = 0$. Direct computations show that this identity holds for every x and y.
It is also not difficult to verify that (BCC1) holds if x, y, z are different.

Hence this algebra is a BCC-algebra.
But we have

$$((2 \cdot 1) \cdot (2 \cdot 3)) \cdot (3 \cdot 1) = (2 \cdot 1) \cdot 3 = 2 \cdot 3 = 1 \neq 0,$$

i.e., the axiom (BCK1) is not satisfied and the algebra defined in Table 1.1.1 is a proper BCC-algebra. □

Example 1.1.6. Let $\mathscr{F}(X)$ be the set of all real valued functions defined on a nonempty set X with the operation:

$$(f \star g)(x) = \begin{cases} 0 & \text{if } \ f(x) \leq g(x), \\ f(x) - g(x) & \text{if } \ g(x) < f(x). \end{cases}$$

Then the algebra $\langle \mathscr{F}(X); \star, 0 \rangle$ is a BCC-algebra. □

Example 1.1.7. Let $\mathscr{P}(X)$ be a power set and the operation $\diamond$ be defined in the following way:

$$A \diamond B = \begin{cases} \emptyset & \text{if } A \subseteq B, \\ A \setminus B & \text{otherwise.} \end{cases}$$

Then the algebra $\langle \mathscr{P}(X); \diamond, \emptyset \rangle$ is a BCC-algebra. □

Example 1.1.8. An interesting example of a BCC-algebra is the set of all ideals on a given principal ideal domain R with the operation $(a):(b)$. □

Example 1.1.9. Let $\mathfrak{X} = \langle X;\ \cdot\,,0\rangle$ be an arbitrary proper weak BCC-algebra and let $\mathbb{Z}$ be the set of all integers. Let us consider an algebra

$$\mathfrak{X} \times \mathfrak{Z} = \langle X \times \mathbb{Z};\ \star\,,(0,0)\rangle$$

with the operation $\star$ defined as follows:

$$(x,m)\star(y,n) = (xy, m-n)$$

for $x, y \in X$ and $m, n \in \mathbb{Z}$.

Then $\mathfrak{X} \times \mathfrak{Z}$ is a weak BCC-algebra.

First, we check the axiom (BCC1).
For $x_1, x_2, x_3 \in X$, $m_1, m_2, m_3 \in \mathbb{Z}$ we have

$$\begin{aligned}
&(((x_1,m_1)\star(x_2,m_2))\star((x_3,m_3)\star(x_2,m_2)))\star((x_1,m_1)\star(x_3,m_3))\\
&= ((x_1x_2, m_1-m_2)\star(x_3x_2, m_3-m_2))\star(x_1x_3, m_1-m_3)\\
&= (x_1x_2\cdot x_3x_2, m_1-m_2-m_3+m_2)\star(x_1x_3, m_1-m_3)\\
&= ((x_1x_2\cdot x_3x_2)\cdot x_1x_3, m_1-m_2-m_3+m_2-m_1+m_3)\\
&= (0,0),
\end{aligned}$$

because $\mathfrak{X}$ is a weak BCC-algebra.

The axiom (BCC2) is also satisfied:

$$(x,m)\star(x,m) = (xx, m-m) = (0,0)$$

for $x \in X$ and $m \in \mathbb{Z}$.

The proof of the axiom (BCC3) is trivial.

It remains the proof of (BCC5).
For $x_1, x_2 \in X$ and $m_1, m_2 \in \mathbb{Z}$ we have

$$\begin{aligned}
(x_1,m_1)\star(x_2,m_2) = (x_2,m_2)\star(x_1,m_1) = (0,0)&\\
\Longleftrightarrow (x_1x_2, m_1-m_2) = (x_2x_1, m_2-m_1) = (0,0)&\\
\Longrightarrow (x_1,m_1) = (x_2,m_2).&
\end{aligned}$$

So, $\mathfrak{X} \times \mathfrak{Z}$ is a proper weak BCC-algebra.

Note that even if $\mathfrak{X}$ is a BCC-algebra, then $\mathfrak{X} \times \mathfrak{Z}$ is not a BCC-algebra, because $0 - m \neq 0$ for $m \in \mathbb{Z} \setminus \{0\}$. □

Example 1.1.10. Let $\{\langle X_\xi, \cdot_\xi, 0_\xi \rangle\}_{\xi \in \Lambda}$ be a nonempty family of proper BCC-algebras where Λ is a totally ordered set. The Cartesian product of this family is a BCC-algebra with the coordinate-wide defined operation and with the constant $0 = (0_\xi)_{\xi \in \Lambda}$. □

More examples of BCC-algebras are given in the next section.

Now we define a partial order that plays a key role in the theory of BCC-algebras.

Definition 1.1.11. In any weak BCC-algebra $\mathfrak{X}$ we define the relation $\leq$ putting

$$x \leq y \Longleftrightarrow xy = 0 \tag{1.1}$$

for $x, y \in X$.

We say that two elements $x, y \in X$ are *comparable* if $x \leq y$ or $y \leq x$. Otherwise, the elements are *incomparable*. The relation $\leq$ is a partial order.

Indeed, the reflexivity and the antisymmetry follow from (BCC2) and (BCC5), respectively. To show the transitivity let us assume, that $x \leq y$ and $y \leq z$. Then

$$xz = xz \cdot 0 = xz \cdot yz = (xz \cdot yz) \cdot 0 = (xz \cdot yz) \cdot xy = 0, \text{ i.e., } x \leq z,$$

by multiple using of (BCC3) and (BCC1).

Definition 1.1.12. A nonempty subset A of a weak BCC-algebra $\mathfrak{X}$ is called a *chain* if it is linearly ordered by the relation $\leq$.
If the whole BCC-algebra $\mathfrak{X}$ is linearly ordered by the relation $\leq$, then it is called a *BCC-chain*.

Similarly, we define a *BCK-chain*.

Thus every weak BCC-algebra $\mathfrak{X}$ can be considered as a partially ordered set $(X; \leq)$ with some additional properties.

Theorem 1.1.13. *An algebra $\langle X; \cdot, \leq, 0 \rangle$ of type $(2, 2, 0)$ where the relation $\leq$ is defined by (1.1) is a weak BCC-algebra if and only if for all $x, y, z \in X$ the following conditions are satisfied:*

(BCC1$'$) $xy \cdot zy \leq xz$,

(BCC2$'$) $x \leq x$,

(BCC3) $x0 = x$,

(BCC5$'$) $x \leq y$ and $y \leq x$ imply $x = y$.

Corollary 1.1.14. *The class of BCC-algebras is selected from the class of weak BCC-algebras by the following condition:*

(BCC4$'$) $0 \leq x$,

which should be satisfied by all elements.

Since two nonisomorphic weak BCC-algebras may have the same partial order, they cannot be investigated as partially ordered sets only.

Example 1.1.15. It is easy to show that algebras with the operations defined by Tables 1.1.2 and 1.1.3 are both proper weak BCC-algebras.

$\cdot$	0	1	2	3
0	0	0	2	2
1	1	0	2	2
2	2	2	0	0
3	3	3	1	0

Table 1.1.2

$\cdot$	0	1	2	3
0	0	0	2	2
1	1	0	3	3
2	2	2	0	0
3	3	3	1	0

Table 1.1.3

They are proper because in both cases

$$((3 \cdot 1) \cdot (3 \cdot 2)) \cdot (2 \cdot 1) = (3 \cdot 1) \cdot 2 = 3 \cdot 2 = 1 \neq 0.$$

This means that (BCK1) is not satisfied.

(BCC4) is also not satisfied, because in both cases $0 \cdot 2 = 2 \neq 0$. They share the same partial order presented by the following diagram:

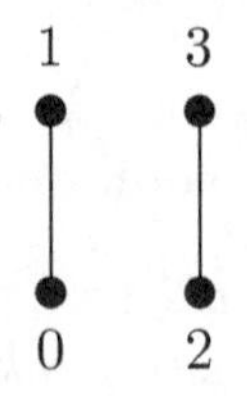

Diagram 1.1.1

In this partial order $0 < 1$, $2 < 3$ and elements 1, 2 and 1, 3 are incomparable. Similarly, 0, 2 and 0, 3. □

Lemma 1.1.16. *In any BCI-algebra $\mathfrak{X}$ the following condition:*

$$(x \cdot xy)y = 0 \tag{1.2}$$

or alternatively the following inequality:

$$x \cdot xy \leq y \tag{1.3}$$

are satisfied for all $x, y \in X$.

Proof. From (BCK3) and (BCK1) we have

$$(x \cdot xy)y = (x0 \cdot xy) \cdot y0 = 0.$$

The alternative form follows from (1.1). □

In weak BCC-algebra $\mathfrak{X}$ the two very helpful implications

$$x \leq y \Longrightarrow xz \leq yz \tag{1.4}$$

$$x \leq y \Longrightarrow zy \leq zx \tag{1.5}$$

are satisfied by all $x, y, z \in X$.

For the proof of (1.4) let us assume $x \leq y$. Then we get

$$xz \cdot yz = (xz \cdot yz)0 = (xz \cdot yz) \cdot xy = 0,$$

i.e., $xz \leq yz$ by respectively (BCC3), the assumption and (BCC1).

For $x \leq y$, the condition (1.5) follows from (BCC3), from the assumption and (BCC1$'$). Indeed, we have

$$zy = zy \cdot 0 = zy \cdot xy \leq zx,$$

i.e., $zy \leq zx$.

The next condition is crucial for the theory of BCK-/BCI-algebras. It is so-called *head-fixed commutative law* and it is satisfied in every BCI-algebra

$$xy \cdot z = xz \cdot y \tag{1.6}$$

This equation was first proven in [77]. Here we give a slightly generalized version of the proof from [30].

Theorem 1.1.17. *Let $\mathfrak{X}$ be a BCI-algebra. Then Equation* (1.6) *is satisfied for any $x, y, z \in X$.*

Proof. Let us observe that from $x \cdot xz \leq z$ by (1.5) it follows that

$$xy \cdot z \leq xy \cdot (x \cdot xz). \tag{1.7}$$

But, by (BCK1)

$$xy \cdot (x \cdot xz) \leq xz \cdot y. \tag{1.8}$$

Combining (1.7) and (1.8) we get

$$xy \cdot z \leq xz \cdot y.$$

Similarly, $xz \cdot y \leq xy \cdot z$.

By (BCK5$'$) we have $xy \cdot z = xz \cdot y$. □

In the following corollary we compare the correlations between the discussed algebras.

Corollary 1.1.18. *The following statements are true:*

(1) *Any BCK-algebra is a BCC-algebra,*

(2) *Any BCC-algebra is a weak BCC-algebra,*

(3) *Any BCK-algebra is a BCI-algebra,*

(4) *Any BCI-algebra is a weak BCC-algebra.*

Proof. (1) Except for (BCK1) and (BCC1) all other axioms are identical. But the condition (1.6) applied to (BCK1) gives (BCC1).

(2) Obvious.

(3) It results from a direct comparison of the axiom systems.

(4) Similar to (1). □

Corollary 1.1.19. *A weak BCC-algebra is a BCI-algebra if and only if it satisfies* (1.6).

Proof. Let $\mathfrak{X}$ be a weak BCC-algebra with the condition (1.6). We have to show only that the axiom (BCK1) is satisfied. Indeed,

$$(xy \cdot xz) \cdot zy = (xy \cdot zy) \cdot xz = 0,$$

by (1.6) and (BCC1).

The converse is obvious. □

Corollary 1.1.20. *A BCC-algebra is a BCK-algebra if and only if it satisfies* (1.6).

Theorem 1.1.21. *The class of all BCK-algebras is equivalent to the class of algebras defined by the independent axiom system:* (BCC1), (BCC3), (BCC4), (BCC5) *and* (1.2).

Proof. From Definition 1.1.1 and from Lemma 1.1.16, it follows that every BCK-algebra satisfies the above conditions.

We prove the converse. Let $\langle X; \cdot, 0\rangle$ be an algebra of type (2,0) in which (BCC1), (BCC3), (BCC4), (BCC5) and (1.2) hold. Then $\langle X; \cdot, 0\rangle$ satisfies also (BCK1) and (BCK2). Indeed, (BCK2) follows from (BCC3) and (1.2), namely for $y = 0$ we have $xx = (x \cdot x0)0 = 0$.

Let us observe that (BCC1), (BCC2), (BCC3) and (BCC5) show that the relation $\leq$ defined by (1.1) is a partial order on X with 0 as the smallest element and of course (BCC1$'$) is satisfied. Moreover, (BCC1) and (1.2) give (1.3) and (1.5).
Thus (BCC1$'$) by (1.5) implies

$$xy \cdot xz \leq xy \cdot (xy \cdot zy),$$

but, by (1.2) we get

$$xy \cdot (xy \cdot zy) \leq zy,$$

which gives

$$xy \cdot xz \leq zy,$$

which, by (1.6), gives (BCK1).
Hence $\langle X; \cdot, 0\rangle$ is a BCK-algebra.

In the sequel we prove that these axioms are independent.

It is clear that the algebra with the operation $\cdot$ defined by Table 1.1.4 satisfies (BCC3), (BCC4) and (BCC5). This algebra satisfies also (BCC1).

Indeed, the cases $x = 0$, $y = 0$ and $y = 2$ are obvious. Similarly, the cases $x = y$, $x = z$ and $y = z$.

We also still consider the case when x, y, z are different and $x \in \{1, 2, 3\}$, $y \in \{1, 3\}$ and $z \in \{0, 1, 2, 3\}$. Direct computations show that (BCC1) holds in such cases.

Since $(2 \cdot (2 \cdot 3)) \cdot 3 = (2 \cdot 1) \cdot 3 = 2 \cdot 3 = 1 \neq 0$, the condition (1.2) is independent.

$\cdot$	0	1	2	3
0	0	0	0	0
1	1	0	0	1
2	2	2	0	1
3	3	3	0	0

Table 1.1.4

The axiom (BCC4) is independent too, because (1.2), (BCC1), (BCC3) and (BCC5) hold in every Boolean group, but (BCC4) holds only in a one-element group.

The algebra defined in Table 1.1.5 satisfies all these axioms with the exception of (BCC5). Indeed, we have $12 = 21 = 0$ but $1 \neq 2$. Thus (BCC5) is independent.

$\cdot$	0	1	2
0	0	0	0
1	1	0	0
2	2	0	0

Table 1.1.5

It is clear that the algebra defined by Table 1.1.6 satisfies (BCC4) and (BCC5). We have to prove that it satisfies also (1.2) and (BCC1). Since the algebra $\mathfrak{G} = \langle\{0, 1, 2\}; \cdot, 0\rangle$ is a BCK-algebra (what can be easily checked), we have to verify these conditions only in the case when at least one of the elements x, y, z is equal to 3.

Direct computations show that in these cases (1.2) holds.

We will now show that the algebra defined in Table 1.1.6 satisfies (BCC1).

Let $x = 3$.

Then the cases in which $y \in \{2, 3\}$ are obvious. The case with $y = 0$ is simple. For $y = 1$ we have $(1 \cdot (z \cdot 1)) \cdot (3 \cdot z)$. Observe that $z \cdot 1$ is equal to 0 for $z \in \{0, 1\}$ or it is equal to 1 for $z \in \{2, 3\}$. The case for $z \in \{2, 3\}$ is obvious. For $z \in \{0, 1\}$ we have $1 \cdot (3 \cdot z)$ which is, as can be easily seen, equal to 0.

Let $y = 3$.

The case for $x = 0$ is obvious.

For $x = 1$ the cases with $z \in \{1, 2\}$ are obvious. For $z = 0$ we have

$$((1 \cdot 3) \cdot (0 \cdot 3)) \cdot (1 \cdot 0) = 1 \cdot 1 = 0.$$

For $z = 3$ we have $((1 \cdot 3) \cdot (3 \cdot 3)) \cdot (1 \cdot 3)$ and it is clearly equal to 0. For $x = 2$ the case is obvious when $z = 2$. For $z = 0$ we have

$$((2 \cdot 3) \cdot (0 \cdot 3)) \cdot (2 \cdot 0) = 2 \cdot 2 = 0.$$

For $z = 1$ we get

$$((2 \cdot 3) \cdot (1 \cdot 3)) \cdot (2 \cdot 1) = (2 \cdot 1) \cdot (2 \cdot 1) = 0$$

and for $z = 3$ we have

$$((2 \cdot 3) \cdot (3 \cdot 3)) \cdot (2 \cdot 3) = 0.$$

Now let $z = 3$.

The cases when $x \in \{0, 3\}$ are obvious. Similarly for $x = 1$ when $y \in \{1, 2\}$. For $y = 0$ we have $(1 \cdot 0) \cdot (3 \cdot 0) = 1 \cdot 2 = 0$. For $x = 2$ we have for $y \in \{0, 1, 2\}$ $2y = 3y$, which satisfies (BCC1).
Hence (BCC1) holds for all $x, y, z \in \{0, 1, 2, 3\}$.

Since $3 \cdot 0 \neq 3$, the axiom (BCC3) is independent.

$\cdot$	0	1	2	3
0	0	0	0	0
1	1	0	0	1
2	2	1	0	2
3	2	1	0	0

Table 1.1.6

Finally we remark that the algebra defined by Table 1.1.7 satisfies (1.2), (BCC3), (BCC4), (BCC5), but

$$((1 \cdot 2) \cdot (0 \cdot 2)) \cdot (1 \cdot 0) = (2 \cdot 0) \cdot 1 = 2 \cdot 1 = 2 \neq 0,$$

i.e., the axiom (BCC1) is independent.

$\cdot$	0	1	2
0	0	0	0
1	1	0	2
2	2	2	0

Table 1.1.7

The proof is completed. $\square$

Theorem 1.1.22. *The class of all BCC-algebras is defined by the independent axiom system:* (BCC1), (BCC3), (BCC4) *and* (BCC5).

Proof. Putting in (BCC1) $y = z = 0$ and using (BCC3) we obtain

$$0 = (x0 \cdot 00) \cdot x0 = x0 \cdot x0 = xx,$$

which shows that (BCC2) follows from (BCC1) and (BCC3).

The independence of the axioms was already shown in Theorem 1.1.21. $\square$

From the above we have the following:

Corollary 1.1.23. *A BCC-algebra satisfying the condition* (1.2) *is a BCK-algebra.*

In other words, a weak BCC-algebra is proper if it does not satisfy (BCC4) and (1.6).

Corollary 1.1.24. *The class of all weak BCC-algebras is defined by the independent axiom system:* (BCC1), (BCC3) *and* (BCC5).

Although a proper weak BCC-algebra $\mathfrak{X}$ is not a BCC-algebra, it contains, as we will see many times, elements of a BCC-algebra or a BCK-algebra forming a subalgebra called the *BCC-part* or the *BCK-part* of $\mathfrak{X}$.

We start with definitions of individual objects.

Definition 1.1.25. Let $\mathfrak{X}$ be a weak BCC-algebra. If $S \subset X$, then $\mathfrak{S} = \langle S; \cdot, 0\rangle$ is called a *subalgebra* of $\mathfrak{X}$, if $\mathfrak{S}$ contains the element 0 and it is closed under the operation $\cdot$, i.e., $xy \in S$ for $x, y \in S$.

A subalgebra $\mathfrak{S}$ of a weak BCC-algebra $\mathfrak{X}$ is called a *BCC-subalgebra* if it satisfies (BCC4). If in a BCC-subalgebra the condition (1.6) holds, then it is called a *BCK*-subalgebra.

Definition 1.1.26. The set

$$B(0) = \{x \in X \mid 0x = 0\} = \{x \in X \mid 0 \leq x\}$$

is called the *BCC-part* of a weak BCC-algebra $\mathfrak{X}$.

If in the BCC-part the condition (1.6) holds, then it is called the *BCK-part*.

Theorem 1.1.27. *Let $\mathfrak{X}$ be a weak BCC-algebra. Then algebra $\mathfrak{B}(0) = \langle B(0); \cdot, 0\rangle$ is the greatest BCC-subalgebra (resp. BCK-subalgebra) due to inclusion contained in $\mathfrak{X}$.*

Proof. First, we prove that $\mathfrak{B}(0)$ is a subalgebra of $\mathfrak{X}$. Obviously $B(0) \subseteq X$ and $0 \in B(0)$. If the elements $x, y \in B(0)$, then $0 \leq x$ and $0 \leq y$, whence, by (1.4), we obtain $0 = 0y \leq xy$, i.e., $xy \in B(0)$. Thus $\mathfrak{B}(0)$ is a subalgebra of $\mathfrak{X}$.

If $\mathfrak{S}$ is another BCC-subalgebra of $\mathfrak{X}$, then for every $x \in S$ we have $0 \leq x$ which means that $x \in B(0)$. Consequently $S \subseteq B(0)$. Thus $\mathfrak{B}(0)$ is the greatest BCC-subalgebra contained in $\mathfrak{X}$.
The proof in the version for the BCK-part is similar. □

Hence every proper weak BCC-algebra is an extension of some BCC-algebra.

Theorem 1.1.28. *The BCC-part (resp. the BCK-part) of a weak BCC-algebra $\mathfrak{X}$ has at least two elements if and only if in X there are at least two comparable elements.*

Proof. Let $x, y \in X$, $x \neq y$. If $x \leq y$, then $0 = yy \leq yx$ by (1.5). Thus $yx \in B(0)$. Obviously $yx \neq 0$ by (BCC5). Hence $B(0)$ has at least two elements. The converse is obvious. □

Corollary 1.1.29. *If in a weak BCC-algebra $\mathfrak{X}$ there exist $x \neq y$ such that $x \leq y$, then there exists also $z \neq 0$ such that $0 \leq z$.*

Theorem 1.1.30. *The relation $\leq$ defined on a BCC-algebra $\mathfrak{X}$ satisfies the following conditions for all $x, y, z, u \in X$:*

(1) $x \cdot yz \leq x \cdot (yu \cdot zu)$,

(2) $(xu \cdot yu) \cdot z \leq xy \cdot z$,

(3) $xy \cdot z \leq xz \leq x \cdot zu$,

(4) $xy \leq x$,

(5) $xy \cdot xz \leq x \cdot xz$,

(6) $xy \cdot z \leq xy \cdot zy$,

(7) $(x \cdot 0z)y \leq x \cdot yz$,

(8) $xy \cdot z \leq x \cdot yz$,

(9) $xy = 0$ *and* $x \neq y$ *implies* $yx \neq 0$,

(10) $xy = z$ *implies* $zx = 0$.

Proof. From (BCC1$'$) we have $yu \cdot zu \leq yz$. By using (1.5) we get (1).

By putting $y = x$ and $z = y$ in the above inequality and by using (1.4) we get (2).

Using (BCC3), (BCC4), (BCC1$'$) and (1.4), we get

$$xy \cdot z = (xy \cdot 0y) \cdot z \leq x0 \cdot z = xz. \tag{1.9}$$

Using (BCC1$'$), (1.5), (BCC4) and (BCC3), we get

$$xz = x \cdot z0 \leq x \cdot (zu \cdot 0u) = x \cdot zu. \tag{1.10}$$

Combining (1.9) and (1.10) we get (3).

By putting $z = 0$ in (3) we get (4).

(5) follows from (4) and (1.4).

(6) follows from (4) and (1.5).

For the proof of the condition (7), we first observe that by (BCC1) and (1.4)

$$yz \cdot 0z \leq y \text{ implies } (yz \cdot 0z)y = 0.$$

The next part of the proof follows from the axioms. We have namely

$$\begin{aligned}
(x \cdot 0z)y &= (x \cdot 0z)y \cdot 0 && \text{from (BCC3)} \\
&= (x \cdot 0z)y \cdot (yz \cdot 0z)y && \text{from the above} \\
&\leq (x \cdot 0z) \cdot (yz \cdot 0z) && \text{from (BCC1}') \\
&\leq x \cdot yz && \text{from (BCC1}').
\end{aligned}$$

For the proof of (8) we know that from (BCC4$'$), (1.5) and (BCC3) $0 \leq z$ implies

$$xy \cdot z \leq xy. \tag{1.11}$$

Then by (4) we have $yz \leq y$ and using (1.5) we get

$$xy \leq x \cdot yz. \tag{1.12}$$

So, by combining (1.11) and (1.12) we get

$$xy \cdot z \leq x \cdot yz,$$

which proves (8).

The implication (9) follows from (BCC5).

(10) follows directly from (4). Indeed, $z = xy \leq x$.

The proof is completed. □

In this book we will also deal with the operation of exponentiation, so now we will give its definition

Definition 1.1.31. Let $\mathfrak{X}$ be a weak BCC-algebra. For $x, y \in X$ and nonnegative integer n we define

$$xy^0 = x, \quad xy^{n+1} = (xy^n)y.$$

Lemma 1.1.32. *In a weak BCC-algebra $\mathfrak{X}$ the following condition holds:*

$$xz^n \cdot yz^n \leq xy \tag{1.13}$$

for all $x, y, z \in X$ and $n \geq 1$.

Proof. For any $x, y, z \in X$ from the above definition and from (BCC1$'$) we have

$$xz^n \cdot yz^n = (xz^{n-1})z \cdot (yz^{n-1})z \leq xz^{n-1} \cdot yz^{n-1} \leq \ldots \leq xz \cdot yz \leq xy.$$

□

1.2 Constructions of BCC-Algebras

In this section we will provide some results concerning the minimum cardinality of proper BCC-algebras and show some interesting methods of constructing them (cf. [36]).

We begin with two construction methods for weak BCC-algebras that are modifications of construction methods for BCI-algebras.

Lemma 1.2.1. *Let $\mathfrak{X} = \langle X; \cdot, 0\rangle$ be a BCC-algebra. If $a \notin X$ then the set $X' = X \cup \{a\}$ with the operation $\star$ defined in the following way:*

$$x \star y = \begin{cases} xy & \text{if} \quad x, y \in X, \\ 0 & \text{if} \quad x = a, \ y = a, \\ a & \text{if} \quad x = a, \ y \in X, \\ a & \text{if} \quad x \in X, \ y = a, \end{cases}$$

is a weak BCC-algebra.

Proof. The axioms (BCC3) and (BCC5) are obviously satisfied. We have to verify (BCC1) only in the case when at least one of x, y, z is equal to a. But the cases are trivial, in which more than one of the elements are equal to a. So we have to verify (BCC1) only when exactly one element is equal to a, and the others belong to X.

Let $x = a$. Then

$$((a \star y) \star (z \star y)) \star (a \star z) = (a \star zy) \star a = a \star a = 0.$$

For $y = a$ we have

$$((x \star a) \star (z \star a)) \star (x \star z) = (a \star a) \star xz = 0 \star xz = 0 \cdot xz = 0,$$

because $\mathfrak{X}$ is a BCC-algebra.

Let now $z = a$. Then

$$((x \star y) \star (a \star y)) \star (x \star a) = (xy \star a) \star a = a \star a = 0.$$

So, the axiom (BCC1) is satisfied. Since $0a = a$, then the axiom (BCC4) does not hold. Hence the algebra $\langle X'; \star, 0\rangle$ is a weak BCC-algebra, which completes our proof. □

Corollary 1.2.2. *If in the above lemma $\mathfrak{X}$ is a BCK-algebra, then $\langle X \cup \{a\}; \star, 0\rangle$ is a BCI-algebra.*

Proof. It is enough to show that condition (1.2) is satisfied when at least one of x, y is equal to a. The proof is very easy. □

Lemma 1.2.3. *Let $\mathfrak{X} = \langle X; \cdot, 0\rangle$ be a BCC-algebra and let $a, b \notin X$. Then the set $X \cup \{a, b\}$ equipped with the operation $\star$ defined in the following way:*

$$x \star y = \begin{cases} xy & \text{if } x, y \in X, \\ 0 & \text{if } x = y = a, \\ a & \text{if } x = a,\ y \in X, \\ a & \text{if } x \in X,\ y = b, \\ a & \text{if } x = b,\ y = a, \\ 0 & \text{if } x = y = b, \\ b & \text{if } x \in X,\ y = a, \\ b & \text{if } x = b,\ y \in X, \\ b & \text{if } x = a,\ y = b, \end{cases}$$

is a weak BCC-algebra.

Proof. In the same way as in the previous lemma, we can prove that $\langle X \cup \{a,b\};\, \star, 0\rangle$ satisfies the axioms (BCC1), (BCC3) and (BCC5). Since $0 \star a = b \neq 0$, it does not satisfy (BCC4). Hence it is a weak BCC-algebra. □

Corollary 1.2.4. *If in the previous lemma $\mathfrak{X}$ is a BCK-algebra, then $\langle X \cup \{a,b\};\, \star, 0\rangle$ is a BCI-algebra.*

Table 1.1.1 provides an example of a proper BCC-algebra. This algebra has four elements. This turns out to be the smallest number of elements in a proper BCC-algebra.

The following results make this statement more precise.

Lemma 1.2.5. *If all elements of a weak BCC-algebra $\mathfrak{X}$ are incomparable, then for $x, y, z \in X$ the following conditions are satisfied:*

(1) $xy \cdot zy = xz,$

(2) $x \neq z \Longrightarrow xy \neq zy,$

(3) $x \neq y \Longrightarrow xy \neq x \quad \text{for} \quad y \neq 0.$

Proof. Since all elements are incomparable, $x \leq y$ implies $x = y$. Thus from (BCC1$'$) it follows (1).

If $xy = zy$ then (BCC2) and (1) imply $0 = xy \cdot zy = xz$. Hence $x \leq z$, i.e., the elements x and z are comparable, which is contrary to our assumption. Hence (2) is true.

To prove (3), observe that for $y \neq 0$, $xy = x$, together with (BCC2) and (1), gives $0x = yy \cdot xy = yx$, which is impossible by (2). Hence (3) holds. □

Theorem 1.2.6. *A proper BCC-algebra has at least four elements.*

Proof. BCC-algebras with the cardinality of 1 and of 2 are obviously not proper.

If a BCC-algebra $\mathfrak{X} = \langle\{0,x,y\};\, \cdot, 0\rangle$ has exactly three elements, then we have two cases:

1^0. At least one element of the word $(xy)z$ is equal to 0. Then using the axioms of BCC-algebra, we can easily check that the identity (1.6) is satisfied.

2^0. At least two elements are equal.
If $x = y$, then the identity (1.6) has the form $(xx)z = (xz)x$. This is equivalent to $0 = (xz)x$. But the right side is, by Theorem 1.1.30(4),

also equal to 0.

The case $x = z$ is similar to the above.

If $y = z$, then the identity (1.6) is clearly satisfied.

Hence, by Corollary 1.1.20, $\mathfrak{X}$ is a BCK-algebra.

But there are BCC-algebras with four elements. For example, the proper BCC-algebra is shown in Example 1.1.5.

The proof is completed. □

From the axiom systems of proper BCC-algebra and proper weak BCC-algebra, it follows that the above theorem is true also for proper weak BCC-algebras. In Tables 1.1.2 and 1.1.3 we have already shown that there exist proper weak BCC-algebras with four elements.

So, we can formulate the following corollary:

Corollary 1.2.7. *A proper weak BCC-algebra has at least four elements.*

But let us see what algebras look like, which are weak BCC-algebras with less than four elements and BCK-part $B(0)$.

Theorem 1.2.8. *A proper weak BCC-algebra $\mathfrak{X}$ with $B(0)$ as its BCK-part has at least four elements.*

Proof. The case is trivial when the cardinality of $\mathfrak{X}$ is 1. For the cardinality of 2 it is not difficult to see that $B(0) = \{0\}$ and $\mathfrak{X}$ is a BCI-algebra.

Let us consider the three-element algebra $\mathfrak{X} = \langle X; \cdot, 0\rangle$, with $X = \{0, 1, 2\}$. We know already that X is partially ordered by the relation $\leq$, and therefore we have to consider the following three cases:

1^0. All elements are incomparable.

In this case, by Theorem 1.1.28, $B(0) = \{0\}$ and it is a trivial BCK-part. In this case every column of the multiplication table of such algebra is, by Lemma 1.2.5(2), a permutation of the set $\{0, 1, 2\}$.

Let us calculate $2 \cdot 1$. It cannot be equal to 0 because all elements are incomparable. It cannot be equal to 2 by Lemma 1.2.5(3), too. So, it remains $2 \cdot 1 = 1$.

Then we calculate $0 \cdot 1$. By (BCC2) and Lemma 1.2.5(1) we have

$$0 \cdot 1 = 0 \cdot (2 \cdot 1) = (1 \cdot 1) \cdot (2 \cdot 1) = 1 \cdot 2.$$

But, by the incomparability of all elements and Lemma 1.2.5(3),

$1 \cdot 2 = 2$. Hence $0 \cdot 1 = 2$.

In a similar way we get $0 \cdot 2 = 1$, which completely determines multiplication table of $\cdot$ (cf. Table 1.2.1). This table coincides with the multiplication table of the algebra $\mathfrak{X}$ defined by $x \cdot y = (x - y)\, mod\, 3$. This algebra is obviously a BCI-algebra.

$\cdot$	0	1	2
0	0	2	1
1	1	0	2
2	2	1	0

Table 1.2.1

2^0. Two elements are comparable.

By Theorem 1.1.28 we have $0 \leq 1$ or $0 \leq 2$.

Let $0 \leq 1$, i.e., $B(0) = \{0, 1\}$ and $\langle B(0); \cdot, 0\rangle$ is by Theorem 1.2.6 a BCK-algebra. Then $2 \cdot 1 \leq 2 \cdot 0 = 2$ and $0 \cdot 2 \leq 1 \cdot 2$, by (1.5), (BCC3) and (1.4). Since (by the assumption) 2 is incomparable with 0 and with 1, $2 \cdot 1 = 2$ and $0 \cdot 2 = 1 \cdot 2 = 1$ or $0 \cdot 2 = 1 \cdot 2 = 2$.

The first case is impossible because, by (BCC1)

$$((2 \cdot 2) \cdot (0 \cdot 2)) \cdot (2 \cdot 0) = (0 \cdot 1) \cdot 2 = 0 \cdot 2 = 1 \neq 0.$$

In the second case we obtain the multiplication table, which, by Corollary 1.2.2, determines a BCI-algebra.

The case $0 \leq 2$ is analogous to the case $0 \leq 1$ and gives a BCI-algebra which is isomorphic to a BCI-algebra obtained in the previous case.

3^0. All elements are comparable.

In this case $B(0) = X$ and it is a BCK-algebra.

Hence a proper weak BCC-algebra with BCK-part has at least four elements.

The proof is completed. □

Theorem 1.2.9. *There are only two proper weak BCC-algebras of order* 4. *Their multiplication tables are given in* Example 1.1.15.

Proof. Let $\mathfrak{X}$ be a weak BCC-algebra with four elements. Without loss of generality, let $X = \{0, 1, 2, 3\}$. We consider the natural partial order $\leq$ induced by the BCC-operation $\cdot$.

1^0. All elements are incomparable.

Therefore, all columns of the multiplication table of this algebra are permutations of elements $0, 1, 2, 3$.

If $0 \cdot 1 = 1$, then $2 \cdot 1 = 3$ and $3 \cdot 1 = 2$. This together with Lemma 1.2.5(1) gives

$$a = 1 \cdot 3 = (0 \cdot 1) \cdot (2 \cdot 1) = 0 \cdot 2,$$

$$b = 1 \cdot 2 = (0 \cdot 1) \cdot (3 \cdot 1) = 0 \cdot 3 \text{ and}$$

$$c = 3 \cdot 2 = (2 \cdot 1) \cdot (3 \cdot 1) = 2 \cdot 3.$$

Obviously, $a \neq b \neq c$ by Lemma 1.2.5(3). Hence the multiplication table of this algebra has the form

$\cdot$	0	1	2	3
0	0	1	a	b
1	1	0	b	a
2	2	3	0	c
3	3	2	c	0

Table 1.2.2

By Lemma 1.2.5 $a \neq 1$, $b \neq 1$, $c = 1$. Thus $a = 2$, $b = 3$, $c = 1$ or $a = 3$, $b = 2$, $c = 1$.

In the first case, we obtain the multiplication table of the group $Z_2 \times Z_2$. In the second case, we obtain the multiplication table of the algebra that coincides with the cyclic group of order 4, i.e., the algebra $\mathfrak{X}$ with the operation $x \cdot y = x - y$, where $\langle X; +, 0\rangle$ is a group isomorphic to the group Z_4. It is easy to see that these algebras are BCI-algebras.

Suppose $0 \cdot 1 = 2$. Then,

$$1 \cdot 2 = (3 \cdot 1) \cdot (0 \cdot 1) = 3 \cdot 0 = 3,$$

$$3 \cdot 2 = (2 \cdot 1) \cdot (0 \cdot 1) = 2 \cdot 0 = 2 \text{ and}$$

$$3 = 2 \cdot 1 = (0 \cdot 1) \cdot (3 \cdot 1) = 0 \cdot 3.$$

Thus $0 \cdot 2 = 1$, $1 \cdot 3 = 2$ and $2 \cdot 3 = 1$ by Lemma 1.2.5. This gives the multiplication table of the BCI-algebra $\mathfrak{X}$ that coincides with the cyclic group of order 4.

Finally let $0 \cdot 1 = 3$. Then $2 \cdot 1 = 1$ and $3 \cdot 1 = 2$ by Lemma 1.2.5. From Lemma 1.2.5(1) we also have

$$0 \cdot 2 = (0 \cdot 1) \cdot (2 \cdot 1) = 3 \cdot 1 = 2$$

and consequently $1 \cdot 2 = 3$, $3 \cdot 2 = 1$.

Similarly,

$$0 \cdot 3 = (0 \cdot 1) \cdot (3 \cdot 1) = 3 \cdot 2 = 1 \ , \ 1 \cdot 3 = 2 \text{ and } 2 \cdot 3 = 3,$$

which completely determines the multiplication table. It is not difficult to see that this table defines the BCI-algebra which coincides with the group Z_4.

Hence a proper weak BCC-algebra $\langle \{0, 1, 2, 3\}; \cdot, 0 \rangle$ in which all elements are incomparable is a BCI-algebra.

2^0. There are comparable elements.

In this case a BCC-part has at least two elements (Theorem 1.1.28). Moreover, without loss of generality, the relation $\leq$ has one of the following diagrams:

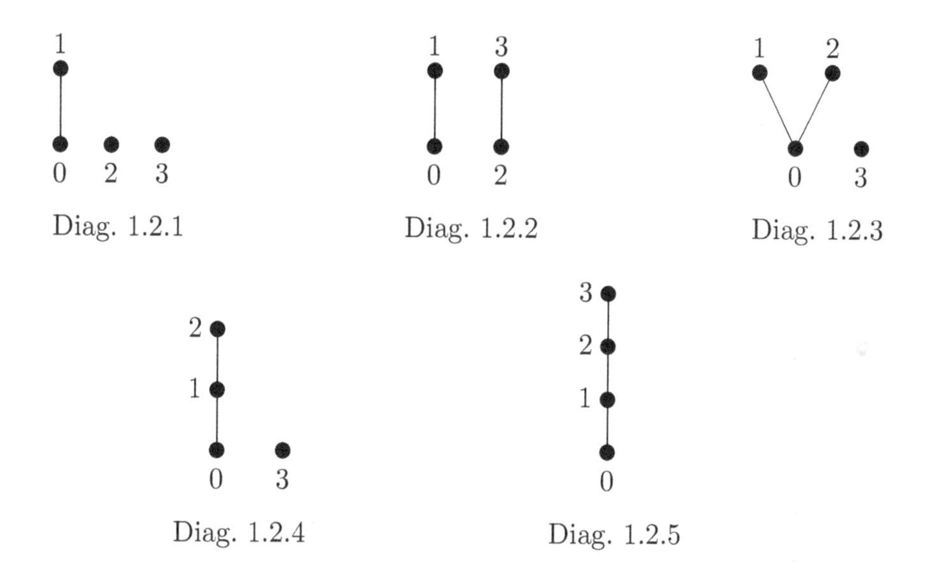

Let $\mathfrak{X}$ be a weak BCC-algebra connected with the first diagram. Then $0 \leq 1$. Since 2 and 3 are incomparable, (1.5) implies $2 \cdot 1 = 2$ and $3 \cdot 1 = 3$. Similarly, (1.4) gives $0 \cdot 2 = 1 \cdot 2 = a$ and $0 \cdot 3 = 1 \cdot 3 = b$. But $(2 \cdot 3) \cdot (1 \cdot 3) = 2$ by (BCC1$'$). Hence $c = 2 \cdot 3 \neq 1 \cdot 3$, i.e., $c \neq b$. Thus the multiplication table of this algebra has the form:

$\cdot$	0	1	2	3
0	0	0	a	b
1	1	0	a	b
2	2	2	0	c
3	3	3	d	0

Obviously $d \neq 0$.

Let $d = 1$. Since $(3 \cdot 2) \cdot (1 \cdot 2) \leq 3$ by (BCC1$'$), $1 \cdot (1 \cdot 2) = 3$ which holds only for $1 \cdot 2 = 1 \cdot 3 = 3$, i.e., for $a = b = 3$. Moreover,

$$3 \cdot (2 \cdot 3) = (1 \cdot 3) \cdot (2 \cdot 3) \leq 1 \cdot 2 = 3 \text{ implies } 2 \cdot 3 = 1.$$

Thus, by (BCC1$'$),

$$3 = 1 \cdot 3 = (2 \cdot 3) \cdot (1 \cdot 3) \leq 2 \cdot 1 = 2,$$

which is a contradiction, because 2 and 3 are incomparable by the assumption. Hence $d \neq 1$.

If $d = 2$, then $2 \cdot a = (3 \cdot 2) \cdot (1 \cdot 2) \leq 3$ is true only for $a = c = 3$. Moreover,

$$2 = 3 \cdot 2 = (1 \cdot 2) \cdot (3 \cdot 2) \leq b \text{ implies } b = 2,$$

which completely determines the above table.

Since the algebra $\langle \{0, 1\}; \cdot, 0 \rangle$ is a BCK-algebra, from Corollary 1.2.4 it follows that the algebra $\langle \{0, 1\} \cup \{2, 3\}; \cdot, 0 \rangle$ is a BCI-algebra.

Let $d = 3$. Then

$$b = 0 \cdot 3 = (2 \cdot 2) \cdot (3 \cdot 2) \leq 2 \cdot 3 = c.$$

But we know already that $b \neq c$. This implies $b = 0$, which is a contradiction.

Thus weak BCC-algebras connected with the first diagram are not proper.

Let $\mathfrak{X}$ be a weak BCC-algebra connected with Diagram 1.2.2. We have $0 \leq 1$ which, by (1.5), implies $2 \cdot 1 \leq 2 \cdot 0 = 2$. Hence $2 \cdot 1 = 2$ because 2 is a minimal element. Similarly, $2 \leq 3$, by (1.4), implies $3 \cdot 2 = 1$ and $0 \cdot 3 \leq 1 \cdot 3 \leq 1 \cdot 2$.

Of course, $0 \leq 1$ implies $3 \cdot 1 \leq 3$.
It is clear that $3 \cdot 1 \leq 3$ implies $3 \cdot 1 = 2$ or $3 \cdot 1 = 3$.

First we consider the case $3 \cdot 1 = 2$. We compute $1 \cdot 3$. Obviously $1 \cdot 3 \neq 0$.

If $1 \cdot 3 = 1$, then $0 \cdot 3 = 1 \cdot 2 = 1$, because $0 \cdot 3 \leq 1 \cdot 3 \leq 1 \cdot 2$ and $0 \cdot 3 \neq 0$. But

$$0 = 1 \cdot 1 = (3 \cdot 2) \cdot (1 \cdot 2) \leq 3 \cdot 1 = 2,$$

which is a contradiction.

Thus $1 \cdot 3 \neq 1$.

If $1 \cdot 3 = 2$, then $0 \cdot 3 = 2$, because $0 \cdot 3 \leq 1 \cdot 3$, and in the consequence

$$0 \cdot 2 = (2 \cdot 3) \cdot (0 \cdot 3) \leq 2.$$

Thus $2 = 0 \cdot 2 \leq 1 \cdot 2$. It is not difficult to verify that for $1 \cdot 2 = 2$ and $1 \cdot 2 = 3$ we obtain two different tables defining operations of two nonisomorphic BCI-algebras. Thus $1 \cdot 3 \neq 2$.

The case $1 \cdot 3 = 3$ is impossible because $2 \leq 3$ implies $3 = 1 \cdot 3 \leq 1 \cdot 2$. So, $1 \cdot 2 = 3$ and $3 = 1 \cdot 3 = (3 \cdot 2) \cdot (1 \cdot 2) \leq 3 \cdot 1 = 2$, which is a contradiction.

Next, let $3 \cdot 1 = 3$. As in the previous case $1 \cdot 3 \neq 0$ and $1 \cdot 3 \neq 1$.

In the same manner we prove that $1 \cdot 3 = 2$ implies $2 = 0 \cdot 3 = 0 \cdot 2 \leq 1 \cdot 2$. Since $1 \cdot 2 = 3$ gives

$$3 = 3 \cdot 1 = (1 \cdot 2) \cdot (3 \cdot 2) \leq 1 \cdot 3 = 2,$$

$1 \cdot 2 \neq 3$. Thus $1 \cdot 2 = 2$. Hence this case yields the proper weak BCC-algebra with the operation $\cdot$ presented in Table 1.1.2.

In the case $1 \cdot 3 = 3$ we have: $2 \leq 3$ imply $1 \cdot 2 = 3$ and $0 \leq 1$ implies $2 \cdot 1 \leq 2 \cdot 0 = 2$ which gives $2 \cdot 1 = 2$. In the consequence,

$$0 \cdot 3 = (2 \cdot 3) \cdot (1 \cdot 3) \leq 2 \cdot 1 = 2 \text{ implies } 0 \cdot 3 = 2,$$

and from $0 \cdot 3 = 2$ and $2 \cdot 3 = 0$ it follows

$$0 \cdot 2 = (2 \cdot 3) \cdot (0 \cdot 3) = 2 \text{ ,i.e., } 0 \cdot 2 = 2.$$

This gives Table 1.1.3 which also determines an operation of a proper weak BCC-algebra.

Let $\mathfrak{X}$ be connected with Diagram 1.2.3. Since 1 and 2 are incomparable, by (BCC3)

$$1 \cdot 2 = (1 \cdot 2) \cdot (0 \cdot 2) \leq 1 \cdot 0 = 1 \text{ implies } 1 \cdot 2 = 1.$$

Analogously,

$$2 \cdot 1 = (2 \cdot 1) \cdot (0 \cdot 1) \leq 2 \text{ gives } 2 \cdot 1 = 2.$$

Similarly, by (1.5), from $0 \leq 1$ it follows $3 \cdot 1 \leq 3 \cdot 0 = 3$. Hence and from Diagram 1.2.3, it follows $3 \cdot 1 = 3$.

Analogously, from $0 \leq 2$ it follows $3 \cdot 2 = 3$.

Now, from $0 \leq 1$, by (1.4), it follows $0 \cdot 3 = 1 \cdot 3$ and from $0 \leq 2$, also by (1.4), it follows $0 \cdot 3 = 2 \cdot 3$, i.e., $0 \cdot 3 = 1 \cdot 3 = 2 \cdot 3$. But $(3 \cdot 3) \cdot (0 \cdot 3) \leq 3 \cdot 0$ gives $0 \cdot (0 \cdot 3) = 3$, which is possible only for $0 \cdot 3 = 3$.

So, we have $x \cdot 3 = 3 \cdot x = 3$ for $x \in \{0, 1, 2\}$. This completely determines the multiplication table of this weak BCC-algebra. It has the form as in Lemma 1.2.1. Since its BCC-part $\mathfrak{B}(0)$ with $B(0) = \{0, 1, 2\}$ (as it is not difficult to verify) is a BCK-algebra, this weak BCC-algebra is, by Corollary 1.2.2, a BCI-algebra.

If $\mathfrak{X}$ is connected with Diagram 1.2.4, then BCC-part $\mathfrak{X}$ with $B(0) = \{0, 1, 2\}$ is, from Theorem 1.1.27, a subalgebra of $\mathfrak{X}$. Thus $2 \cdot 1 = 1$ or $2 \cdot 1 = 2$. Since, by (1.4), $0 \cdot 3 \leq 1 \cdot 3 \leq 2 \cdot 3$ and by (BCC1$'$) $(3 \cdot 3) \cdot (0 \cdot 3) \leq 3 \cdot 0$, $0 \cdot (0 \cdot 3) = 3$, which holds only for $0 \cdot 3 = 3$. Hence $x \cdot 3 = 3$ for $x \in B(0)$.

Similarly, from $0 \leq 1$ and $0 \leq 2$, by (1.5), it follows $3 \cdot 1 \leq 3 \cdot 0 = 3$ and $3 \cdot 2 \leq 3 \cdot 0 = 3$. Hence $3 \cdot x = 3$ for $x \in B(0)$.

This proves that with this diagram are connected two nonisomorphic algebras: the first with $2 \cdot 1 = 1$ and the second with $2 \cdot 1 = 2$. Both have the form as in Lemma 1.2.1. But (as it is not difficult to verify) their $\mathfrak{B}(0)$ are BCK-algebras. Hence, by Corollary 1.2.2, these algebras are BCI-algebras.

With the last diagram are connected only weak BCC-algebras which are not proper. Indeed, we have $0 \leq x$ for all $x \in \{0, 1, 2, 3\}$.

Thus there are only two proper weak BCC-algebras of order 4.
The proof is completed. □

The next important result that will be presented in this section is that for any cardinal $n \geq 4$ there exists a proper weak BCC-algebra with the order n (cf. [30]). We will show the methods of its construction, too.

Lemma 1.2.10. *Let $\mathfrak{X} = \langle X; \cdot, 0\rangle$ be a BCC-algebra and let $a \notin X$. Then the algebra $\mathfrak{X}' = \langle X \cup \{a\}; \star, 0\rangle$ with the operation $\star$ defined as follows:*

$$x \star y = \begin{cases} xy & \text{if} \quad x, y \in X, \\ a & \text{if} \quad x = a, \; y = 0, \\ 0 & \text{if} \quad x = a, \; y \neq 0, \\ x & \text{if} \quad x \in X, \; y = a, \end{cases}$$

is a BCC-algebra. Moreover, a BCC-algebra $\mathfrak{X}'$ is proper if an only if $\mathfrak{X}$ is proper.

Proof. We verify only the axiom (BCC1). The verification of other axioms is very easy. For $x, y, z \in X$ (BCC1) is satisfied.

Let $x = a$, then we have to check two following cases:
1^0. $z = 0$. In this case we have

$$((a \star y) \star (0 \star y)) \star (a \star 0) = (a \star y) \star a = \begin{cases} 0 \star a = 0 & \text{if} \quad y \neq 0, \\ a \star a = 0 & \text{if} \quad y = 0. \end{cases}$$

2^0. Let $z \neq 0$, then we have

$$((a \star y) \star (z \star y)) \star (a \star z) = (a \star y) \star (z \star y) = \begin{cases} 0 \cdot zy = 0 & \text{if} \quad y \neq 0, \\ a \star z = 0 & \text{if} \quad y = 0. \end{cases}$$

It follows that (BCC1) is satisfied for $x = a$.

For $y = a$ we have

$$((x \star a) \star (z \star a)) \star (x \star z) = (x \star z) \star (x \star z) = xz \cdot xz = 0.$$

Thus (BCC1) holds for $y = a$, too.

Let $z = a$, then

$$((x \star y) \star (a \star y)) \star (x \star a) = \begin{cases} xy \cdot x = 0 & \text{if} \quad y \neq 0, \\ (x \star a) \star (x \star a) = 0 & \text{if} \quad y = 0. \end{cases}$$

Hence (BCC1) holds for $z = a$.

Simple computations show that in the case $x = a$ or $y = a$ the condition

(1.2) holds. Hence a BCC-algebra $\langle X'; \star, 0\rangle$ satisfies (1.2) if and only if (1.2) holds for all $x, y \in X$.

So $\mathfrak{X}'$ is proper if and only if $\mathfrak{X}$ is proper, which completes the proof. □

Lemma 1.2.11. *Any weak BCC-algebra can be extended to a weak BCC-algebra containing one more element.*

Proof. Let $\langle X; \cdot, 0\rangle$ be a weak BCC-algebra and $\mathfrak{X}' = \langle X'; \star, 0\rangle$ be an algebra with the base set $X' = X \cup \theta$, where $\theta \notin X$, equipped with the operation $\star$ defined as follows:

$$x \star y = \begin{cases} xy & \text{if} \quad x, y \in X, \\ \theta & \text{if} \quad x = \theta,\ y = 0, \\ 0y & \text{if} \quad x = \theta,\ y \neq 0, \\ 0 & \text{if} \quad x = \theta,\ y = \theta, \\ x & \text{if} \quad x \in X,\ y = \theta. \end{cases}$$

We will show that $\mathfrak{X}'$ is a weak BCC-algebra.

As it can be easily seen, the axioms (BCC3) and (BCC5) are satisfied. From the assumption the axiom (BCC1) is satisfied for all $x, y, z \in X$, so it has to be verified only when at least one of x, y, z is equal to θ.

Let $x = \theta$. Then the left side of (BCC1) takes the form

$$((\theta \star y) \star (z \star y)) \star (\theta \star z) \tag{1.14}$$

and we have the following two cases:

1^0. $y = 0$. Then (1.14) has the form

$$((\theta \star 0) \star (z \star 0)) \star (\theta \star z) = (\theta \star z) \star (\theta \star z) = \begin{cases} 0z \cdot 0z = 0 & \text{if} \quad z \neq 0, \\ \theta \star \theta = 0 & \text{if} \quad z = 0. \end{cases}$$

2^0. For $y \neq 0$, (1.14) is equal to

$$(0y \star zy) \star (\theta \star z) = \begin{cases} (0y \cdot zy) \cdot 0z = 0 & \text{if} \quad z \neq 0, \\ 0 \star \theta = 0 & \text{if} \quad z = 0. \end{cases}$$

The condition for $z \neq 0$ is due to assumption that $\mathfrak{X}$ is a weak BCC-algebra.

So, for $x = \theta$, (BCC1) is satisfied.

For $y = \theta$, (BCC1) is also satisfied. Indeed,

$$((x \star \theta) \star (z \star \theta)) \star (x \star z) = (x \star z) \star (x \star z) = xz \cdot xz = 0.$$

Now, let $z = \theta$. Then the left side of (BCC1) has the form

$$((x \star y) \star (\theta \star y)) \star x \tag{1.15}$$

Here, too, we have two simple cases depending on the value of y.

1^0. $y = 0$,

$$((x \star 0) \star (\theta \star 0)) \star x = (x \star \theta) \star x = x \star x = xx = 0.$$

2^0. $y \neq 0$,

$$((x \star y) \star (\theta \star y)) \star x = (xy \cdot 0y)x \leq xx = 0,$$

by (1.4). In this case (BCC) is also satisfied.

Let $x = y = \theta$. Then

$$((\theta \star \theta) \star (z \star \theta)) \star (\theta \star z) = (0 \star z) \star (\theta \star z) = \begin{cases} 0 \star \theta & \text{if} \quad z = 0, \\ 0z \star 0z & \text{if} \quad z \neq 0. \end{cases}$$

It is easy to see that in both cases, the result of the operation $\star$ is equal to 0. The remaining cases are trivial.

The proof is completed. □

Since the proof depends only on the type of algebra, we can formulate the following corollary:

Corollary 1.2.12. *Any BCI-algebra (BCK-algebra) can be extended to a BCI-algebra (BCK-algebra) containing one more element.*

Theorem 1.2.13. *Any weak BCC-algebra can be embedded into a BCC-algebra.*

Proof. Let $\langle X; \cdot, 0\rangle$ be a weak BCC-algebra and let $X' = X \cup \{\theta\}$, where $\theta \notin X$. Then, as it is not difficult to see, $\langle X'; \star, \theta\rangle$ with the operation

$$x \star y = \begin{cases} xy & \text{if} \quad xy \neq 0, \\ \theta & \text{if} \quad xy = 0, \\ \theta & \text{if} \quad x = \theta,\ y \in X', \\ x & \text{if} \quad x \in X',\ y = \theta, \end{cases}$$

is a BCC-algebra. □

Example 1.2.14. Using the last construction, we can extend the weak BCC-algebra $\langle\{0,1,2,3,4\}\ \cdot, 0\rangle$ with the operation $\cdot$ defined by Table 1.2.3 (in fact it is a BCI-algebra (cf. [70])) into the following BCC-algebra:

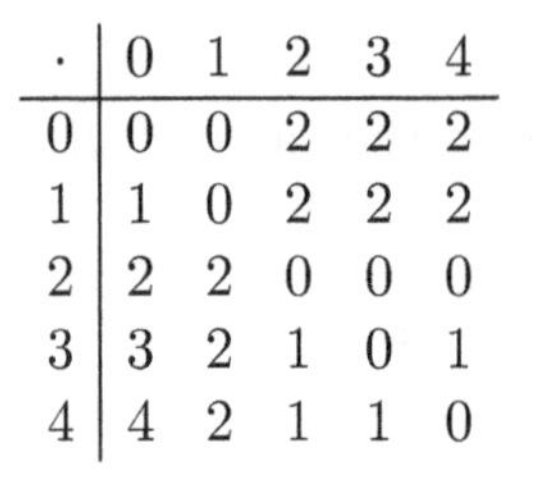

$\cdot$	0	1	2	3	4
0	0	0	2	2	2
1	1	0	2	2	2
2	2	2	0	0	0
3	3	2	1	0	1
4	4	2	1	1	0

Table 1.2.3

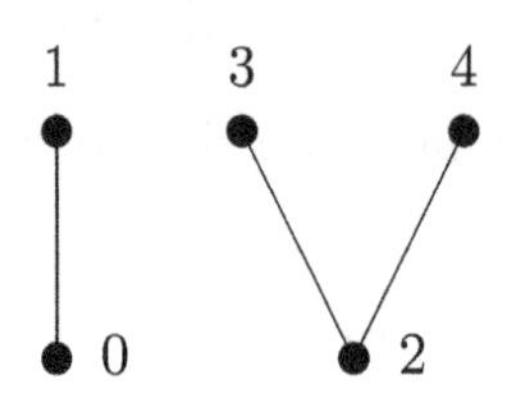

Diagram 1.2.6

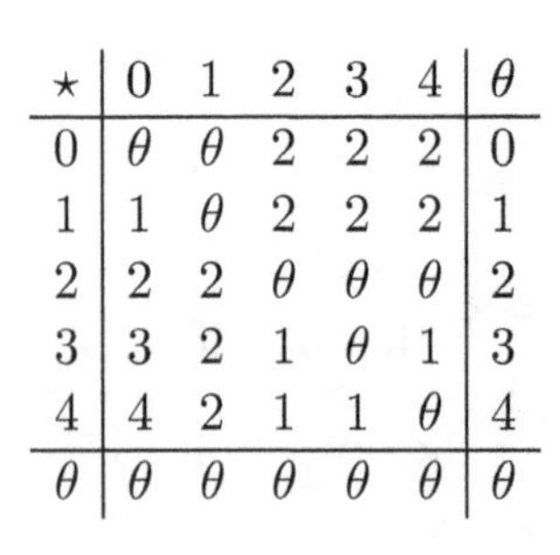

$\star$	0	1	2	3	4	θ
0	θ	θ	2	2	2	0
1	1	θ	2	2	2	1
2	2	2	θ	θ	θ	2
3	3	2	1	θ	1	3
4	4	2	1	1	θ	4
θ	θ	θ	θ	θ	θ	θ

Extended Table 1.2.3

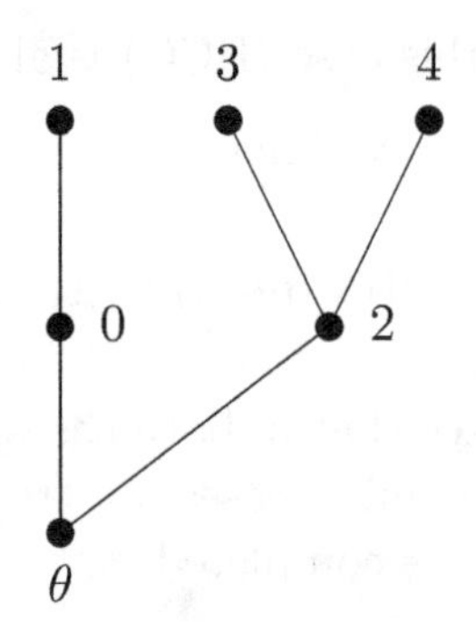

Extended Diagram 1.2.6 □

We summarize the above results in the following theorem:

Theorem 1.2.15. *For any cardinal $n \geq 4$ there exists a proper BCC-algebra which has the order n.*

In the next example we show a construction made for BCK-algebras and generalized to BCC-algebras, whose correctness is not difficult to prove.

Example 1.2.16. If $\{\langle X_\xi, \cdot_\xi, 0_\xi\rangle\}_{\xi\in\Lambda}$ is a nonempty family of BCC-algebras where Λ is a totally ordered set with distinguished element α, then the set $X = \bigcup_{\xi\in\Lambda} X_\xi$ with the operation $\star$ defined by the formula

$$x \star y = \begin{cases} x \cdot_\alpha y & \text{if} \quad x, y \in X_\alpha, \\ x \cdot_\xi y & \text{if} \quad x, y \in X_\xi, x \cdot_\xi y \neq 0_\xi,\ \alpha \neq \xi, \\ 0_\alpha & \text{if} \quad x, y \in X_\xi, x \cdot_\xi y = 0_\xi, \alpha \neq \xi, \\ 0_\alpha & \text{if} \quad x \in X_\xi, y \in X_\eta, \xi < \eta, \\ x & \text{if} \quad x \in X_\eta, y \in X_\xi, \xi < \eta, \end{cases}$$

is a BCC-algebra. □

The natural order $\leq$, induced by this operation, coincides with the original order of each X_ξ. Obviously, $x \leq y$ for all $x \in X_\xi$ and $y \in X_\eta$, where $\xi < \eta$. Moreover, if at least one of the BCC-algebras $\langle X_\eta;\ \cdot_\eta, 0_\eta\rangle$ is proper, then the BCC-algebra $\langle X;\ \star, 0_\alpha\rangle$ is proper, too.

Let us note that this construction is not a generalization of a construction from Lemma 1.2.10.

Finally we show some examples of BCC-algebras. We begin with the direct product of a family of BCC-algebras.

Example 1.2.17. Let $\{X_t\}_{t\in T}$ be a nonempty family of BCC-algebras. We consider the direct product $\prod X_t$. For any fixed $t \in T$ let a_t be an element of $\prod X_t$ such that $a_t(s) = 0$ for any $s \neq t$ and $a_t(t) = a \in X_t$. Then for each $t \in T$, the set $\overline{X}_t = \{a_t \,|\, a \in X_t\}$ is a subalgebra of $\prod X_t$ isomorphic to X_t. If $t \neq s$, then $a_t a_s = a_t$ and $\overline{X_t} \cap \overline{X_s} = \{0\}$. Hence

$$\bigcup_{t\in T} \overline{X_t} = \sum_{t\in T} \overline{X_t}$$

and $\bigcup_{t\in T} \overline{X_t}$ is a subalgebra of $\prod X_t$. Moreover,

$$p_K(\bigcup_{t\in T} \overline{X_t}) = X_K$$

for any projection p_K.

Then

$$\bigcup_{t\in T} \overline{X_t} \cong \sum_{t\in T} X_t$$

and $\sum_{t\in T} X_t$ is a subdirect product of $\{X_t\}_{t\in T}$.

If we identify each X_t with $\overline{X_t}$, then $\sum_{t\in T} X_t$ is the smallest subalgebra of $\prod X_t$ containing all X_t. □

Example 1.2.18. Let $\{X_t\}_{t\in T}$ be a nonempty family of BCC-algebras such that $X_i \cap X_j = \{0\}$ for any distinct $i, j \in T$. We assume that any set of indexes has at least two elements. In the union $\bigcup_{t\in T} X_t$ we define an operation $\cdot$, which coincides with the original BCC-operation in A_t for any $t \in T$.
If x and y belong to distinct BCC-algebras, then we define $xy = x$. The union $\bigcup_{t\in T} X_t$ is a BCC-algebra with respect to the operation $\cdot$. This BCC-algebra is denoted by

$$\sum_{t\in T} X_t = \langle \bigcup_{t\in T} X_t; \cdot, 0\rangle.$$

In the general case, where $\{X_t\}_{t\in T}$ is an arbitrary nonempty family of BCC-algebras, we consider $\{X_t \times \{t\}\}_{t\in T}$ and identify all $(0_t, t)$. By identifying each X_t with $X_t \times \{t\}$, the assumption of the definition mentioned above is satisfied. Consequently, we can define the union $\sum_{t\in T} X_t$. □

Example 1.2.19. Let $\mathfrak{X} = \langle X; \cdot_X, 0_X\rangle$ and $\mathfrak{Y} = \langle Y; \cdot_Y, 0_Y\rangle$ be BCC-algebras and let $X \cap Y = \emptyset$. We define the operation $\cdot_{X\cup Y}$ in the following way:

$$x \cdot_{X\cup Y} y = \begin{cases} x \cdot_X y & \text{if} \quad x, y \in X, \\ x & \text{if} \quad x \in X, y \in Y, \\ 0_X & \text{if} \quad x \in Y,\ y \in X, \\ x \cdot_Y y & \text{if} \quad x, y \in Y,\ xy \neq 0_Y, \\ 0_X & \text{if} \quad x, y \in Y,\ xy = 0_Y. \end{cases}$$

It is easy to verify that the algebra $\langle X \cup Y; \cdot_{X\cup Y}, 0_X\rangle$ is a BCC-algebra whose natural order coincides with the original order on X and Y. Moreover, $x \leq_{\cdot_{X\cup Y}} y$ for any $x \in X$ and $y \in Y$. □

The purpose of the next part of this section is to present a theorem that says that every BCC-algebra of a finite order contains a subalgebra containing one element less (cf. [31]). Moreover, we will show all nonisomorphic BCC-algebras of order 4.

We define the matrix A_n as follows:

$$A_n = \begin{pmatrix} a_1a_1 & a_1a_2 & \dots & a_1a_n \\ a_2a_1 & a_2a_2 & \dots & a_2a_n \\ \vdots & \vdots & & \vdots \\ a_na_1 & a_na_2 & \dots & a_na_n \end{pmatrix}.$$

Let $\mathfrak{X}_n = \langle \{a_1, a_2, ..., a_n\}; \cdot, 0\rangle$ be a BCC-algebra such that $a_i \neq a_j$ for $i \neq j$. Without loss of generality we can put $a_1 = 0$. Then the matrix A_n, by (BCC2), (BCC3) and (BCC4), has the form

$$A_n = \begin{pmatrix} 0 & 0 & \dots & 0 \\ a_2 & 0 & \dots & a_2a_n \\ \vdots & \vdots & & \vdots \\ a_n & a_na_2 & \dots & 0 \end{pmatrix}.$$

If in the i-th row of this matrix we have $a_ia_j = 0$ if and only if $i = j$, then this row is called *proper*.

Lemma 1.2.20. *For a BCC-algebra $\mathfrak{X}_n$ (*for $n \geq 2$*) the matrix A_n has at least one proper row.*

Proof. The method of the proof is an induction on n. For $n = 2$, $0 \neq a_2$ and A_n has the form

$$A_2 = \begin{pmatrix} a_1a_1 & a_1a_2 \\ a_2a_1 & a_2a_2 \end{pmatrix} = \begin{pmatrix} 0 & 0 \\ a_2 & 0 \end{pmatrix}.$$

As it can be easily seen, for $n = 2$ the lemma is true.

We assume that the lemma is true for $n - 1$, i.e., the following matrix:

$$A_{n-1} = \begin{pmatrix} a_1a_1 & a_1a_2 & \dots & a_1a_{n-1} \\ a_2a_1 & a_2a_2 & \dots & a_2a_{n-1} \\ \vdots & \vdots & & \vdots \\ a_{n-1}a_1 & a_{n-1}a_2 & \dots & a_{n-1}a_{n-1} \end{pmatrix}$$

$$= \begin{pmatrix} 0 & 0 & \dots & 0 \\ a_2 & 0 & \dots & a_2a_{n-1} \\ \vdots & \vdots & & \vdots \\ a_{n-1} & a_{n-1}a_2 & \dots & 0 \end{pmatrix}$$

has at least one proper row.

Without loss of generality, we can assume that the second row of A_{n-1} is proper, i.e., $a_2 a_i \neq 0$ for $i = 1, 3, ..., n-1$.

If $a_2 a_n \neq 0$, then the second row of A_n is proper.

Let us assume that the second row of A_n is not proper, i.e., $a_2 a_n = 0$. Then by Theorem 1.1.30(9) $a_n a_2 \neq 0$, because, from definition of $\mathfrak{X}_n$, $a_2 \neq a_n$.

But in this case $a_n a_i \neq 0$ for all $i = 1, 2, ..., n-1$. Indeed, if $a_n a_i = 0$ for some $i = 1, 3, ..., n-1$, then we have $a_2 \leq a_n$ and $a_n \leq a_i$. Hence, by the transitivity of $\leq$, we get $a_2 \leq a_i$, i.e., $a_2 a_i = 0$, which is a contradiction.

This proves that in A_n the second or the n-th row is proper.

The proof is completed. □

Theorem 1.2.21. *Every BCC-algebra of the order $n+1$, where $n \geq 1$, contains a subalgebra of order n* (for $n \geq 1$)*.*

Proof. Let $\mathfrak{X}$ with $X = \{0, a_1, a_2, ..., a_n\}$ be a BCC-algebra of the order $n+1$. Then by Lemma 1.2.20, the matrix A_n has at least one proper row. Without loss of generality, we can assume that the n-th row is proper, i.e., $a_n a_i \neq 0$ for all i= 1, 2, ... , n-1. Then $\mathfrak{S}$ with $S = \{0, a_1, a_2, ..., a_{n-1}\}$ is a subalgebra of $\mathfrak{X}$. Indeed, if $\mathfrak{S}$ is not a subalgebra, then it is not closed under multiplication, i.e., there exist $a_i, a_j \in S$ such that $a_i a_j = a_n$. Hence by Theorem 1.1.30(10) $a_n a_i = 0$, which is impossible because the n-th row is proper. □

Corollary 1.2.22. *Every BCC-algebra of the order $n+1$, where $n \geq 1$, contains at least one subalgebra of the order $i = 1, 2, ..., n-1$.*

It can be easily seen that there are only three nonisomorphic BCK-algebras of order 3. Their multiplication tables are as follows:

·	0	1	2
0	0	0	0
1	1	0	0
2	2	1	0

Table 1.2.4

·	0	1	2
0	0	0	0
1	1	0	0
2	2	2	0

Table 1.2.5

·	0	1	2
0	0	0	0
1	1	0	1
2	2	2	0

Table 1.2.6

On the other hand, by Theorem 1.2.6, every subalgebra of a proper BCC-algebra of order 4 is a BCK-algebra. Moreover, from Theorem 1.2.21 immediately follows that a BCC-algebra of order 4 has at least one subalgebra with three elements. Obviously this subalgebra is a BCK-algebra isomorphic to one of BCK-algebras given above. Hence every BCC-algebra of order 4 is isomorphic to one-element extension of these BCK-algebras.

Direct but very long computations show that there exist 45 distinct proper BCC-algebras defined on the set $S = \{0, 1, 2, 3\}$. These BCC-algebras are isomorphic to one of the following:

$\cdot$	0	1	2	3
0	0	0	0	0
1	1	0	0	0
2	2	1	0	0
3	3	2	2	0

Table 1.2.7

$\cdot$	0	1	2	3
0	0	0	0	0
1	1	0	0	0
2	2	2	0	0
3	3	2	2	0

Table 1.2.8

$\cdot$	0	1	2	3
0	0	0	0	0
1	1	0	0	0
2	2	2	0	0
3	3	3	1	0

Table 1.2.9

$\cdot$	0	1	2	3
0	0	0	0	0
1	1	0	0	0
2	2	2	0	1
3	3	3	3	0

Table 1.2.10

$\cdot$	0	1	2	3
0	0	0	0	0
1	1	0	0	1
2	2	2	0	1
3	3	3	3	0

Table 1.2.11

$\cdot$	0	1	2	3
0	0	0	0	0
1	1	0	0	1
2	2	1	0	1
3	3	3	3	0

Table 1.2.12

$\cdot$	0	1	2	3
0	0	0	0	0
1	1	0	1	0
2	2	2	0	0
3	3	3	1	0

Table 1.2.13

$\cdot$	0	1	2	3
0	0	0	0	0
1	1	0	1	0
2	2	2	0	1
3	3	3	1	0

Table 1.2.14

The BCC-algebra defined by Table 1.2.14 is isomorphic to three BCC-algebras, defined on the set $S = \{0, 1, 2, 3\}$. Each of the other BCC-algebras is isomorphic to six BCC-algebras. Moreover, in BCC-algebras given by Tables 1.2.7, 1.2.8 and 1.2.9, any two elements are comparable.

We will now look at bounded BCC-algebras and show a method and examples of their construction.

First, let us remind that a BCC-algebra $\mathfrak{X}$ is called *bounded* if there is an element $1 \in X$ such that $x \leq 1$ for all $x \in X$, i.e., 1 is the greatest element of X.

Since $0 \leq 1$, 1 belongs to the BCC-part of $\mathfrak{X}$. This means that a bounded weak BCC-algebra is not proper. Indeed, let $x \in X$. Then $x \leq 1$ implies

$$0 = x1 \cdot 01 \leq x$$

by (BCC1$'$). So, we have $0 \leq x$ and x belongs to BCC-part $B(0)$ of $\mathfrak{X}$. Hence $\mathfrak{B}(0) = \mathfrak{X}$. Thus $\mathfrak{X}$ is a bounded BCC-algebra.

The BCC-algebras with their operations defined by Tables 1.2.7, 1.2.8, 1.2.9 and 1.2.13 are examples of bounded BCC-algebras.

Corollary 1.2.23. *If* $\mathfrak{X} = \langle X; \cdot, 0\rangle$ *is a BCC-algebra and* $1 \notin X$, *then the algebra* $\mathfrak{X}' = \langle X'; \star, 0\rangle$, *where* $X' = X \cup \{1\}$ *with the operation* $\star$ *defined in the following way:*

$$x \star y = \begin{cases} xy & \text{if } \; x, y \in X, \\ 0 & \text{if } \; x \in X', \; y = 1, \\ 1 & \text{if } \; x = 1, \; y \in X, \end{cases}$$

is a BCC-algebra with 1 as the greatest element. This BCC-algebra $\mathfrak{X}'$ *is proper if and only if* $\mathfrak{X}$ *is proper.*

Proof. Since for every $x \in X'$ we have $x \star 1 = 0$, 1 is the greatest element of X'. We show only that the algebra $\langle X'; \star, 0\rangle$ satisfies the axiom (BCC1). It can be easily checked that the other axioms are satisfied.

For $x, y, z \in X$ (BCC1), of course, holds.

If $x = 1$, then the left side of (BCC1) takes the form

$$((1 \star y) \star (z \star y)) \star (1 \star z) = (1 \star (z \star y)) \star 1 = (1 \star zy) \star 1 = 1 \star 1 = 0.$$

Thus (BCC1) is, in this case, satisfied.

For $y = 1$ (BCC1) is satisfied because $x \star 1 = z \star 1 = 0$.

Now, let $z = 1$, then

$$((x \star y) \star (1 \star y)) \star (x \star 1) = xy \star 1 = 0.$$

The other cases are very simple.

So, (BCC1) is satisfied in $\mathfrak{X}'$ and $\mathfrak{X}'$ is a BCC-algebra.

It can be easily shown that in both cases, when $x = 1$ and when $y = 1$, condition (1.2) is satisfied. Indeed, we have

$$(1 \star (1 \star y)) \star y = (1 \star 1) \star y = 0 \star y = 0y = 0$$

and

$$(x \star (x \star 1)) \star 1 = (x \star 0) \star 1 = x0 \star 1 = x \star 1 = 0.$$

So, we can conclude that $\mathfrak{X}^{'}$ is proper if and only if $\mathfrak{X}$ is proper.

The proof is completed. □

Theorem 1.2.24. *Let $\langle X; \cdot, 0\rangle$ be a bounded BCC-algebra with x_0 as the greatest element. If $a \notin X$, then the set $X^{'} = X \cup a$ with the operation $\star$ defined in the following way:*

$$x \star y = \begin{cases} xy & \text{if} \quad x, y \in X, \\ 0 & \text{if} \quad x \in X^{'},\ y = a, \\ a & \text{if} \quad x = a,\ y = 0, \\ x_0 & \text{if} \quad x = a,\ y \in X \setminus \{0\}, \end{cases}$$

is a proper BCC-algebra with a as the greatest element.

Proof. We verify only the axiom (BCC1). For $x, y, z \in X$ the axiom is satisfied because $\langle X; \cdot, 0\rangle$ is a BCC-algebra.

Also in the case when $y = a$ or $z = a$, (BCC1) holds.

If $x = a$, then for $y \in X \setminus \{0\}$, we have

$$((a \star y) \star (z \star y)) \star (a \star z) = (x_0 \star (z \star y)) \star x_0 = (x_0 \cdot zy)x_0 = 0,$$

by Theorem 1.1.30(4).

The other cases are very simple.

Thus (BCC1) is satisfied for all $x, y, z \in X^{'}$.

Since for $y \in X \setminus \{0, x_0\}$,

$$(a \star (a \star y)) \star y = (a \star x_0) \star y = x_0 \star y \neq 0,$$

$\langle X^{'}; \star, 0\rangle$ is a proper BCC-algebra by Corollary 1.1.23, which finishes the proof. □

Corollary 1.2.25. *For any natural $n \geq 4$ there exists a bounded proper BCC-algebra with n elements.*

Proof. This corollary is a simple consequence of the above theorem and examples given by Tables 1.2.4 and 1.2.5. □

Theorem 1.2.26. *Let $\langle X;\cdot,0\rangle$ be a BCC-algebra with an element $x_0 \neq 0$ such that $x_0 \leq x$ for all $x \in X\setminus\{0\}$. If $a \notin X$, then the set $X' = X \cup \{a\}$ with the operation $\star$ defined as follows:*

$$x \star y = \begin{cases} xy & \text{if} \quad x, y \in X, \\ 0 & \text{if} \quad x = 0,\ y = a, \\ x_0 & \text{if} \quad x \in X\setminus\{0\},\ y = a, \\ 0 & \text{if} \quad x = y = a, \\ a & \text{if} \quad x = a,\ y \in X, \end{cases}$$

is a proper BCC-algebra.

Proof. In the similar way as in the proof of Theorem 1.2.24 we verify that $\langle X';\star,0\rangle$ is a BCC-algebra. This BCC-algebra is proper. Indeed, for $x \neq 0$ we have

$$(x \star (x \star a)) \star a = (x \star x_0) \star a = xx_0 \star a = x_0 \neq 0,$$

because $xx_0 \neq 0$. □

It can be easily shown that every poset (X, ρ), with the element 0 of X such that $(x, 0) \in \rho$ for all $x \in X$, can be thought of as a BCK-algebra $\mathfrak{X} = \langle X;\cdot,0\rangle$. The operation $\cdot$ we define as follows:

$$x \cdot y = \begin{cases} 0 & \text{if } (x,y) \in \rho, \\ x & \quad \text{otherwise.} \end{cases}$$

Note, that the relation ρ coincides with the natural partial order $\leq$ defined by the BCK-operation.

As it is well-known, every BCK-algebra satisfies the identity

$$x(x \cdot xy) = xy, \tag{1.16}$$

which is a consequence of the inequalities (1.2) and (1.5). It is called the *absorptance* of elements. It holds in some proper BCC-algebras, too. For example, it holds in proper BCC-algebras defined by Tables 1.2.7, 1.2.8 and 1.2.12, but it is not satisfied in the BCC-algebras defined by other tables.

It can be easily seen that if a BCC-algebra $\langle X;\cdot,0\rangle$ satisfies (1.16), then a new BCC-algebra $\langle X';\star,0\rangle$, constructed by the method given in Lemma 1.2.10, Corollary 1.2.23 and Theorem 1.2.24, also satisfies

(1.16). Moreover, if (1.16) holds in all BCC-algebras $\{(X_\xi, \cdot_\xi, 0_\xi)\}_{\xi\in\Lambda}$, then it holds in a BCC-algebra $\langle X; \star, 0_\alpha\rangle$ constructed in Example 1.2.16.

So we get the following corollary:

Corollary 1.2.27. *For any cardinal $n \geq 4$ there exists a proper BCC-algebra satisfying* (1.16) *which has n elements.*

1.3 Estimating the Number of Subalgebras

In this section we give a method for estimating the number of BCC-subalgebras with i-elements in proper BCC-algebras with n-elements. W.A. Dudek explored this subject in his works (cf. [31]). In [30], he posed two problems concerning a full characterization of finite BCC-algebras such that the number of subalgebras is equal to the number of ways for selecting $i-1$ elements from $n-1$ nonzero elements and of proper BCC-algebras having only improper BCC-algebras (BCK-algebras) as subalgebras.

The problems were partially solved in [41] and finally in [13].

Subalgebras of BCC-algebras are of course BCC-algebras, but there are proper BCC-algebras in which all (or only some) subalgebras are not proper.

Example 1.3.1. It is not difficult to verify that the set $X = \{0, a, b, c, 1\}$ with the operation $\cdot$ defined by the following table:

$\cdot$	0	a	b	c	1
0	0	0	0	0	0
a	a	0	a	0	0
b	b	b	0	0	0
c	c	b	a	0	0
1	1	c	c	c	0

Table 1.3.1

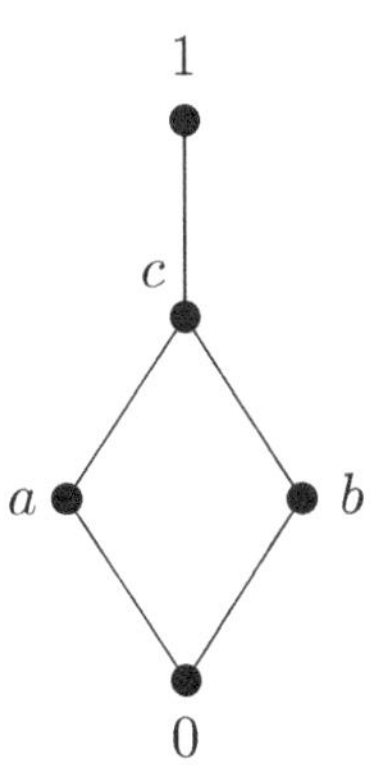

Diagram 1.3.1

is a BCC-algebra. It is proper since $1b \cdot a \neq 1a \cdot b$. In this BCC-algebra all proper subalgebras are BCK-algebras. □

Let $N(i)$ denote the *number of subalgebras* of the order i of $\mathfrak{X}$. Let us recall that the formula for an n-element set, the number of possible combinations of subsets with i elements, where $0 \leq i \leq n$ is the following:

$$C_n^i = \frac{n!}{i!(n-i)!}.$$

Example 1.3.2. The set $X = \{0, 1, 2, 3, 4\}$ with the multiplication defined by Table 1.3.2 is a proper BCC-algebra because $(2{\cdot}1){\cdot}4 \neq (2{\cdot}4){\cdot}1$.

These are the subalgebras of this BCC-algebra:
$\{0\}, \{0,1\}, \{0,2\}, \{0,3\}, \{0,4\}, \{0,1,2\}, \{0,1,4\}, \{0,1,2,3\}, \{0,1,2,4\}$, but only the last is not a BCK-algebra. It is isomorphic to a proper BCC-algebra with an operation defined by Table 1.2.10.

Hence $N(2) = 4 = C_{5-1}^{2-1}$, $N(3) = 2 < C_{5-1}^{3-1}$, $N(4) = 2 < C_{5-1}^{4-1}$.

·	0	1	2	3	4
0	0	0	0	0	0
1	1	0	0	0	0
2	2	2	0	0	1
3	3	2	1	0	1
4	4	4	4	4	0

Table 1.3.2 □

Since for $1 \leq i \leq n$ a subalgebra of the order i contains 0 and $i-1$ nonzero elements, by Theorem 1.2.21, we have $1 \leq N(i) \leq C_{n-1}^{i-1}$, where C_{n-1}^{i-1} denotes the number of ways for selecting $i-1$ elements from $n-1$ nonzero elements.

It is clear that every nonzero element of a BCC-algebra together with 0 forms a subalgebra of order 2. Hence a BCC-algebra of order $n \geq 2$ contains $N(2) = C_{n-1}^{2-1} = n-1$ subalgebras of order 2.

On the other hand, for the BCC-algebra defined in Example 1.3.1 we have $N(3) < C_{5-1}^{3-1}$ and $N(4) < C_{5-1}^{4-1}$. The situation is similar for the BCC-algebra defined in Example 1.3.2. But in the BCC-algebra $\mathfrak{X}_n = \langle\{0, 1, 2, ..., n-1\}; \star, 0\rangle$ with the operation $\star$ defined as follows:

$$x \star y = \begin{cases} x & \text{if } x > y, \\ 0 & \text{otherwise} \end{cases}$$

every subset containing 0 is a subalgebra. Thus in this BCC-algebra we have $N(i) = C_{n-1}^{i-1}$ for every $i = 1, 2, ..., n-1$.

A BCC-algebra with the operation $\star$ defined above is called a BCC-algebra with *trivial structure.*

Theorem 1.3.3. *If for some fixed $i \geq 2$ every subset of a BCC-algebra $\mathfrak{X}$ containing 0 and i nonzero elements is a subalgebra, then every subset of X containing 0 is a subalgebra.*

Proof. Let $M = \{0, a_1, a_2, ..., a_{i+1}\}$ be an arbitrary subset of X, where $\mathfrak{X} = \langle X; \cdot, 0\rangle$ is a BCC-algebra with i satisfying the assumption. Then
$S_1 = \{0, a_2, a_3, ..., a_{i+1}\}$,
$S_2 = \{0, a_1, a_3, ..., a_{i+1}\}$ and
$S_3 = \{0, a_1, a_2, a_4, ..., a_{i+1}\}$
are subalgebras. Thus for all $x, y \in M = S_1 \cup S_2 \cup S_3$ we have $xy \in M$, which proves the M is a subalgebra. Hence, by induction, every subset containing 0 and j nonzero elements for $j \geq i$ is a subalgebra.

All subsets containing 0 and $j < i$ nonzero elements are subalgebras, too. Indeed, if some $S_j = \{0, a_1, a_2, ..., a_j\}$ is not a subalgebra, then there exist $x, y \in S_j$ such that $xy = z \neq a_k$ for any $a_k \in S_j$.

Thus $M = S_j \cup (\{a_{j+1}, ..., a_i\} \setminus \{z\})$ containing 0 and i nonzero elements is not a subalgebra, which as a contradiction. □

Corollary 1.3.4. *In a BCC-algebra all subsets containing 0 are subalgebras if and only if every subset containing 0 and two nonzero elements is a subalgebra.*

Corollary 1.3.5. *In a BCC-algebra of the order $n > 3$ for every $i = 3, ..., n$ we have either $N(i) = C_{n-1}^{i-1}$ or $N(i) < C_{n-1}^{i-1}$.*

As a consequence of the above results we obtain the following theorem:

Theorem 1.3.6. *A finite BCC-algebra of order n in which for some $3 \leq i \leq n$ the condition $N(i) = C_{n-1}^{i-1}$ holds, is a BCK-algebra.*

Proof. Indeed, if $N(i) = C_{n-1}^{i-1}$ for some i, then for $x, y \in X$ every set of the form $\{0, x, y\}$ is a subalgebra of $\mathfrak{X}$. By Theorem 1.2.6 it is a BCK-algebra. This means that (1.2) holds for all x, y. Hence this BCC-algebra is a BCK-algebra. □

Theorem 1.3.7. *Let $\mathfrak{X}$ be a proper BCC-algebra of order $n \geq 5$. If each proper subalgebra of $\mathfrak{X}$ is a BCK-algebra, then*

$$N(i) \leq C_{n-1}^{i-1} - C_{n-4}^{i-4}$$

for all $4 \leq i < n$.

Proof. Let, by contrary, there exists $4 \leq i' < n$ such that

$$N(i') > C_{n-1}^{i'-1} - C_{n-4}^{i'-4}.$$

Since $\mathfrak{X}$ is a proper BCC-algebra, (1.6) is not satisfied for some $a, b, c \in X$. Let us consider two following families of sets:

$A' = \{A \mid A \subseteq X,\ Card(A) = i',\ 0 \in A$ and A includes at most two elements of the set $\{a, b, c\}\}$ and
$B' = \{B \mid B \subseteq X,\ Card(B) = i'$ and $\{0, a, b, c\} \subseteq B\}$.

By assumptions $n \geq 5$ and $i' < n$. So, $A' \neq \emptyset$ and $B' \neq \emptyset$. Moreover, $A' \bigcap B' = \emptyset$.

The following equation is obvious:

$$Card(B') = C_{n-4}^{i'-4}.$$

This implies

$$C_{n-1}^{i'-1} = Card(A' \bigcup B') = Card(A') + Card(B') = Card(A') + C_{n-4}^{i'-4}.$$

So,

$$Card(A') = C_{n-1}^{i'-1} - C_{n-4}^{i'-4} < N(i').$$

This means that there exists a subalgebra $\mathfrak{S}$ with $Card(S) = i'$ such that $S \in B'$. It is because $S \in A' \bigcup B'$ and $S \notin A'$. So, from the definition of B' it follows that the elements $a, b, c \in S$. Since the subalgebra $\mathfrak{S}$ has to be a BCK-algebra, Equation (1.6) is satisfied. This is a contradiction to the assumption. Therefore, for all $4 \leq i \leq n$, we have $N(i) \leq C_{n-1}^{i-1} - C_{n-4}^{i-4}$.

The proof is completed. □

The converse of Theorem 1.3.7 is not true. We show it in the following example:

Example 1.3.8. Let us consider the algebra $\mathfrak{X} = \langle\{0,1,2,3,4\}; \cdot, 0\rangle$ with the operation $\cdot$ defined by Table 1.3.2.
$\mathfrak{X}$ is a proper BCC-algebra. Of course, $4 \leq i < 5$ means $i = 4$. It is easy to show that the only subalgebras of order $i = 4$ are $A = \{0,1,2,3\}$ and $B = \{0,1,2,4\}$. Hence

$$N(i) = N(4) = 2 \leq 3 = C_{5-1}^{4-1} - C_{5-4}^{4-4} = C_{n-1}^{i-1} - C_{n-4}^{i-4},$$

for all $4 \leq i < 5$. But the subalgebra $\mathfrak{B} = \langle\{0,1,2,4\}; \cdot, 0\rangle$ is not a BCK-algebra, because $0 = (2 \cdot 1) \cdot 4 \neq (2 \cdot 4) \cdot 1 = 1$. Therefore, the converse of Theorem 1.3.7 is not true in general. □

In the next corollary, we summarize the obtained results.

Corollary 1.3.9. *Let $\mathfrak{X}$ be a proper BCC-algebra of order $n \geq 5$ such that each proper subalgebra of $\mathfrak{X}$ is a BCK-algebra. Then,*

$$N(i) : \begin{cases} = 1 & \text{if} \quad i = 1, \\ = n - 1 & \text{if} \quad i = 2, \\ \leq C_{n-1}^{i-1} - 1 & \text{if} \quad i = 3, \\ \leq C_{n-1}^{i-1} - C_{n-4}^{i-4} & \text{if} \quad 4 \leq i \leq n-1, \\ = 1 & \text{if} \quad i = n. \end{cases}$$

Theorem 1.3.10. *If in a bounded BCC-algebra $\mathfrak{X}$ for some fixed $i \geq 2$ all subsets of the form $\{0, a_1, a_2, ..., a_i, 1\}$ are subalgebras, then every subset of X containing $0, 1$ and at least two elements is a subalgebra.*

Proof. The proof is a simple modification of the proof of Theorem 1.3.3. □

Definition 1.3.11. A subalgebra $\mathfrak{S}$ of a bounded BCC-algebra $\mathfrak{X}$ is called *extremal* if it contains 1.

It is easy to see that an extremal subalgebra has at least two elements, namely 0 and 1.

The BCC-algebra in Example 1.3.1 has only two extremal subalgebras: $\{0,1\}$ and $\{0,c,1\}$. Indeed, the subsets $S_1 = \{0,a,c,1\}$ and $S_2 = \{0,b,c,1\}$ are not subalgebras, because for $c, a \in S_1$ the product $ca = b \notin S_1$ and for $c, b \in S_2$ the product $cb = a \notin S_2$.

Since $1a = 1b = c$, the subsets $\{0,a,1\}$ and $\{0,b,1\}$ also are not subalgebras.

The algebra from the Example 1.3.2, on the other hand, has no extremal subalgebras because it is not bounded.

Let $N_e(i)$ denote the number of extremal subalgebras of the order $i \geq 2$. Since every such subalgebra contains 0 and 1, $N_e(i) \leq C_{n-2}^{i-2}$ for all bounded BCC-algebras of the order $n \geq 2$.

Corollary 1.3.12. *In a BCC-algebra of the order $n \geq 3$, for every $2 \leq i \leq n$ we have either $N_e(i) = C_{n-2}^{i-2}$ or $N_e(i) < C_{n-2}^{i-2}$.*

Let us note that in the BCC-algebra from Example 1.3.1 we have $N_e(3) = 1 < C_{5-2}^{3-2}$ and $N_e(4) = 0$, but in the above-defined BCC-algebra $\mathfrak{X}_n$ we have $N_e(i) = C_{n-2}^{i-2}$ for all $i = 2, 3, ..., n$.

On the other hand, the set $X = \{0, a, b, c, 1\}$ with the multiplication $\cdot$ defined by the following table:

$\cdot$	0	a	b	c	1
0	0	0	0	0	0
a	a	0	0	a	0
b	b	b	0	a	0
c	c	c	c	0	0
1	1	1	c	a	0

Table 1.3.3

is an example of a BCC-algebra in which $N_e(i) = 1$ for every $i = 2, 3, ..., n$.

2

Special Objects in Weak BCC-Algebras

The purpose of this chapter is to describe so-called solid weak BCC-algebras.

We start out by giving some results about the atom. Next, we go to the initial elements, which are determinants of the branches described later.

Then, starting with the description of the *BCC*-ideals and their connections with congruences, we characterize them and show that their equivalence classes are branches.

It turns out that the set of initial elements creates the greatest group-like BCC-algebra in given weak BCC-algebra. In the section devoted to this interesting subclass of BCC-algebras, we describe its properties and show that they are determined by groups. We also give an interesting result showing that a weak BCC-algebra induced by a commutative group is not proper, it is a BCI-algebra, but if a weak BCC-algebra is induced by a noncommutative group, then it is proper.

Then, having all the components needed to describe solid BCC-algebras, we devote a whole section to them.

2.1 Atoms

We start with the definition of atom in the sense of [59].

Definition 2.1.1. A nonzero element $a \in X$ is called an *atom* of a BCC-algebra $\mathfrak{X}$ if $x \leq a$ implies $x = 0$ or $x = a$ for all $x \in X$.

The set of all atoms of a BCC-algebra $\mathfrak{X}$ is denoted by $A(X)$.

Lemma 2.1.2. *If $a \neq b$ are atoms of a BCC-algebra $\mathfrak{X}$, then $ab = a$.*

Proof. Let $a, b \in X$ be atoms of $\mathfrak{X}$. If $a = 0$ or $b = 0$, then obviously $ab = a$. If $a \neq b$ are nonzero atoms, then $0 \leq b$ and (1.5) imply $ab \leq a$. Since a is an atom of $\mathfrak{X}$, we have $ab = 0$ or $ab = a$. In the first case we obtain $a \leq b$, which is impossible. So, $ab = a$ for all $a, b \in A(X)$, $a \neq b$. □

Since in the above lemma in the equation for $a = 0$ we get the axiom (BCC4), the set $A(X)$ is included in the BCC-algebra.

As a simple consequence of Lemma 2.1.2 we obtain

Corollary 2.1.3. *For any cardinal $n \geq 2$ there exists only one BCC-algebra in which all nonzero elements are atoms.*

For a BCC-algebra $\mathfrak{X}$ we consider the set of head-fixed commutative elements of X, i.e.,

$$HF(X) = \{z \in X \mid zx \cdot y = zy \cdot x \text{ for all } x, y \in X\}.$$

Obviously, $0 \in HF(X)$.

Lemma 2.1.4. *If $a \in HF(X)$, then $a \cdot ax \leq x$ for all $x \in X$.*

Proof. Indeed, $(a \cdot ax)x = ax \cdot ax = 0$. □

Theorem 2.1.5. *$\mathfrak{HF}(X)$ is a BCK-subalgebra of a BCC-algebra $\mathfrak{X}$.*

Proof. Let $\mathfrak{X}$ be a BCC-algebra and let $a, b \in HF(X)$, $x, y \in X$. Then by (BCC1$'$) and Lemma 2.1.4

$$ab \cdot (ab \cdot y) = ab \cdot (ay \cdot b) \leq a \cdot ay \leq y, \text{ i.e., } ab \cdot (ab \cdot y) \leq y.$$

This, by (1.5) and again by (BCC1$'$) gives

$$\begin{aligned}(ab \cdot x)y &\leq (ab \cdot x)(ab \cdot (ab \cdot y)) \\ &= (ax \cdot b)(a(ab \cdot y) \cdot b) \\ &\leq ax \cdot a(ab \cdot y) \\ &= a(a(ab \cdot y)) \cdot x \\ &\leq (ab \cdot y)x,\end{aligned}$$

since $a(a(ab \cdot y)) \leq ab \cdot y$, by Lemma 2.1.4.
Thus

$$(ab \cdot x)y \leq (ab \cdot y)x,$$

which by symmetry gives

$$(ab \cdot y)x \leq (ab \cdot x)y,$$

i.e.,

$$(ab \cdot x)y = (ab \cdot y)x.$$

Hence $ab \in HF(X)$, i.e., $\mathfrak{HF}(X)$ is a BCC-subalgebra of $\mathfrak{X}$. It is also a BCK-subalgebra because by the definition of $HF(X)$, the condition (1.6) is satisfied in $\mathfrak{HF}(X)$. □

Theorem 2.1.6. $\mathfrak{A}(X)$ *is a BCK-subalgebra contained in* $\mathfrak{HF}(X)$.

Proof. $A(X)$ is nonempty because $0 \in A(X)$. By Lemma 2.1.2 it is closed with respect to the BCC-operation.

We prove that $A(X) \subset HF(X)$.

Let $a \in A(X)$. Then for $x, y \in X$ we have the following four cases:

1^0. $a \leq x$ and $a \leq y$,

2^0. $a \leq x$ and $not(a \leq y)$,

3^0. $not(a \leq x)$ and $a \leq y$,

4^0. $not(a \leq x)$ and $not(a \leq y)$.

First, observe that:

(a) If $a \leq x$, then $ax \cdot y = 0$ by (1.1) and (BCC4), for every $y \in X$.

(b) If $a > x$, then $ax = a$, because $ax \leq a$ (by Theorem 1.1.30(4)) and $ax \neq 0$.

Thus in the case 1^0., by (a), we obtain $ax \cdot y = 0 = ay \cdot x$, i.e., $a \in HF(X)$.

In the case 2^0. we have, by (a), $ax \cdot y = 0$ and, by (b) and assumption, $ay \cdot x = ax = 0$, which gives $ax \cdot y = ay \cdot x$.

The case 3^0. is analogous.

In the case 4^0. we have, using (b), $ax \cdot y = ay = a$. Similarly, $ay \cdot x = a$. So, we have $ax \cdot y = ay \cdot x$.

Hence, in any case we obtain $ax \cdot y = ay \cdot x$, which proves that $A(X)$ is contained in $HF(X)$ and, by Lemma 2.1.2, closed under $\cdot$.

Since $\mathfrak{A}(X)$ is a BCK-algebra, it is a BCK-subalgebra of $\mathfrak{HF}(X)$. □

Corollary 2.1.7. *If all nonzero elements of a given BCC-algebra are atoms, then it is a BCK-algebra.*

Corollary 2.1.8. *We proved the existence of the following chain:* $\mathfrak{A}(X)$ *is a BCK-subalgebra of* $\mathfrak{HF}(X)$ *which is a BCK-subalgebra of a BCC-algebra* $\mathfrak{X}$.

Finally, we note that if a BCK-algebra $\mathfrak{X}$ has at least one non-atom element, then $A(X) \neq HF(X) = X$. On the other hand, in some BCC-algebras $A(X) = HF(X) \neq X$. As an example we consider the algebra $\mathfrak{X} = \langle\{0,1,2,3,4,5\};\ \cdot, 0\rangle$ with the operation $\cdot$ defined by the table

·	0	1	2	3	4	5
0	0	0	0	0	0	0
1	1	0	0	0	0	1
2	2	2	0	0	1	1
3	3	2	1	0	1	1
4	4	4	4	4	0	1
5	5	5	5	5	5	0

Table 2.1.1

We know that $\mathfrak{S} = \langle\{0,1,2,3,4\};\ \cdot, 0\rangle$ is a proper BCC-algebra defined by Table 1.3.2. Thus $\mathfrak{X}$ is, as the extension of $\mathfrak{S}$ shown in Theorem 1.2.26, a proper BCC-algebra. In this BCC-algebra $A(X) = HF(X) = \{0,1,5\}$, but in a subalgebra $\mathfrak{S}$ we have $A(S) = \{0,1\} \neq HF(S) = \{0,1,4\}$.

Let us note also that $HF(S)$ is not the greatest BCK-subalgebra of $\mathfrak{S}$ since it does not contain the BCK-subalgebra $\mathfrak{B}(0) = \langle\{0,1,2\};\ \cdot, 0\rangle$ defined in Table 1.2.5.

2.2 Branches

In the investigations of weak BCC-algebras, an important role plays a mapping denoted by φ, which was formally introduced in [57] for decomposition of BCH-algebras, but it was earlier used to investigate various classes of BCI-algebras (cf. [25] and [27]). It was also applied in studies of some subsets of weak BCC-algebras known as *branches*, which are covered later in this section. This function was denoted as $'$ for BZ-algebras in, for example, [165].

We begin with the definition of this map.

Definition 2.2.1. In a weak BCC-algebra $\mathfrak{X}$ we define a mapping φ as $\varphi(x) = 0x$ for every $x \in X$.

Generally this is not the case, although, as we will show later in the book, in some subclasses of weak BCC-algebras the mapping φ is an endomorphism.

The main properties of this mapping, in the case of weak BCC-algebras, are collected in the following lemma.

Lemma 2.2.2. *Let $\mathfrak{X}$ be a weak BCC-algebra. Then*

(1) $\varphi(xy) \leq yx$,
(2) $\varphi^2(x) \leq x$,
(3) $\varphi(x) \cdot yx = \varphi(y)$,
(4) $\varphi(xy)\varphi(x) = \varphi^2(y)$,
(5) $\varphi^3(x) = \varphi(x)$,
(6) $\varphi^2(xy) = \varphi^2(x)\varphi^2(y)$,
(7) $x \leq y \Longrightarrow \varphi(x) = \varphi(y)$,
(8) $\varphi^2(xy) = \varphi(yx)$,
(9) $(xy \cdot z)\varphi(z) \leq x(yz \cdot \varphi(z))$

for all $x, y, z \in X$.

Proof. (1) follows from (BCC2) and (BCC1$'$). Indeed, we have

$$\varphi(xy) = 0 \cdot xy = yy \cdot xy \leq yx.$$

(2) follows from (BCC2), (BCC1$'$) and (BCC3). We have

$$\varphi^2(x) = 0 \cdot 0x = xx \cdot 0x \leq x.$$

To prove (3) let us observe, that from (BCC1$'$) the two conditions follow:

$$\varphi(x) \cdot yx = 0x \cdot yx \leq 0y = \varphi(y), \text{ i.e.,}$$

$$\varphi(x) \cdot yx \leq \varphi(y) \tag{2.1}$$

and

$$yx \cdot \varphi(x) = yx \cdot 0x \leq y, \text{ i.e.,}$$

$$yx \cdot \varphi(x) \leq y.$$

Moreover, $xy \leq z$ implies $uy \cdot z \leq ux$, because, by (1.5) and (BCC1$'$), we have

$$uy \cdot z \leq uy \cdot xy \leq ux.$$

So, we get

$$yx \cdot \varphi(x) \leq y \Longrightarrow \varphi(y) = \varphi(x)\varphi(x) \cdot y \leq \varphi(x) \cdot yx, \text{ i.e.,}$$

$$\varphi(y) \leq \varphi(x) \cdot yx. \tag{2.2}$$

Then (2.1) and (2.2) imply $\varphi(x) \cdot yx = \varphi(y)$, which proves (3).

(4) is a consequence of (3). Indeed, by putting $x = xz$ in (3), we get

$$\varphi(xz) \cdot (y \cdot xz) = \varphi(y).$$

Next, let us put $y = \varphi(y)$, then

$$\varphi(xz) \cdot (\varphi(y) \cdot xz) = \varphi^2(y).$$

And, by putting $z = y$, we get

$$\varphi(xy) \cdot (\varphi(y) \cdot xy) = \varphi^2(y).$$

From (3) we have

$$\varphi(xy)\varphi(x) = \varphi^2(y),$$

which proves (4).

To prove (6), we put $x = \varphi(x)$ in (3) and we get

$$\varphi^2(x) \cdot y\varphi(x) = \varphi(y). \tag{2.3}$$

By putting $y = \varphi(xy)$ in (2.3) we have

$$\varphi^2(x) \cdot \varphi(xy)\varphi(x) = \varphi^2(xy),$$

which, by (4), implies

$$\varphi^2(x)\varphi^2(y) = \varphi^2(xy),$$

which proves (6).

Another consequence of (3) is (7). Indeed, for $yx = 0$ we have $\varphi(y) \cdot 0 = \varphi(x)$. Hence $\varphi(y) = \varphi(x)$.

(5) follows from (2) and (7). We have

$$\varphi^2(x) \leq x \Longrightarrow \varphi(\varphi^2) = \varphi(x) \Longrightarrow \varphi^3(x) = \varphi(x).$$

(8) Using (BCC1$'$), (1.5) we have

$$\varphi^2(xy) = \varphi(0 \cdot xy) = \varphi(yy \cdot xy) \geq \varphi(yx).$$

Next, by (7) and (5)

$$\varphi^2(yx) = \varphi(\varphi^2(xy)) = \varphi^3(xy) = \varphi(xy),$$

which gives (8).

To prove (9) let us observe, that from (BCC1$'$) for any $v \in X$ we have

$$vz \cdot 0z \leq v \Longleftrightarrow vz \cdot \varphi(z) \leq v.$$

Putting $v = xy$, we get

$$(xy \cdot z)\varphi(z) \leq xy. \tag{2.4}$$

On the other hand, from (BCC1$'$) it follows

$$yz \cdot 0z \leq y \Longrightarrow yz \cdot \varphi(z) \leq y,$$

which, by (1.5), implies

$$xy \leq x(yz \cdot \varphi(z)). \tag{2.5}$$

Combining (2.4) with (2.5), we get (9).

The proof is completed. □

Corollary 2.2.3. *The mapping φ^2 is always an endomorphism.*

The next object we will present are *branches*. They play a key role in the research of many subclasses of weak BCC-algebras. We provide their definition and the most important properties.

We begin with the following definition:

Definition 2.2.4. Let $\mathfrak{X}$ be a weak BCC-algebra. By an *initial element* of X we mean an element $a \in X$ such that $x \leq a \Longrightarrow x = a$ for every $x \in X$.

The set of all initial elements of X is denoted by $I(X)$, i.e.,

$$I(X) = \{a \in X \mid x \leq a \Longrightarrow x = a, \text{ for every } x \in X\}.$$

Obviously $0 \in I(X)$.

The set $I(X)$ coincides with the set of all minimal elements of the poset $(X;\leq)$. In this book, however, we will continue using the name *initial element* because it better reflects the role of these elements.

As it can be easily seen,

$$I(X) \cap A(X) = \{0\},$$

where $A(X)$ is the set of all atoms of $\mathfrak{X}$.

Theorem 2.2.5. *Let $\mathfrak{X}$ be a weak BCC-algebra, then*

$$I(X) = \{a \in X \mid \varphi^2(a) = a\}.$$

Proof. Indeed, if a is an initial element of X, then from Lemma 2.2.2(2) we have $\varphi^2(a) \leq a$ which, by Definition 2.2.4, implies $\varphi^2(a) = a$.

Conversely, let $\varphi^2(a) = a$ for some $a \in X$. Assume that $x \leq a$ for some $x \in X$. Then $xa = 0$ and, by (BCC1) and (BCC3),

$$ax = \varphi^2(a)x = (0 \cdot 0a)x = (xa \cdot 0a) \cdot x0 = 0.$$

Thus $ax = xa = 0$. This, by (BCC5), implies $x = a$. So, $a \in I(X)$. □

Theorem 2.2.6. *$I(X) = \varphi(X)$ for any weak BCC-algebra $\mathfrak{X}$.*

Proof. Indeed, if $x \in \varphi(X)$, then $x = \varphi(y)$ for some $y \in X$. Thus, by Lemma 2.2.2(5),

$$\varphi^2(x) = \varphi^3(y) = \varphi(y) = x.$$

Hence $\varphi^2(x) = x$. This means that $x \in I(X)$, by Theorem 2.2.5.
So, $\varphi(X) \subseteq I(X)$.

Conversely, for $x \in I(X)$ we have

$$x = \varphi^2(x) = \varphi(\varphi(x)) = \varphi(y),$$

where $y = \varphi(x) \in X$. Thus $I(X) \subseteq \varphi(X)$, which completes the proof. □

Corollary 2.2.7. *Let $\mathfrak{X}$ be a weak BCC-algebra. For every element $x \in X$ there exists exactly one initial element $a \in I(X)$.*

Theorem 2.2.8. *In a weak BCC-algebra $\mathfrak{X}$ the following conditions are equivalent:*

(1) *a is an initial element,*

(2) *$a = \varphi(x)$ for some x,*

(3) *$\varphi^2(a) = a$,*

(4) *$\varphi(xa) = ax$,*

(5) *$\varphi^2(ax) = ax$,*

(6) *ax is an initial element,*

(7) *$y \leq z$ implies $ay = az$*

for $x, y, z \in X$.

Proof. (1) $\Longleftrightarrow$ (2) It follows from Theorem 2.2.6.

(1) $\Longleftrightarrow$ (3) It follows from Theorem 2.2.5.

(3) $\Longrightarrow$ (4) If $\varphi^2(a) = a$, then for any $x \in X$, by Lemma 2.2.2(8), (6), (2) and (1.5), we have

$$\varphi(xa) = \varphi^2(ax) = \varphi^2(a)\varphi^2(x) = a\varphi^2(x) \geq ax$$

and $ax \leq \varphi(xa)$.

On the other hand, $\varphi(xa) \leq ax$, by Lemma 2.2.2(1). So,

$$\varphi(xa) = ax.$$

(4) $\Longrightarrow$ (5) Indeed, using (4) we obtain

$$\varphi^2(ax) = \varphi(\varphi(ax)) = \varphi(xa) = ax.$$

(5) $\Longrightarrow$ (6) $\varphi^2(ax) = ax$ means, by Theorem 2.2.5, that $ax \in I(X)$. Hence ax is an initial element.

(6) $\Longrightarrow$ (7) For $y \leq z$ we have, by (1.5), $az \leq ay$. But ay is, by the assumption, an initial element for $y \in X$. This implies $ay = az$.

(7) $\Longrightarrow$ (1) $\varphi^2(a) \leq a$ according to Lemma 2.2.2(2). Whence, applying (7), we obtain

$$a\varphi^2(a) = aa = 0.$$

So,

$$a \leq \varphi^2(a).$$

Thus $\varphi^2(a) = a$. Theorem 2.2.5 completes the proof. □

Corollary 2.2.9. *Let $\varphi^2(A) = \{\varphi^2(x) \mid x \in A\}$, where A is a nonempty subset of the base set X of a weak BCC-algebra $\mathfrak{X}$. Then $\varphi^2(A) \subseteq I(X)$ and $\varphi^2(X) = I(X)$.*

Proof. Let us note that for any $x \in \varphi^2(A)$, there exists $y \in A$ such that $x = \varphi^2(y)$. Then, by Lemma 2.2.2(5),

$$\varphi^2(x) = \varphi^4(y) = \varphi^2(y) = x,$$

which, by Theorem 2.2.8(3), implies $\varphi^2(A) \subseteq I(X)$.
Consequently $\varphi^2(X) \subseteq I(X)$. □

Definition 2.2.10. In a weak BCC-algebra $\mathfrak{X}$ the set

$$B(a) = \{x \in X \,|\, a \leq x\},$$

where $a \in X$ is fixed, is called a *branch* of X initiated by a, if for every $b \leq a$ the condition $B(b) = B(a)$ holds. A branch containing only initial element is called *trivial.*

A totally ordered branch is called a *linear branch.* A branch which has at least two incomparable elements is called *expanded.*

A branch $B(a) \subset X$ is called *improper* if there exists a branch $B(b) \subset X$ such that $B(a) \subset B(b)$, i.e., if there exists $b < a$.

In the sequel of the book, we will investigate only *proper* branches, i.e., branches $B(a)$ for a being an initial element. That means only $B(a)$ for $a \in I(X)$.

The branch initiated by 0:

$$B(0) = \{x \in X \,|\, 0 \leq x\},$$

coincides with the BCC-part of a weak BCC-algebra $\mathfrak{X}$.

It is easy to see that the BCC-part is the kernel of the map φ, namely

$$ker\,\varphi = \{x \in X \,|\, \varphi(x) = 0\}.$$

Theorem 2.2.11. *In any weak BCC-algebra* $\mathfrak{X}$

$$B(0) = ker\,\varphi = ker\,\varphi^2.$$

Proof. Of course, $B(0) = ker\,\varphi$. If $x \in ker\,\varphi^2$, then $0 = \varphi^2(x) \leq x$, by Lemma 2.2.2(2). Thus $x \in B(0)$. Consequently, $ker\,\varphi^2 \subseteq B(0)$.

To prove the converse inclusion, observe that for $x \in B(0)$ we have $0 = 0x = \varphi(x)$. This implies $\varphi^2(x) = 0$. So, $B(0) \subseteq ker\,\varphi^2$.
Therefore, $ker\,\varphi^2 = B(0)$. □

We know that each endomorphism h of a weak BCC-algebra $\mathfrak{X}$ saves its element 0, i.e., we have $h(0) = 0$. Thus $x \leq y$ implies $h(x) \leq h(y)$. Moreover, for $a \in I(X)$, by Theorem 2.2.8(3), we have

$$h(a) = h(0 \cdot 0a) = 0 \cdot 0h(a).$$

This, by Theorem 2.2.5, means that

$$h(I(X)) \subset I(X).$$

Consequently,

$$h(B(a)) \subset B(h(a)) \quad \text{and} \quad h(B(0)) \subset B(0).$$

We now present an important result about a some special subalgebra.

Let, for a weak BCC–algebra $\mathfrak{X}$, an algebra $\mathfrak{I}(\mathfrak{X})$ be defined as follows:

$$\mathfrak{I}(\mathfrak{X}) = \langle I(X); \cdot, 0\rangle.$$

Then we have the following:

Theorem 2.2.12. *$\mathfrak{I}(\mathfrak{X})$ is a weak BCC-subalgebra of $\mathfrak{X}$.*

Proof. Obviously $I(X) \subseteq X$ and $0 \in I(X)$. Let $a, b \in I(X)$. Then, from Theorem 2.2.8(3) it follows $a = \varphi^2(a)$ and $b = \varphi^2(b)$. Hence, by Lemma 2.2.2(6),

$$ab = \varphi^2(a) \cdot \varphi^2(b) = \varphi^2(ab).$$

Which means that $ab \in I(X)$. □

Corollary 2.2.13. *Let $\mathfrak{X}$ be a weak BCC-algebra. If $a \in I(X)$, then $x \in B(a)$ if and only if $\varphi(x) = \varphi(a)$.*

Proof. If $\varphi(x) = \varphi(a)$ holds for some $a \in I(X)$ and $x \in X$, then

$$a = \varphi^2(a) = \varphi^2(x) \leq x,$$

from Theorem 2.2.8(3) and Lemma 2.2.2(2). Therefore, $x \in B(a)$.

Conversely, the proof follows directly from Lemma 2.2.2(7). □

Corollary 2.2.14. *Let $\mathfrak{X}$ be a weak BCC-algebra. Then*

$$\varphi(a) = a \text{ if and only if } \varphi(x) \leq x \text{ for every } x \in B(a).$$

Proof. Let $\varphi(a) = a$ for some $a \in X$. Then $a \in I(X)$. Hence for every $x \in B(a)$ we have $a \leq x$. From this, applying Lemma 2.2.2(7), we obtain

$$\varphi(x) = \varphi(a) = a \leq x, \text{ i.e., } \varphi(x) \leq x.$$

Conversely, let $\varphi(x) \leq x$ for any $x \in B(a)$. Then, by Lemma 2.2.2(7),

$$\varphi^2(x) = \varphi(x) \Longleftrightarrow \varphi(a) = a.$$

□

This means that the first row of the Cayley table determining a weak BCC-algebra contains only initial elements.

According to Corollary 2.2.14, each element satisfying the condition $\varphi(a) = a$ is initial, but this condition is not characteristic for initial elements.

Example 2.2.15. Using a computer, one can check that the following table defines an operation of a weak BCC-algebra:

$\cdot$	0	1	2	3	4	5
0	0	0	0	4	3	4
1	1	0	1	4	3	4
2	2	2	0	4	3	4
3	3	3	3	0	4	0
4	4	4	4	3	0	3
5	5	3	5	1	4	0

Table 2.2.1

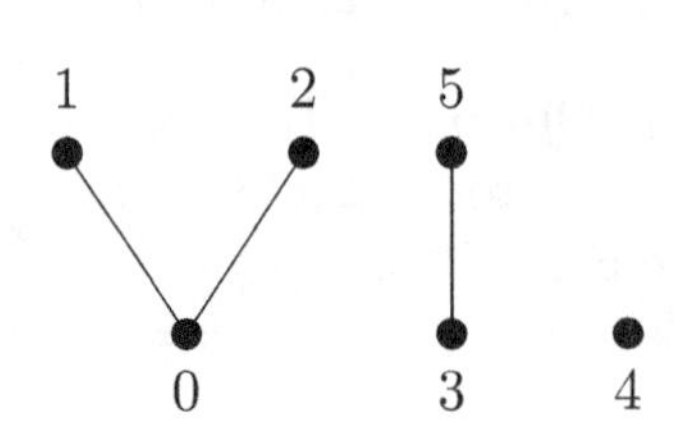

Diagram 2.2.1

This weak BCC-algebra has three initial elements: 0, 3, 4. But $\varphi(3) \neq 3$ and $\varphi(4) \neq 4$. □

Corollary 2.2.16. *Branches initiated by two different initial elements are disjoint.*

Corollary 2.2.17. *In any weak BCC-algebra $\mathfrak{X}$, its base set X is a set-theoretic union of disjoint branches initiated by elements from $I(X)$.*

Example 2.2.18. Let us consider two weak BCC-algebras with their operations defined in Table 1.1.2 and Table 1.1.3. In these algebras we have $B(0) = \{0,1\}$, $B(1) = \{1\}$, $B(2) = \{2,3\}$, $I(X) = \{0,2\}$. The branches $B(0)$ and $B(2)$ are proper, the branch $B(1)$ is improper. $\mathfrak{I}(\mathfrak{X}) = \langle (I(X); \cdot, 0\rangle$ is, of course, a subalgebra of those algebras. □

As it can be easily seen, each branch including a BCC-algebra, is linear or expanded. Let a chain with the smallest element a be denoted by $C(a)$.

Example 2.2.19. Let us consider a set $X = \{0, 1, 2, 3\}$ with the operation $\cdot$ defined by Table 2.2.2.

The algebra $\langle X; \cdot, 0\rangle$ is a proper BCC-algebra. In this algebra $B(0) = X$ and $I(X) = \{0\}$. It is a set-theoretic union of two chains: $C_1(0) = \{0, 1, 2\}$ and $C_2(0) = \{0, 1, 3\}$. As elements 2 and 3 are incomparable, it is an expanded branch. There are no other proper branches. An improper expanded branch $B(1) = \{1, 2, 3\}$ is a union of chains $C_1(1) = \{1, 2\}$ and $C_2(1) = \{1, 3\}$.

$\cdot$	0	1	2	3
0	0	0	0	0
1	1	0	0	0
2	2	2	0	1
3	3	3	1	0

Table 2.2.2

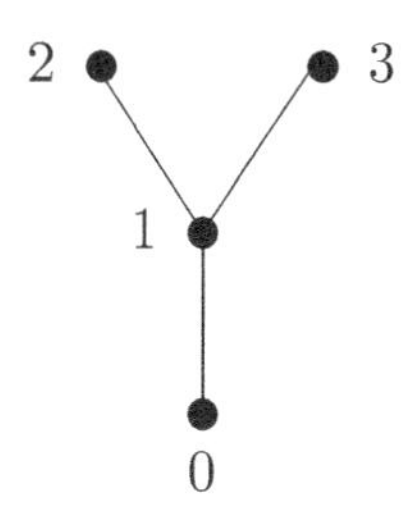

Diagram 2.2.2 □

Example 2.2.20. Let us consider a set $X = \{0, 1, 2, 3\}$ with the operation $\cdot$ defined by the following table:

$\cdot$	0	1	2	3
0	0	0	3	2
1	1	0	3	2
2	2	2	0	3
3	3	3	2	0

Table 2.2.3

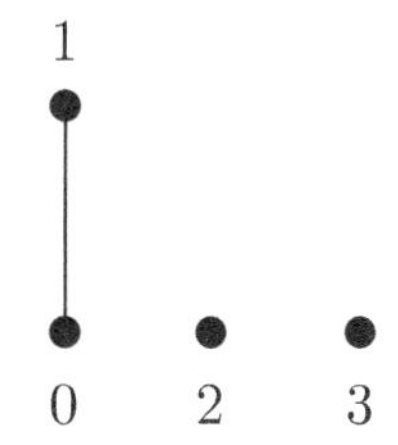

Diagram 2.2.3

It is not difficult to see that $\langle X; \cdot, 0\rangle$ is a weak BCC-algebra. In this case $I(X) = \{0, 2, 3\}$. Two branches are trivial. □

Example 2.2.21. Let us consider the set $X = \{0, 1, 2, 3, 4\}$ with the operation given in Table 1.2.3.
Then $\langle X; \cdot, 0\rangle$ is a weak BCC-algebra with two nontrivial branches $B(0)$ and $B(2)$. □

2.3 BCC-Ideals

In many algebras some special subsets uniquely determine congruences. For example, normal subgroups determine the congruences of groups. In the case of rings, ideals determine the congruences of rings. The situation is similar in BCC-algebras. Here, too, there are strong connections between ideals and congruences.

In this section, we will describe ideals before looking at these connections in the next one. We will use the research method in weak BCC-algebras developed and first published in [61].

K. Iseki and S. Tanaka observed in the mid-1970s that any subset A of X containing 0 and such that $y \in A$, $xy \in A$ imply $x \in A$ determines some congruence on X.

On the other hand, the kernel

$$\rho(0) = \{x \in X \mid (x, 0) \in \rho\}$$

of a congruence ρ on a BCK-algebra X has the following property:

$$y \in \rho(0), xy \in \rho(0) \text{ imply } x \in \rho(0).$$

This means that some special subsets of BCI- and BCK-algebras, so-called BCK-ideals, determine congruences.

For weak BCC-algebras the situation is more complex. In this case, some congruences are not determined by BCK-ideals. Hence the need arises to create an ideal better suited to BCC-algebras, the so-called BCC-ideal.

In the sequel, we consider the BCC-ideals and show that congruences and their relations to BCK-ideals determine them. We also present their properties, the knowledge of which is necessary to understand the following sections of the chapter.

We begin with the definition of the BCK-ideals.

Definition 2.3.1. Let $\mathfrak{X}$ be a weak BCC-algebra. A nonempty subset A of X is called *a BCK-ideal* of $\mathfrak{X}$ if

$(\mathrm{I1})_{BCK}$ $0 \in A$,

$(\mathrm{I2})_{BCK}$ $xy \in A$ and $y \in A \Longrightarrow x \in A$.

For a BCK-ideal A of $\mathfrak{X}$ we define a relation $\theta_{BCK} \subseteq X^2$ in the following way:

$$(x, y) \in \theta_{BCK} \iff xy,\ yx \in A. \tag{2.6}$$

It is well known in the theory of BCK-algebras, that if A is a BCK-ideal of a BCK-algebra $\mathfrak{X}$, then the relation θ_{BCK} is a congruence (cf. [77]).

This result is not true for BCC-algebras. We show it in the following example:

Example 2.3.2. Let $\mathfrak{X} = \langle X; \cdot, 0\rangle$, where $X = \{0, 1, 2, 3, 4\}$ and the binary operation $\cdot$ is defined by Table 2.3.1.

$\cdot$	0	1	2	3	4
0	0	0	0	0	0
1	1	0	1	0	0
2	2	2	0	0	0
3	3	3	1	0	0
4	4	3	4	3	0

Table 2.3.1

We prove that this algebra is a BCC-algebra. It is clear that such algebra satisfies (BCC3), (BCC4) and (BCC5). We prove (BCC1).

If x, y, z are not different, then obviously (BCC1) holds.

For different x, y, z we verify only the case when one of elements x, y, z is equal to 4 because $S = \{0, 1, 2, 3\}$ is a BCC-algebra, see Table 1.2.13.

Since for $y \in S$ we have $4y \in \{3, 4\}$ and $u3 = u4 = 0$ for all $u \in S$, (BCC1) holds for $z = 4$.

For $y = 4$ it holds, too.

For $x = 4$ the axiom (BCC1$'$) has the form $4y \cdot zy \leq 4z$, which for $z = 4$ is clearly satisfied. So, we have to check (BCC1$'$) only the cases when $z \in S$.

For $y = 1$ it can be easily checked that $(4 \cdot 1) \cdot z1 \leq 4z$. For $y = 2$ we have $4 \cdot z2 \leq 4$ for all $z \in S$. If $y = 3$, then $(4 \cdot 3) \cdot z3 = 3 \leq 4$.

The other cases are very simple.

This completes the proof that $\mathfrak{X}$ is a BCC-algebra.

It is not difficult to verify that $A = \{0, 1\}$ is a BCK-ideal of this BCC-algebra, but the relation θ_{BCK} defined by this ideal is not a congruence.

Indeed, $(4, 4) \in \theta_{BCK}$ and $(2, 3) \in \theta_{BCK}$. But $(4 \cdot 2, 4 \cdot 3) \notin \theta_{BCK}$ because $(4 \cdot 2) \cdot (4 \cdot 3) = 3 \notin A$. □

In connection with this, W.A. Dudek and X.H. Zhang introduced a new concept of ideals in weak BCC-algebras.

Definition 2.3.3. Let $\mathfrak{X}$ be a weak BCC-algebra. A nonempty subset A of X is called *a BCC-ideal* if the following conditions are satisfied:

(I1) $0 \in A$,

(I2) $xy \cdot z \in A$ and $y \in A \Longrightarrow xz \in A$ for all $x, y, z \in X$.

Theorem 2.3.4. *In a weak BCC-algebra any BCC-ideal is a BCK-ideal.*

Proof. Indeed, putting $z = 0$ in (I2) we obtain $(I2)_{BCK}$. □

Theorem 2.3.5. *In a BCK-algebra any BCK-ideal is a BCC-ideal.*

Proof. Let $\mathfrak{X}$ be a BCK-algebra and A be a BCK-ideal. Then $0 \in A$. Now, let $xy \cdot z \in A$ and $y \in A$. Then from (1.6) $xy \cdot z = xz \cdot y$ and from assumption $xz \in A$. So, (I2) is satisfied and A is a BCC-ideal. □

Theorem 2.3.6. *In a BCC-algebra any BCK-ideal is a BCC-subalgebra.*

Proof. Let A be a BCK-ideal of a BCC-algebra $\mathfrak{X}$. Then $0 \in A$ and from Theorem 1.1.30(4) we have $xy \leq x \Longrightarrow xy \cdot x = 0$. Thus for $x, y \in A$ we have $xy \cdot x \in A$, which by $(I2)_{BCK}$ implies $xy \in A$. Therefore A forms a subalgebra of $\mathfrak{X}$. □

Theorem 2.3.7. *The BCC-part $B(0)$ of a weak BCC-algebra $\mathfrak{X}$ is a BCC-ideal.*

Proof. Obviously $0 \in B(0)$. Let $xy \cdot z$, $y \in B(0)$. Then $0 \leq xy \cdot z$ and from Theorem 1.1.30(3) we obtain $xy \cdot z \leq xz$, which by transitivity of $\leq$ implies $0 \leq xz$ and consequently $xz \in B(0)$. Therefore (I2) is satisfied and $B(0)$ is a BCC-ideal. □

Corollary 2.3.8. *Any BCC-ideal of a BCC-algebra is its BCC-subalgebra.*

The following example shows that a *BCC*-ideal is not a BCK-subalgebra, in general.

Example 2.3.9. Let $\mathfrak{X} = \langle X; \cdot, 0\rangle$, where $X = \{0,1,2,3,4,5\}$ and the binary operation $\cdot$ is defined by Table 2.1.1.

It is easy to see that $S = \{0,1,2,3,4\}$ is a *BCC*-ideal of $\mathfrak{S}$, but it is not a BCK-algebra since $(2\cdot 1)\cdot 4 \neq (2\cdot 4)\cdot 1$, i.e., the condition (1.6) is not satisfied.

On the other hand, in Example 2.3.2 the set $S = \{0,1,2,3\}$ is the base set of the BCC-subalgebra $\mathfrak{S}$, but S is not a *BCK*-ideal, because $4\cdot 3 = 3 \in S$, but $4 \notin S$.

Similarly, $G = \{0,1,2\}$ is the base set of the BCK-subalgebra $\mathfrak{G}$ shown in Table 1.2.6 and G is not a *BCK*-ideal since $3\cdot 2 \in G$, but $3 \notin G$.

Thus in BCC-algebras *BCC*-ideals, *BCK*-ideals and BCK-subalgebras are independent concepts. □

In the next results, we will show the conditions that must be met for a subalgebra to be a *BCC*- or a *BCK*-ideal.

Theorem 2.3.10. *Let $\mathfrak{X}$ be a BCC-algebra. Then a BCC-subalgebra $\mathfrak{A}$ of $\mathfrak{X}$ is a BCC-ideal if and only if $x \in A, yz \notin A$ imply $yx \cdot z \notin A$.*

Proof. If a BCC-subalgebra $\mathfrak{A}$ is a *BCC*-ideal, then

$$x \in A, yz \notin A \Longrightarrow yx \cdot z \notin A.$$

If not, then

$$yx \cdot z \in A, x \in A \Longrightarrow yz \in A,$$

which is a contradiction.

Conversely, let $\mathfrak{A}$ be a BCC-subalgebra in which

$$x \in A, yz \notin A \Longrightarrow yx \cdot z \notin A.$$

Then obviously $0 \in A$.

Moreover,

$$x \in A, yx \cdot z \in A \Longrightarrow yz \in A,$$

because for $yz \notin A$ we have (by assumption)

$$yx \cdot z \notin A.$$

Hence A is a *BCC*-ideal. □

Putting $z = 0$ in the condition of the above theorem, we obtain:

Corollary 2.3.11. *Let $\mathfrak{X}$ be a BCC-algebra. Then a BCC-subalgebra $\mathfrak{A}$ of $\mathfrak{X}$ is a BCK-ideal if and only if $x \in A, y \notin A$ imply $yx \notin A$.*

Theorem 2.3.12. *If A is a BCC-ideal and $\mathfrak{S}$ is a subalgebra of a weak BCC-algebra $\mathfrak{X}$ then $A \cap S$ is a BCC-ideal of $\mathfrak{S}$.*

Proof. Obviously $0 \in A \cap S$. Let $x, y, z \in S$. Then $xy \cdot z \in A \cap S$ and $y \in A \cap S$ imply $xz \in A$ because A is a BCC-ideal, and $xz \in S$ because $\mathfrak{S}$ is a subalgebra. So, $xz \in A \cap S$. □

The next lemma says that any ideal containing some element of a given branch contains also its initial element.

Lemma 2.3.13. *Let A be a BCC-ideal of a weak BCC-algebra $\mathfrak{X}$. Then A is a BCK-ideal in the sense of ordered structures, i.e., the following condition holds:*

$$y \in A \wedge x \leq y \Longrightarrow x \in A. \tag{2.7}$$

Proof. If A is a BCC-ideal then, from Theorem 2.3.4, we know that it is also a BCK-ideal and the implication (2.7) follows from the definition of the relation $\leq$ and from $(\mathrm{I}2)_{BCK}$. □

Lemma 2.3.14. *Let $\mathfrak{X}$ be a weak BCC-algebra, then $\varphi^2(A) = A \cap I(X)$ for any BCK-ideal A of $\mathfrak{X}$.*

Proof. If A is a BCK-ideal of $\mathfrak{X}$ and $y \in \varphi^2(A)$, then $y = \varphi^2(x)$ for some $x \in A$. By Lemma 2.2.2(2) we have $y \leq x$ and by Lemma 2.3.13 $y \in A$. Now by Lemma 2.2.2(5), we have

$$\varphi^2(y) = \varphi^4(x) = \varphi^2(x) = y.$$

So, by Theorem 2.2.5, $y \in I(X)$. Therefore, $\varphi^2(A) \subseteq A \cap I(X)$.

The converse inclusion is obvious. □

Lemma 2.3.15. *A nonzero element $a \in X$ is an atom of a BCC-algebra $\mathfrak{X}$ if $\{0, a\}$ is a BCK-ideal of $\mathfrak{X}$.*

Proof. The proof follows directly from Definition 2.1.1 and from Lemma 2.3.13. □

The converse is not true. Indeed, in a proper BCC-algebra defined by Table 1.2.7 an element 1 is an atom, but $2 \cdot 1 = 1$ belongs to $\{0, 1\}$ and 2 does not belong to $\{0, 1\}$, so $\{0, 1\}$ is not a BCK-ideal.

Lemma 2.3.16. *Let $\mathfrak{X} = \langle X; \cdot, 0\rangle$ be a BCC-algebra. If every nonzero element of X is an atom, then any subalgebra of $\mathfrak{X}$ is a BCK-ideal.*

Proof. Let $\mathfrak{S}$ be a subalgebra of $\mathfrak{X}$ and let $x, yx \in S$. Then, by Theorem 1.1.30(4), $yx \leq y$ for all $x, y \in X$ and y is an atom of $\mathfrak{X}$. Hence, $yx = 0$ or $y = yx \in S$. If $yx = 0$, then $y \leq x$. Which gives $y = 0$ or $y = x$. Thus $y \in S$, which completes the proof. □

From the above two lemmas, we obtain:

Theorem 2.3.17. *A BCC-algebra $\mathfrak{X}$ contains only atoms if and only if every subalgebra of $\mathfrak{X}$ is a BCK-ideal.*

The next corollary follows directly from Theorem 2.3.17.

Corollary 2.3.18. *Every subalgebra of a BCK-algebra $\mathfrak{X} = \langle X; \cdot, 0\rangle$ is a BCK-ideal if and only if every nonzero element of X is an atom.*

Theorem 2.3.19. *Let A be a BCK-ideal of a BCC-algebra $\mathfrak{X}$. If B is a BCK-ideal such that $B \subseteq A$, then it is a BCK-ideal of $\mathfrak{X}$.*

Proof. B is a BCK-ideal and $B \subseteq A$, so $0 \in B$. Let $y, xy \in B$ for some $x \in \mathfrak{X}$. Then $y, xy \in A$ and $x \in A$ because $B \subseteq A$ and A is a BCK-ideal of $\mathfrak{X}$. Thus $xy, y \in B$ imply $x \in B$.

This completes the proof. □

Corollary 2.3.20. *Let A be a BCC-ideal of a BCC-algebra $\mathfrak{X}$. If B is a BCK-ideal of A, then B is a BCK-ideal of $\mathfrak{X}$.*

Now we will answer the question: when is $I(X)$ a BCC-ideal ?

Example 2.3.21. In the proper weak BCC-algebras, say $\mathfrak{X}$ and $\mathfrak{Y}$, with the operations defined in Table 1.1.2 and Table 1.1.3 in Example 1.1.15 the subset $\{0, 2\}$ is a BCK-ideal but not a BCC-ideal.
Indeed, $(3 \cdot 2) \cdot 1 = 0 \in \{0, 2\}$, $2 \in \{0, 2\}$ but $3 \cdot 1 = 3 \notin \{0, 2\}$.
As it can be easily seen, the subset $\{0, 2\}$ is equal to I(X) and I(Y). Therefore, we can say that for weak BCC-algebra $\mathfrak{X}$, I(X) is not a BCC-ideal, in general. □

Theorem 2.3.22. *I(X) is a BCC-ideal of a weak BCC-algebra $\mathfrak{X}$ if and only if the following conditions are satisfied:*

(1) $y \in I(X)$ *implies* $x \cdot xy \in I(X)$, *for* $x \in X$,

(2) $xb = ab$ *implies* $x = a$, *for* $a, b \in I(X)$.

Proof. Assume that $I(X)$ is a BCC-ideal of $\mathfrak{X}$ and consider an arbitrary element $y \in I(X)$. Then $xy \cdot xy = 0 \in I(X)$, which implies $x \cdot xy \in I(X)$. Thus (1) holds.

Let now $xb = ab$ for some $a, b \in I(X)$. Since, by Theorem 2.2.12, $\mathfrak{I}(\mathfrak{X})$ is a subalgebra and from assumption it is a BCC-ideal, then $xb = ab \in I(X)$, whence $x \in I(X)$ by Theorem 2.3.4 and $\mathrm{I2}_{BCK}$. Applying Theorem 2.2.8(3), Lemma 2.2.2(4) and (BCC1), we obtain

$$x = \varphi^2(x) = \varphi(ax)\varphi(a) = ((ab \cdot xb) \cdot ax)\varphi(a) = 0\varphi(a) = \varphi^2(a) = a.$$

This proves (2).

To prove the converse statement, let us assume that $I(X)$ satisfies (1) and (2). Clearly, $0 \in I(X)$.

If $xy \in I(X)$, $y \in I(X)$, then, by Theorem 2.2.8(3) and Lemma 2.2.2(6), we have

$$xy = \varphi^2(xy) = \varphi^2(x)\varphi^2(y) = \varphi^2(x)y,$$

which, by (2), implies $x = \varphi^2(x)$ and hence $x \in I(X)$.

Thus $I(X)$ is a BCK-ideal.

In fact, it is a BCC-ideal. Indeed, for $xy \cdot z \in I(X)$, $y \in I(X)$, from (BCC1)

$$(xz \cdot (xy \cdot z))(x \cdot xy) = 0 \in I(X)$$

and, by(1) and $\mathrm{I2}_{BCK}$, we get $xz \cdot (xy \cdot z) \in I(X)$.

Now, again by $\mathrm{I2}_{BCK}$ we deduce $xz \in I(X)$.

This completes the proof. □

Corollary 2.3.23. *Let $\mathfrak{X}$ be a weak BCC-algebra. A subalgebra $\mathfrak{S}$ with $S \subseteq I(X)$ is a BCC-ideal of $\mathfrak{X}$ if and only if the following conditions are satisfied:*

(1) $y \in S$ *implies* $x \cdot xy \in S$,

(2) $xb = ab$ *implies* $x = a$, *for* $a, b \in S$.

There is also a generalized form of BCC-ideals called $BCC(m, n)$-ideals. We show some examples and results concerning these objects.

We begin with the following definition:

Definition 2.3.24. [97] Let $\mathfrak{X}$ be a weak BCC-algebra. A nonempty subset A of X is called a *BCC(m, n)-ideal* of $\mathfrak{X}$ if there are positive integers m and n such that the following conditions are satisfied for all $x, y, z \in X$:

$$(I1) \qquad 0 \in A,$$
$$(I2)_{(m,n)} \quad xy \cdot z^m \in A \text{ and } y \in A \Longrightarrow xz^n \in A.$$

Note that a $BCC(1,1)$-ideal is a BCC-ideal.

Example 2.3.25. Let us consider the following weak BCC-algebra: $\mathfrak{X} = \langle\{0, 1, 2, 3, 4\}; \cdot, 0\rangle$ with the operation $\cdot$ defined by the following table:

$\cdot$	0	1	2	3	4
0	0	0	2	2	0
1	1	0	2	2	1
2	2	2	0	0	2
3	3	2	1	0	2
4	4	4	2	2	0

Table 2.3.2

In this algebra the subset $A = \{0, 4\}$ is a $BCC(2,4)$-ideal and a $BCC(2,10)$-ideal of $\mathfrak{X}$. But A is neither a $BCC(1,1)$-ideal nor a $BCC(1,2)$-ideal of $\mathfrak{X}$. □

Example 2.3.26. The algebra $\mathfrak{X} = \langle\{0, 1, 2, 3, 4, 5\}; \cdot, 0\rangle$ with the operation $\cdot$ defined by the following table is a weak BCC-algebra:

$\cdot$	0	1	2	3	4	5
0	0	2	1	3	4	5
1	1	0	2	4	5	3
2	2	1	0	5	3	4
3	3	4	5	0	2	1
4	4	5	3	1	0	2
5	5	3	4	2	1	0

Table 2.3.3

It can be checked that the set $A = \{0, 1, 2\}$ is a $BCC(3,1)$-ideal of $\mathfrak{X}$. It is a $BCC(4,2)$-ideal and $BCC(1,1)$-ideal of $\mathfrak{X}$, too. But A is neither a $BCC(2,1)$-ideal nor a $BCC(3,2)$-ideal of $\mathfrak{X}$. □

Theorem 2.3.27. *Let $m, n \in \mathbb{N}$. If A is a $BCC(m,n)$-ideal of $\mathfrak{X}$, then for $x, y \in X$*

$$xy^{m+1} \in A, y \in A \Longrightarrow xy^n \in A.$$

Proof. By putting $z = y$ in $(\mathrm{I2})_{(m,n)}$ we get the statement of the theorem. □

Finally, we prove Theorem 2.3.4 in the version for $BCC(m,n)$-ideals.

Theorem 2.3.28. *In a weak BCC-algebra $\mathfrak{X}$ every $BCC(m,n)$-ideal is a BCK-ideal, for any $m, n \in \mathbb{N}$.*

Proof. By putting $z = 0$ in $(\mathrm{I2})_{(m,n)}$ we obtain $(I2)_{BCK}$. □

2.4 Congruences

In this part of the chapter, we will describe congruences on BCC-algebras and show their strong connections with BCC-ideals and branches.

We start with the following definition:

Definition 2.4.1. For a BCC-ideal A of a weak BCC-algebra $\mathfrak{X}$ we define a relation $\theta \subseteq X^2$ in the following way:

$$(x, y) \in \theta \iff xy,\ yx \in A. \tag{2.8}$$

Theorem 2.4.2. *If A is a BCC-ideal of a weak BCC-algebra $\mathfrak{X}$, then the relation θ, defined by* (2.8), *is a congruence on X.*

Proof. It is clear that the relation θ is reflexive and symmetric.
It is also transitive because

$$(x, y) \in \theta \text{ and } (y, z) \in \theta \text{ imply } xy, yx, yz, zy \in A$$

and, by (BCC1), $(xz \cdot yz) \cdot xy = 0 \in A$. Hence by (I2) $xz \cdot xy \in A$, which, by Theorem 2.3.4, gives $xz \in A$.

Similarly,

$$(zx \cdot yx) \cdot zy = 0 \in A \text{ implies } zx \in A.$$

Thus $(x, z) \in \theta$ and θ is an equivalence relation.

Let now $(x, u) \in \theta$ and $(y, v) \in \theta$. Then $(xy \cdot uy) \cdot xu = 0 \in A$ and $xu \in A$, which implies

$$xy \cdot uy \in A.$$

Similarly,

$$uy \cdot xy \in A.$$

Hence we have

$$(xy, uy) \in \theta. \tag{2.9}$$

On the other hand,

$$(uy \cdot vy) \cdot uv = 0 \in A \text{ and } vy \in A \text{ imply } uy \cdot uv \in A.$$

In the same manner from

$$(uv \cdot yv) \cdot uy = 0 \in A \text{ and } yv \in A$$

we obtain

$$uv \cdot uy \in A.$$

Similarly,

$$uy \cdot uv \in A.$$

Thus

$$(uy, uv) \in \theta. \tag{2.10}$$

Since θ is transitive, from (2.9) and (2.10) it follows $(xy, uv) \in \theta$, which proves that the relation θ has the substitution property, so it is a congruence. □

We say that the congruence θ is *induced* by the ideal A.

Let ρ be a congruence on a weak BCC-algebra $\mathfrak{X}$. Then an equivalence class of ρ containing an element $x \in X$ is denoted by C_x^ρ, i.e.,

$$C_x^\rho = \{y \in X \mid (x, y) \in \rho\}.$$

For $x, y \in X$, we define

$$C_x^\rho \cdot C_y^\rho = C_{x \cdot y}^\rho \tag{2.11}$$

Since ρ has the substitution property, the operation $\cdot$ is well defined. The quotient algebra is denoted as $\mathfrak{X}/\rho = \langle X/\rho\,;\, \cdot\,, C_0^\rho \rangle$, where $X/\rho = \{C_x^\rho \mid x \in X\}$.

For simplicity and without loss of generality, we will mostly denote C_x instead of C_x^ρ.

Corollary 2.4.3. *For the congruence θ defined by (2.8) we have*

$$X/\theta = X/A \text{ and } C_0 = A.$$

Lemma 2.4.4. *If ρ is a congruence on a BCC-algebra $\mathfrak{X}$, then C_0 is a BCC-ideal.*

Proof. Obviously $0 \in C_0$. If $xy \cdot z, y \in C_0$, then $(xy \cdot z, 0) \in \rho$ and $(y, 0) \in \rho$. From the substitution property and by (BCC3) we get

$$\begin{aligned} ((x,x) \in \rho \text{ and } (z,z) \in \rho) &\Longrightarrow (xy, x0) \in \rho \\ &\Longrightarrow (xy \cdot z, x0 \cdot z) \in \rho \\ &\Longleftrightarrow (xy \cdot z, xz) \in \rho. \end{aligned}$$

Thus, since ρ is symmetric and transitive, we have

$$\begin{aligned} &(xy \cdot z, 0) \in \rho \text{ and } (xy \cdot z, xz) \in \rho \\ &\Longleftrightarrow (0, xy \cdot z) \in \rho \text{ and } (xy \cdot z, xz) \in \rho \\ &\Longrightarrow (0, xz) \in \rho, \end{aligned}$$

i.e., $xz \in C_0$, which completes the proof. □

Since $C_0 = A$ for a congruence defined by (2.8), the following important corollary is a consequence of the above results:

Corollary 2.4.5. *Any BCC-ideal is determined by some congruence.*

Theorem 2.4.6. *The lattice of all congruences of a BCC-algebra $\mathfrak{X}$ is complete. The least congruence is induced by a BCC-ideal $\{0\}$, the greatest by $A = X$.*

Each BCC-ideal of a BCC-algebra $\mathfrak{X}$ is determined by some congruence on $\mathfrak{X}$, and conversely, each BCC-ideal of $\mathfrak{X}$ induces by (2.8) some congruence on $\mathfrak{X}$.

We will show that, similarly to BCK-algebras, infinitive BCC-algebras include congruences that are not induced by BCC-ideals. On the other hand, in finite BCC-algebras all congruences are induced by BCC-ideals. Let us take a closer look at it.

As it can be easily seen, $\langle X/\rho;\ \cdot, C_0\rangle$ satisfies all the axioms of a BCC-algebra except (BCC5). This axiom is satisfied only in some cases.

It is also not satisfied in the case of BCK-algebras (cf. [78] and [137]).

Definition 2.4.7. The congruence ρ defined on a BCC-algebra $\mathfrak{X}$ is called *regular* if and only if $C_xC_y = C_yC_x = C_0$ implies $C_x = C_y$.

Corollary 2.4.8. *The quotient algebra $\mathfrak{X}/\rho$ is a BCC algebra if the relation θ is regular.*

The following lemma is obvious.

Lemma 2.4.9. *A congruence ρ on a BCC-algebra $\mathfrak{X}$ is regular if and only if for all $x, y \in X$ from $(xy, 0) \in \rho$ and $(yx, 0) \in \rho$ it follows $(x, y) \in \rho$.*

Corollary 2.4.10. *The congruence relation θ defined by (2.8) and induced by a BCC-ideal A is a regular congruence on a weak BCC-algebra $\mathfrak{X}$.*

Regular congruences are characterized by BCC-ideals.

Theorem 2.4.11. *A congruence is regular if and only if it is induced by some BCC-ideal.*

Proof. Let ξ be a congruence induced by a BCC-ideal A. Then $C_0 = A$ and $C_xC_y = C_0 = C_yC_x$ imply $xy, yx \in A$, which shows that $(x, y) \in \theta$ and $C_x = C_y$ for θ defined by (2.8). Hence a congruence induced by a BCC-ideal is regular.

Let ρ be an arbitrary regular congruence. If $(x, y) \in \rho$, then from $(y, y) \in \rho$ and from the substitution property it follows that $(xy, yy) \in \rho \Longleftrightarrow (xy, 0) \in \rho$. In a similar way $(yx, 0) \in \rho$. Therefore, $C_{xy} = C_0 = C_{yx}$ and $xy, yx \in C_0$ and $A = C_0$ is, by Lemma 2.4.4, a BCC-ideal. Hence $\rho \subseteq \xi$.

Conversely, if $(x, y) \in \xi$ then $xy, yx \in A = C_0$ and $C_xC_y = C_0 = C_yC_x$, which by the regularity of ξ implies $C_x = C_y$. Thus $(x, y) \in \rho$ and $\xi \subseteq \rho$. Hence $\rho = \xi$.

The proof is completed. □

Corollary 2.4.12. *All congruences of a finite BCC-algebra are regular.*

Example 2.4.13. The relation $\sim$ defined on a weak BCC-algebra $\mathfrak{X}$ by

$$(x,y) \in \sim \Longleftrightarrow \varphi(x) = \varphi(y) \tag{2.12}$$

is an equivalence on $\mathfrak{X}$.

We can interpret the condition (2.12) as follows: x and y belong to the relation $\sim$ if and only if they are in the same branch.

Moreover, if $(x,y) \in \sim$ and $(u,v) \in \sim$, then $\varphi(x) = \varphi(y)$ and $\varphi(u) = \varphi(v)$. Hence, by Lemma 2.2.2(8), (6), we obtain

$$\varphi(ux) = \varphi^2(xu) = \varphi^2(x)\varphi^2(u) = \varphi^2(y)\varphi^2(v) = \varphi^2(yv) = \varphi(vy),$$

which implies $(ux, vy) \in \sim$. Thus $\sim$ is a congruence.

It can be easily seen that the corresponding quotient algebra

$$\mathfrak{X}/{\sim} = \langle X/{\sim}\,;\,\cdot\,,0\rangle,$$

where $X/{\sim} = \{C_x \,|\, x \in X\}$ and the operation $\cdot$ is defined by (2.11), satisfies the axioms (BCC1) and (BCC3).
Indeed, by (2.11) we have

$$(C_xC_y \cdot C_zC_y) \cdot C_xC_z = C_{(xy\cdot zy)\cdot xz} = C_0.$$

The proof of (BCC3) is similar.

Moreover, if for some $C_x, C_y \in X/{\sim}$ the equalities $C_xC_y = C_yC_x = C_0$ hold, then $\varphi(xy) = \varphi(yx) = \varphi(0) = 0$. This, by Lemma 2.2.2, implies

$$\varphi^2(y)\varphi^2(x) = \varphi^2(yx) = \varphi(xy) = 0 = \varphi(yx) = \varphi^2(xy) = \varphi^2(x)\varphi^2(y).$$

Therefore $\varphi^2(x) = \varphi^2(y)$, and consequently

$$\varphi(x) = \varphi^3(x) = \varphi(\varphi^2(x)) = \varphi(\varphi^2(y)) = \varphi^3(y) = \varphi(y).$$

Thus $C_x = C_y$ and (BCC5) is satisfied.

Then $\sim$ is a regular congruence and, by Corollary 1.1.24, $\mathfrak{X}/{\sim}$ is a weak BCC-algebra. □

Theorem 2.4.14. *Let $\sim$ and θ defined by (2.12) and (2.8) are congruences on weak BCC-algebra $\mathfrak{X}$. Then the congruence $\sim$ coincides with the congruence θ induced by $B(0)$.*

Proof. If $(x, y) \in \sim$, then $\varphi(x) = \varphi(y)$ and, by Lemma 2.2.2(8), (6),

$$\varphi(xy) = \varphi^2(yx) = \varphi^2(y)\varphi^2(x) = 0,$$

i.e., $0 \leq xy$. Hence $xy \in B(0)$.

Similarly, $yx \in B(0)$. Thus $(x, y) \in \theta$.

Conversely, let $(x, y) \in \theta$ with $A = B(0)$. Then $xy, yx \in B(0)$ and consequently $\varphi(xy) = \varphi(yx) = 0$. Thus

$$0 = \varphi^2(xy) = \varphi^2(x)\varphi^2(y) \text{ and } 0 = \varphi^2(yx) = \varphi^2(y)\varphi^2(x),$$

so,

$$\varphi^2(x)\varphi^2(y) = \varphi^2(y)\varphi^2(x) = 0,$$

which, by (BCC5), implies

$$\varphi^2(x) = \varphi^2(y).$$

Therefore,

$$\varphi(x) = \varphi^3(x) = \varphi(\varphi^2(x)) = \varphi(\varphi^2(y)) = \varphi^3(y) = \varphi(y),$$

which proves $(x, y) \in \sim$.

The proof is completed. □

Theorem 2.4.15. *In a weak BCC-algebra $\mathfrak{X}$, for $x \in X$ the class $C_x^\sim$ coincides with the branch containing x.*

Proof. Indeed, let $x \in B(a)$ and $y \in C_x^\sim$. Then $a \leq x$ and $\varphi(y) = \varphi(x)$. Thus, by (1.5), $\varphi(x) \leq \varphi(a)$, which implies $\varphi(y) \leq \varphi(a)$ and

$$\varphi^2(a) \leq \varphi^2(y). \tag{2.13}$$

Consequently, since a is an initial element, by Theorem 2.2.8(3), (2.13) and Lemma 2.2.2(2), we have

$$a = \varphi^2(a) \leq \varphi^2(y) \leq y.$$

This shows that $y \in B(a)$. So, $C_x^\sim \subseteq B(a)$.

Let now $x, y \in B(a)$. Then $a \leq y$ and, by Lemma 2.2.2(7), $\varphi(a) = \varphi(y)$. Similarly, $\varphi(a) = \varphi(x)$. Thus

$$\varphi(y) = \varphi(a) = \varphi(x).$$

So, we have $(x, y) \in \sim$. Hence $y \in C_x^\sim$ which means $B(a) \subseteq C_x^\sim$.

Consequently, $C_x^\sim = B(a)$ for $a = \varphi(x)$, which completes the proof. □

Corollary 2.4.16. *Branches of a weak BCC-algebra $\mathfrak{X}$ coincide with the equivalence classes of a congruence induced by its BCC-part $B(0)$. This means $B(a) = C_a^{\theta}$ for any $a \in I(X)$.*

Proof. The proof follows from Theorem 2.4.14. □

Since we have $B(a) = C_a^{\sim} = C_a^{\theta}$, we will denote it simply as $B(a) = C_a$.

Corollary 2.4.17. *Let $\mathfrak{X}$ be a weak BCC-algebra and let $a, b \in I(X)$. Then*

$$B(a)B(b) = B(ab).$$

Theorem 2.4.18. *If $x \in B(a)$, $y \in B(b)$, for some $a, b \in I(X)$, then $xy \in B(ab)$.*

Proof. Let $x \in B(a)$, $y \in B(b)$. We obtain $(a, x) \in \sim$ and $(b, y) \in \sim$, which implies $(xy, ab) \in \sim$. This gives $xy \in C_{ab} = B(ab)$. □

Lemma 2.4.19. $xy \in B(0) \Longleftrightarrow yx \in B(0)$.

Proof. Indeed, by (BCC1$'$), from $xy \in B(0)$ it follows

$$0 = 0 \cdot xy = yy \cdot xy \leq yx, \text{ i.e., } yx \in B(0).$$

Similarly, $yx \in B(0)$ implies $xy \in B(0)$, which completes the proof. □

Theorem 2.4.20. *Elements $x, y \in X$ are in the same branch if and only if $xy \in B(0)$.*

Proof. If $x, y \in B(a)$, then $a \leq x$ and $a \leq y$, which, by (1.4), gives $0 = ay \leq xy$. Hence $xy \in B(0)$.

Conversely, if $xy \in B(0)$, then, by Lemma 2.4.19, $yx \in B(0)$. Thus

$$
\begin{aligned}
& (x, y) \in \theta \\
& \Longrightarrow (x, y) \in \rho && \text{from Theorem 2.4.14} \\
& \Longleftrightarrow \varphi(x) = \varphi(y) && \text{by (2.12).}
\end{aligned}
$$

Let $I(X) \ni \varphi(x) = \varphi(y) = a$. By Theorem 2.2.8(3), $\varphi^2(a) = a$, which implies $x, y \in B(\varphi(a))$.

Thus x, y belong to the same branch.

This completes the proof. □

Corollary 2.4.21. *Comparable elements are in the same branch.*

Example 2.4.22. The converse statement to the above corollary is not true. In a weak BCC-algebra $\mathfrak{X}$ with the operation $\cdot$ defined by Table 2.4.1 elements 1 and 2 belong to the same branch but they are incomparable.

$\cdot$	0	1	2	3	4	5
0	0	0	0	0	4	4
1	1	0	1	0	4	4
2	2	2	0	0	4	4
3	3	3	3	0	5	4
4	4	4	4	4	0	0
5	5	5	5	4	3	0

Table 2.4.1 □

As a consequence of Lemma 2.4.19 and Theorem 2.4.20 we also obtain:

Corollary 2.4.23. *In a weak BCC-algebra $\mathfrak{X}$ for any element $a \in I(X)$, from $x \in B(a)$ and $y \notin B(a)$ it follows xy, $yx \notin B(0)$.*

Corollary 2.4.24. *Let $\mathfrak{X}$ be a weak BCC-algebra with $I(X) \subsetneq X$. Then X has at least two nontrivial branches.*

Proof. Since $I(X) \neq X$, there exists $x \in X$ and $x \notin I(X)$. Let $x \in B(a)$ for any $a \in I(X)$. Then, from Theorem 2.4.20, $xa \in B(0)$. So, the branches $B(0)$ and $B(a)$ are nontrivial. □

Theorem 2.4.25. *Let $\mathfrak{X}$ be a weak BCC-algebra. The sum of all branches $B(a)$ of $\mathfrak{X}$ such that $a \in A \subset I(X)$ is a subalgebra of $\mathfrak{X}$ if and only if $\mathfrak{A}$ is a subalgebra of $\mathfrak{I}(\mathfrak{X})$.*

Proof. Let S be the set-theoretic sum of all branches $B(a)$ of $\mathfrak{X}$ such that $a \in A$. Let $\mathfrak{S}$ be a subalgebra of $\mathfrak{X}$. Clearly, $B(0) \subset S$ and hence $0 \in A$. For $a, b \in A$ we have $a \in B(a)$ and $b \in B(b)$. From definition of S it follows $B(a) \subset S$ and $B(b) \subset S$. So, $a, b \in S$. From the assumption about $\mathfrak{S}$, it follows that $ab \in S$.

On the other hand, from the fact that $A \subseteq I(X)$ and from Theorem 2.2.12 it follows that $ab \in I(X)$. So, $ab \in S \cap I(X) = A$.

This proves that $\mathfrak{A}$ is a subalgebra of $\mathfrak{I}(\mathfrak{X})$.

Now, let $\mathfrak{A}$ be a subalgebra of $\mathfrak{I}(\mathfrak{X})$. From $0 \in A$ it follows $B(0) \subset S$.

Let $x, y \in S$. Then $x \in B(a) \subset S$ and $y \in B(b) \subset S$ and $a, b \in A$. Hence $ab \in A$ and $B(ab) \subset S$. But $xy \in B(ab)$. So, $xy \in S$.

This proves that $\mathfrak{S}$ is a subalgebra of $\mathfrak{X}$.

The proof is completed. □

Theorem 2.4.26. *Let $\mathfrak{X}$ be a weak BCC-algebra. The sum of all branches $B(a)$ of $\mathfrak{X}$ such that $a \in A \subset I(X)$ is a BCC-ideal of $\mathfrak{X}$ if and only if $\mathfrak{A}$ is a BCC-ideal of $\mathfrak{I}(\mathfrak{X})$.*

Proof. Let S be the set-theoretic sum of all branches $B(a)$ of $\mathfrak{X}$ such that $a \in A$. Let $\mathfrak{S}$ be a BCC-ideal of $\mathfrak{X}$. From $B(0) \subset S$ it follows $0 \in A$. Let $ab \cdot c$ and b belong to A. Then, as in the proof of the previous theorem, $B(ab \cdot c)$ and $B(b)$ are included in S. Let $x \in B(a)$, $y \in B(b)$ and $z \in B(c)$. Then, since S is a BCC-ideal, $xz \in S$. So, $xz \in B(ac) \subset S$ and hence $ac \in A$.

We have proven that A is a BCC-ideal.

Let now A be a BCC-ideal. Of course, the branch $B(0)$ is included in S. Let $xy \cdot z$ and y belong to S. Let $x \in B(a), y \in B(b)$ and $z \in B(c)$. Then $B(a)B(b) \cdot B(c) = B(ab \cdot c) \subset S$ and $B(b) \subset S$. Hence $ab \cdot c \in A$ and $b \in A$. But A is a BCC-ideal. Hence $ac \in A$ and $B(ac) \subset S$. Therefore, S is a BCC-ideal of $\mathfrak{X}$. □

If $\mathfrak{X}/\mathfrak{A}$ is a BCC-algebra, then the canonical mapping $f : X \longrightarrow X/A$ defined by $f(x) = C_x$ is an epimorphism $\mathfrak{X}$ onto $\mathfrak{X}/\mathfrak{A}$. Since the kernel $ker\, f = f^{-1}(C_0)$ of any BCC-homomorphism is a BCC-ideal (Theorem 2.3.7), from Universal Algebra we get the following results:

Theorem 2.4.27. *If f is an epimorphism from a BCC-algebra $\mathfrak{X}$ onto a BCC-algebra $\mathfrak{Y}$, then the quotient BCC-algebra $\mathfrak{X}/ker\, f$ is isomorphic to $\mathfrak{Y}$.*

Theorem 2.4.28. *Let $\mathfrak{X}, \mathfrak{Y}, \mathfrak{Z}$ be BCC-algebras and let $h : X \longrightarrow Y$ be an epimorphism $\mathfrak{X}$ onto $\mathfrak{Y}$, and let $g : X \longrightarrow Z$ be a homomorphism $\mathfrak{X}$ into $\mathfrak{Z}$.*

If $ker\, h \subset ker\, g$, then there exists a unique homomorphism $f : Y \longrightarrow Z$ of BCC-algebra $\mathfrak{Y}$ into a BCC-algebra $\mathfrak{Z}$ such that $f \circ h = g$.

Corollary 2.4.29. *Let $\mathfrak{X}, \mathfrak{Y}, \mathfrak{Z}$ be BCC-algebras and let θ be a regular congruence on a BCC-algebra $\mathfrak{X}$ defined by a BCC-ideal A, and let h be a canonical mapping from $\mathfrak{X}$ onto $\mathfrak{Y} = \mathfrak{X}/\mathfrak{A}$. If $C_0 \subset ker\, g$, where $g : X \longrightarrow Z$ is a homomorphism $\mathfrak{X}$ into $\mathfrak{Z}$ then there exists a unique homomorphism $f : X/A \longrightarrow Z$ such that $f \circ h = g$.*

Corollary 2.4.30. *Let h be a homomorphism from a BCC-algebra $\mathfrak{X}$ onto a BCC-algebra $\mathfrak{Y}$. Then the inverse image of a BCC-ideal, a BCC-subalgebra and a BCK-subalgebra of $\mathfrak{Y}$ is a BCC-ideal, a BCC-subalgebra and a BCK-subalgebra of $\mathfrak{X}$, respectively.*

We say that two congruences ρ and σ on a BCC-algebra $\mathfrak{X}$ *commute*, if the condition

$$\rho \circ \sigma = \sigma \circ \rho$$

is satisfied.

The following theorem is a transcription of a result known from Universal Algebra.

Theorem 2.4.31. *The composition $\rho \circ \sigma$ of two congruences on a BCC-algebra $\mathfrak{X}$ is a congruence on $\mathfrak{X}$ if and only if these congruences commute.*

Corollary 2.4.32. *Let A and B be BCC-ideals of a weak BCC-algebra $\mathfrak{X}$ and ρ_A and ρ_B be a congruences induced by A and B, respectively. If congruences ρ_A and ρ_B commute, then*

$$\bigcup_{a \in A} C_a^{\rho_B} = \bigcup_{b \in B} C_b^{\rho_A}$$

is a BCC-ideal.

Proof. Let $\rho_A \circ \rho_B = \rho$. It follows from Theorem 2.4.31 that ρ is a congruence on $\mathfrak{X}$. Then we have

$$\begin{aligned} \bigcup_{a \in A} C_a^{\rho_B} &= \{x \in B \mid (x, a) \in \rho_B \text{ for some } a \in A\} \\ &= \{x \in X \mid (x, a) \in \rho_B \text{ and } (a, 0) \in \rho_A\} \\ &= \{x \in X \mid (x, 0) \in \rho\} = C_0. \end{aligned}$$

Lemma 2.4.4 implies $\bigcup_{a \in A} C_a^{\rho_B}$ is a BCC-ideal. Since ρ_A and ρ_B commute,

$$\bigcup_{a \in A} C_a^{\rho_B} = \bigcup_{b \in B} C_b^{\rho_A},$$

which completes the proof. □

2.5 Group-Like Weak BCC-Algebras

One of the most important classes of weak BCC-algebras is the class of the *group-like weak BCC-algebras* also called *anti-grouped BZ-algebras.*

The group-like weak BCC-algebras were first described by W.A. Dudek, while X.H. Zhang characterized the anti-grouped BZ-algebras. We will prove that any weak BCC-algebra has a group-like weak BCC-subalgebra. We will also show the difference between group-like weak BCC-algebras and group-like BCI-algebras. In the main results, we will indicate the strong connections between group-like weak BCC-algebras and groups.

Definition 2.5.1. A weak BCC-algebra is *group-like* if all its branches are trivial.

This means that the order $\leq$ on a group-like weak BCC-algebra is a discrete (all elements are incomparable).

We begin with theorem that characterizes the class of weak BCC-algebras.

Theorem 2.5.2. *In a weak BCC-algebra $\mathfrak{X}$, the following conditions are equivalent:*

(1) *$\mathfrak{X}$ is group-like,*

(2) $\varphi^2(x) = x$,

(3) $xy \cdot zy = xz$,

(4) $xy = 0$ implies $x = y$ for all $x, y \in X$,

(5) $\varphi(xy) = yx$ for all $x, y \in X$,

(6) $B(0) = \{0\}$,

(7) $\varphi(x) = y$ implies $\varphi(y) = x$.

Proof. (1)$\Longleftrightarrow$(2) From Definition 2.5.1 it follows that for every $x \in X$, there is $y \in X$ such that $x = \varphi(y)$. So, by Lemma 2.2.2(5) we have

$$\varphi^2(x) = \varphi^3(y) = \varphi(y) = x.$$

Conversely, if the condition (2) is true, then, according to Theorem 2.2.5, it follows $X = I(X)$ and, by Definition 2.5.1, $\mathfrak{X}$ is group-like.

(2)$\Longleftrightarrow$(3) Let us assume that (2) holds. Then,

$$\begin{aligned} x &= \varphi^2(x) \\ &= \varphi(\varphi(y)\cdot xy) && \text{by Lemma 2.2.2(3)} \\ &\leq xy\cdot\varphi(y) && \text{by Lemma 2.2.2(1)} \end{aligned}$$

Hence $x \leq xy\cdot\varphi(y)$, which, by (1.4), gives

$$xz \leq (xy\cdot\varphi(y))z$$

and, by Theorem 1.1.30(7),

$$(xy\cdot\varphi(y))z \leq xy\cdot zy.$$

So, we get

$$xz \leq xy\cdot zy. \tag{2.14}$$

On the other hand, by (BCC1$'$), we have $xy\cdot zy \leq xz$.

Combining the last inequality with (2.14), we can see that (3) is satisfied.

To prove the converse, observe that by putting $y = x$ and $z = 0$ in (3) we obtain $0\cdot 0x = x \Longleftrightarrow \varphi^2(x) = x$ for $x \in X$, which means that (3) implies (2).

(1)$\Longrightarrow$(4) Obvious.

(4)$\Longrightarrow$(3) From (BCC1), by (4), we get (3).

(3)$\Longrightarrow$(5) From (3) we immediately get

$$\varphi(xy) = 0\cdot xy = yy\cdot xy = yx.$$

(5) $\Longrightarrow$(2) Putting $x = 0$ in (5) we get (2).

(6)$\Longleftrightarrow$(1) Let us assume (6) and let $B(a), a \in I(X)$ be nontrivial. Then there exists $x \neq a$ in B(a). Then $xa \neq 0$ and, by Theorem 2.4.20, $xa \in B(0)$, but, by assumption, $B(0) = \{0\}$. Hence $x \leq a$ and $x = a$. This means that any branch has only one element. By Definition 2.5.1, $\mathfrak{X}$ is group-like.

The converse is obvious.

(7) $\Longleftrightarrow$ (2) Let us assume (2). Let $\varphi(x) = y$, then $\varphi(y) = \varphi^2(x) = x$.

Conversely, $\varphi^2(x) = \varphi(y)$ implies $\varphi^2(x) = x$.

The proof is completed. □

As a consequence of the above theorem, we obtain

Corollary 2.5.3. *A weak BCC-algebra $\mathfrak{X}$ is group-like if and only if $X = I(X)$ or if and only if $X = \varphi(X)$.*

Corollary 2.5.4. *$\varphi(\mathfrak{X}) = \langle \varphi(X);\, \cdot\, , 0\rangle$ is the greatest group-like subalgebra of each weak BCC-algebra $\mathfrak{X}$.*

Proof. By Theorem 2.2.12 and Theorem 2.2.6, $\varphi(\mathfrak{X}) = \mathfrak{I}(\mathfrak{X})$ is a subalgebra. By Corollary 2.5.3, it is group-like. To prove it is the greatest, let us consider an arbitrary group-like subalgebra $\mathfrak{A}$ of $\mathfrak{X}$. Then, from Theorem 2.5.2(2) it follows that for any $x \in A$ we have $x = \varphi^2(x) = \varphi(\varphi(x))$, which means that $x \in \varphi(X)$. Thus $A \subseteq \varphi(X)$ for any group-like subalgebra $\mathfrak{A}$ of $\mathfrak{X}$. Hence $\varphi(X)$ is the greatest group-like subalgebra of $\mathfrak{X}$. □

Corollary 2.5.5. *$\mathfrak{I}(\mathfrak{X})$ is the greatest group-like subalgebra of each weak BCC-algebra $\mathfrak{X}$.*

Proof. The proof follows from Theorem 2.2.6 and Corollary 2.5.4. □

Theorem 2.5.6. *A weak BCC-algebra $\mathfrak{X}$ is group-like if and only if it satisfies the condition*

$$xz^n \cdot yz^n = xy \tag{2.15}$$

for all $x, y, z \in X$ and for $n \geq 1$.

Proof. By Lemma 1.1.32, $\mathfrak{X}$ satisfies $xz^n \cdot yz^n \leq xy$. But it is group-like with discrete order $\leq$, by Definition 2.5.1. Therefore, (2.15) is satisfied.

Conversely, from Lemma 1.1.32 and from (2.15), it follows

$$xz^n \cdot yz^n = (xz^{n-1})z \cdot (yz^{n-1})z = xz^{n-1} \cdot yz^{n-1} = \ldots = xz \cdot yz = xy.$$

Therefore, the condition (3), by Theorem 2.5.2, is satisfied. Hence $\mathfrak{X}$ is a group-like weak BCC-algebra. □

Group-like weak BCC-algebras are strongly connected with groups. The connection is characterized in the next theorem.

Theorem 2.5.7. *A weak BCC-algebra $\mathfrak{X} = \langle X;\, \cdot\, , 0\rangle$ is group-like if and only if $\langle X; *, e\rangle$, where $e = 0$ and $x * y = x \cdot 0y$, is a group. Moreover, in this case $xy = x * y^{-1}$.*

Proof. Let $\langle X, \cdot, 0\rangle$ be a group-like weak BCC-algebra. We define

$$x * y = x \cdot 0y.$$

We will prove that $\langle X; *, e\rangle$ is a group. First, we show the associativity. We have

$$\begin{aligned}
(x * y) * z &= (x \cdot 0y) \cdot 0z \\
&= (x \cdot \varphi(y))\varphi(z) \\
&= (x \cdot \varphi(y))\varphi^3(z) && \text{Lemma 2.2.2(5)} \\
&= (x \cdot \varphi(y))\varphi^2(\varphi(z)) \\
&= (x \cdot \varphi(y))(\varphi(y \cdot \varphi(z))\varphi(y)) && \text{Lemma 2.2.2(4)} \\
&= x \cdot \varphi(y \cdot \varphi(z)) && \text{by (BCC1)} \\
&= x * (y * z).
\end{aligned}$$

We now show that e is the identity element. From (BCC3) it follows

$$x * e = x \cdot 00 = x0 = x.$$

On the other hand, by Theorem 2.5.2(2), we have

$$e * x = 0 \cdot 0x = \varphi^2(x) = x.$$

It remains to show that $0x$ is the inverse element of x.

By Theorem 2.5.2(2), we have

$$x * (0x) = x(0 \cdot 0x) = x\varphi^2(x) = xx = 0.$$

On the other hand

$$(0x) * x = 0x \cdot 0x = 0.$$

We proved that $\langle X; *, e\rangle$ is a group.

To prove the converse, let $\langle X; *, e\rangle$, where $e = 0$, be a group.

We prove the axiom (BCC1):

$$\begin{aligned}
&xy \cdot zy \\
&= (x * y^{-1})(z * y^{-1}) \\
&= (x * y^{-1}) * (z * y^{-1})^{-1} \\
&= (x * y^{-1}) * (y * z^{-1}) \\
&= x * y^{-1} * y * z^{-1} \\
&= x * z^{-1} \\
&= xz.
\end{aligned}$$

Hence (BCC1) is satisfied.

(BCC3) is also satisfied because $e = 0$ is the identity element in $\langle X; *, e\rangle$.

For the proof of (BCC5), we assume that

$$xy = yx = 0, \text{ i.e. } x * y^{-1} = y * x^{-1}.$$

In a group we get $x = y$.

It can be easily shown that Theorem 2.5.2(5) is satisfied. Indeed, we have

$$0 \cdot xy = (x * y^{-1})^{-1} = y * x^{-1} = yx.$$

The weak BCC-algebra $\langle X; \cdot, 0\rangle$ is group-like.

This completes the proof. □

We call the group in the last theorem *corresponding* to the weak BCC-algebra. The weak BCC-algebra is then *induced* by the group.

Theorem 2.5.8. *If the corresponding group $\langle X; *, e\rangle$ is commutative, then the induced weak BCC-algebra $\mathfrak{X} = \langle X; \cdot, 0\rangle$ is a BCI-algebra.*

Proof. Indeed, we have namely

$$\begin{aligned}
&xy \cdot z \\
&= (x * y^{-1}) * z^{-1} \\
&= x * y^{-1} * z^{-1} \\
&= x * z^{-1} * y^{-1}
\end{aligned}$$

$$= (x * z^{-1}) * y^{-1}$$

$$= xz \cdot y.$$

From Corollary 1.1.19 it follows that $\mathfrak{X}$ is a BCI-algebra. □

Theorem 2.5.8 implies two simple corollaries:

Corollary 2.5.9. *A group-like weak BCC-algebra $\mathfrak{X}$ is proper if and only if it is induced by a noncommutative group.*

Corollary 2.5.10. *ρ is a congruence of a group-like weak BCC-algebra $\mathfrak{X}$ if and only if it is a congruence of the corresponding group.*

Example 2.5.11. A weak BCC-algebra $\mathfrak{X}$ with the condition

$$xy \cdot zu = xz \cdot yu \tag{2.16}$$

for $x, y, z, u \in X$, is called *medial.*

This class of algebras was introduced by W.A. Dudek in [26] and investigated in many papers. It is not proper, because for $u = 0$ we get (1.6).
It satisfies Theorem 2.5.2(5). Indeed,

$$xy = xy \cdot 0 = xy \cdot xx = xx \cdot yx = 0 \cdot yx = \varphi(yx).$$

So, a medial weak BCC-algebra is a group-like BCI-algebra.

Conversely, let $\mathfrak{X}$ be a group-like BCI-algebra. Then, by Theorem 2.5.2(5) and (1.6), we have

$$x \cdot yz = 0 \cdot (yz \cdot x) = 0 \cdot (yx \cdot z) = z \cdot yx.$$

So, we proved that

$$x \cdot yz = z \cdot yx. \tag{2.17}$$

By (2.17), we have

$$xy \cdot zu = u(z \cdot xy) = u(y \cdot xz) = xz \cdot yu,$$

and the condition (2.16) is satisfied. Therefore, a weak BCC-algebra $\mathfrak{X}$ is medial if and only if it is a group-like BCI-algebra. □

Example 2.5.12. Let $\mathfrak{X}$ be a BCH-algebra, i.e., an algebra $\langle X; \cdot, 0\rangle$ of type (2,0) that satisfies the conditions: (BCC2), (BCC5) and (1.6). For details about this class of algebras, refer to [57].

Also in the case of BCH-algebras one can prove that $\mathfrak{X}$ is a medial BCH-algebra if and only if it is a group-like BCI-algebra. Moreover, $\varphi(X)$ is the greatest medial BCI-algebra included in a BCH-algebra $\mathfrak{X}$. □

Therefore, group-like BCI-algebras are a significant class of subalgebras determined by groups, and at the same time, as we will see in the next chapter, being the core on which all BCH-, BCI- and weak BCC-algebras are based.

It is worth mentioning that group-like BCI-algebras are sometimes called *p-semisimple* BCI-algebras (cf. [146]).

The next result is a characterization of a general version of the condition (2) from Theorem 2.5.2.

Corollary 2.5.13. *A weak BCC-algebra $\mathfrak{X}$ satisfies the identity*

$$x \cdot xy = y \tag{2.18}$$

for any $x, y \in X$ if and only if it is a group-like BCI-algebra.

Proof. In a weak BCC-algebra $\mathfrak{X} = \langle X; \cdot, 0\rangle$ satisfying the identity (2.18) we have $\varphi^2(y) = y$ for every $y \in X$. This means that this weak BCC-algebra is group-like. Thus $xy = x * y^{-1}$ for some group $\langle X; *, 0\rangle$. Hence $x \cdot xy = x * y * x^{-1} = y$, which implies the commutativity of $\langle X; *, 0\rangle$. From Theorem 2.5.8, it follows that $\mathfrak{X}$ is a BCI-algebra.

Conversely, we have $(x \cdot xy)y = 0$, by (1.2).
On the other hand,

$$y(x \cdot xy) = 0 \cdot (x \cdot xy)y = 00 = 0,$$

by Theorem 2.5.2(5). Hence, from (BCK5), it follows (2.18). □

From Theorem 2.5.2, it follows that the operation $\cdot$ of a group-like weak BCC-algebra is *right-cancellative.*

Indeed, let $xy = zy$. Then $0 = xy \cdot zy \le xz$. Hence $x \le z$ and, by Theorem 2.5.2(4), we have $x = z$, i.e.,

$$xy = zy \Longrightarrow x = z. \tag{2.19}$$

However, $\cdot$ is not *left-cancellative* in a group-like weak BCC-algebra, while in any group-like BCI-algebra it is.

Indeed, let $xy = xz$. Then

$$0 = (xy \cdot zy) \cdot xz = (xy \cdot xz) \cdot zy.$$

But, in the above computations, we have used the condition (1.6) and the axiom (BCK1). Therefore, the proof cannot be applied to group-like weak BCC-algebras.

As shown in [27], in a group-like BCI-algebra, each of the equations $ax = b$ and $ya = b$ has a unique solution $x = ab$ and $y = b \cdot 0a$. So a group-like BCI-algebra is a quasi-group.

A proper group-like weak BCC-algebra is not a quasi-group, because the solution $x = ab$ of $ax = b$ holds, by (2.18), only in group-like BCI-algebra.

It is interesting that in a finite group-like weak BCC-algebra in the Cayley table, each element occurs exactly once in every row and in every column - like in the case of groups.

We can now reformulate Theorem 2.3.22. We have the following corollary:

Corollary 2.5.14. *$I(X)$ is a BCC-ideal of a weak BCC-algebra $\mathfrak{X}$ if and only if the following condition is satisfied:*

$$y \in I(X) \Longrightarrow x \cdot xy \in I(X), \tag{2.20}$$

for $x \in X$.

Proof. Since $\mathfrak{I}(\mathfrak{X})$ is a group-like weak BCC-algebra, the right-cancellativity is satisfied. Therefore, the second condition in Theorem 2.3.22 is satisfied. So, $I(X)$ is a BCC-ideal of $\mathfrak{X}$.

The converse is obvious. □

Corollary 2.5.15. *Any group-like BCI-algebra is a BCC-ideal.*

Proof. In any group-like BCI-algebra $X = I(X)$ and, by (2.18), the condition (2.20) holds. So, it is a BCC-ideal. □

Theorem 2.5.16. *A weak BCC-algebra $\mathfrak{X}$ is a semigroup if and only if it is induced by a Boolean group.*

Proof. Let a weak BCC-algebra $\mathfrak{X} = \langle X; \cdot, 0\rangle$ be a semigroup. Then, by (BCC2) and from associativity, we get

$$0x = 00 \cdot x = 0 \cdot 0x,$$

which, by Lemma 2.2.2(2), gives $0x \leq x$.

On the other hand, by associativity and (BCC3), we get

$$x \cdot 0x = x0 \cdot x = xx = 0$$

and $x \leq 0x$.

Hence $x = 0x$ for every $x \in X$.

This means that $X = I(X)$ and from Corollary 2.5.3 it follows that $\mathfrak{X}$ is group-like. From Theorem 2.5.7 it follows that there exists a corresponding group $\langle X; *, 0\rangle$.

Since $0x = x$, we have

$$x * x = x \cdot 0x = xx = 0.$$

Therefore, the corresponding group to $\mathfrak{X}$ is a boolean group.

The converse is obvious. □

Theorem 2.5.17. *Any group induces one weak group-like BCC-algebra and one weak BCC-algebra which is not group-like.*

Proof. Let $\langle X; \cdot, e\rangle$ be a group. Then, by Theorem 2.5.7, $\langle X, *, e\rangle$ with the operation $x * y = xy^{-1}$ is a weak group-like BCC-algebra.

By the construction proposed in Lemma 1.2.11, $X' = X \cup a$, where $a \notin X$, with the operation

$$x \star y = \begin{cases} xy^{-1} & for \quad x, y \in X, \\ x & for \quad x \in X, \ y = a, \\ y^{-1} & for \quad x = a, \ y \in X \setminus \{e\}, \\ a & for \quad x = a, \ y = e, \\ e & for \quad x = a, \ y = a \end{cases}$$

is a weak BCC-algebra. It is not group-like since $e \star a = e \star e = e$. This means B(e) ={e, a}, which contradicts Theorem 2.5.2(6). □

2.6 Solid Weak BCC-Algebras

The part of basic results proven for various classes of BCK-algebras can be proven for analogous classes of BCI-algebras. However, in many cases, these analogous classes of BCI-algebras must be defined in a different way. Unfortunately, the results proven for BCI-algebras cannot be transferred to weak BCC-algebras since, in the proofs of these results, a crucial role plays the identity

$$xy \cdot z = xz \cdot y.$$

M.A. Chaudhry observed (cf. [17], [18] and [19]) that some results proven for BCI-algebras that satisfy some identities are also valid when these identities are satisfied only by elements belonging to the same branch. In this case, the proofs must be new.

Many years ago, W.A. Dudek observed that some classical results proven for BCI-algebras can be transferred to weak BCC-algebras in which Equation (1.6) is valid only for elements belonging to the same branch. In [42], he investigated weak BCC-algebras in which the above equation is valid only when at least two elements are in the same branch. Such algebras are called *solid.* It turned out that such weak BCC-algebras have properties similar to BCI-algebras, but proofs of these properties are completely different.

In the previous sections, we collected many results concerning objects needed to investigate solid weak BCC-algebras. Now, with the necessary tools, we can take a closer look at this subclass of weak BCC-algebras.

We start with its definition.

Definition 2.6.1. A weak BCC-algebra $\mathfrak{X} = \langle X; \cdot, 0\rangle$ is called *left solid* (shortly: *solid*) if Equation (1.6)

$$xy \cdot z = xz \cdot y$$

is valid for all x, y belonging to the same branch and arbitrary $z \in X$. If it is valid for y, z belonging to the same branch and arbitrary $x \in X$, then we say that a weak BCC-algebra X is *right solid.* A left and right solid weak BCC-algebra is called *super solid.*

All BCI-algebras and all BCK-algebras are solid weak BCC-algebras. A solid weak BCC-algebra containing only one branch is a BCK-algebra.

Example 2.6.2. The proper weak BCC-algebra $\mathfrak{X}$ with the operation given by Table 1.1.2 is the smallest right solid proper weak BCC-algebra due to cardinality of its base set X. It it the only such nonisomorphic algebra. □

Example 2.6.3. Let us take a proper weak BCC-algebra $\mathfrak{X} = \langle\{0,1,2,3,4\};\ \cdot,0\rangle$ with the operation $\cdot$ defined by the following table:

$\cdot$	0	1	2	3	4
0	0	0	0	3	3
1	1	0	0	3	3
2	2	1	0	4	4
3	3	3	3	0	0
4	4	3	3	1	0

Table 2.6.1

Then $\mathfrak{X}$ is the smallest solid proper weak BCC-algebra due to cardinality of its base set X. □

Example 2.6.4. Let us take a proper weak BCC-algebra $\mathfrak{X} = \langle\{0,1,2,3,4,5\};\ \cdot,0\rangle$ with the operation $\cdot$ defined by the following table:

$\cdot$	0	1	2	3	4	5
0	0	0	3	2	2	0
1	1	0	3	2	2	0
2	2	2	0	3	3	2
3	3	3	2	0	0	3
4	4	3	2	1	0	3
5	5	1	4	2	2	0

Table 2.6.2

Then $\mathfrak{X}$ is the smallest super solid proper weak BCC-algebra due to cardinality of its base set X. □

The proper weak BCC-algebras with the operations defined by Table 1.1.2 and Table 1.1.3 are not solid, because in both of these cases $(3 \cdot 2) \cdot 1 \neq (3 \cdot 1) \cdot 2$. Moreover, the operation defined in Table 1.1.3 is not right solid because $(1 \cdot 3) \cdot 2 \neq (1 \cdot 2) \cdot 3$.

The next example shows solid weak BCC-algebras divided into branches.

Example 2.6.5. Using a computer, it is not difficult to verify that the following two tables define the operations $\cdot$ of proper solid weak BCC-algebras $\mathfrak{X} = \langle \{0, 1, 2, 3, 4, 5, \}; \cdot, 0 \rangle$:

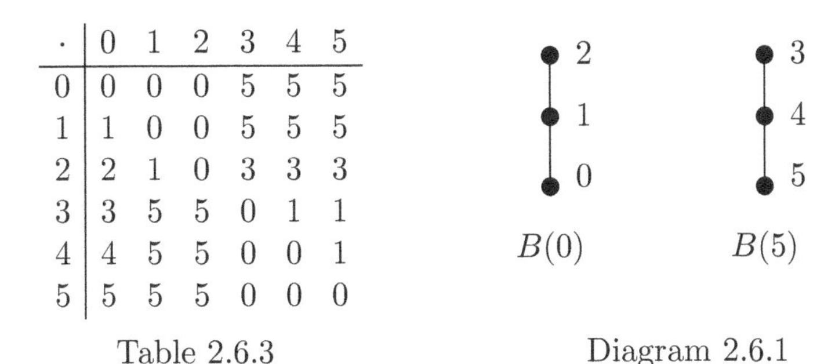

$\cdot$	0	1	2	3	4	5
0	0	0	0	5	5	5
1	1	0	0	5	5	5
2	2	1	0	3	3	3
3	3	5	5	0	1	1
4	4	5	5	0	0	1
5	5	5	5	0	0	0

Table 2.6.3

Diagram 2.6.1

The weak BCC-algebra with the operation $\cdot$ defined by Table 2.6.3 has two branches: $B(0) = \{0, 1, 2\}$ and $B(5) = \{3, 4, 5\}$.

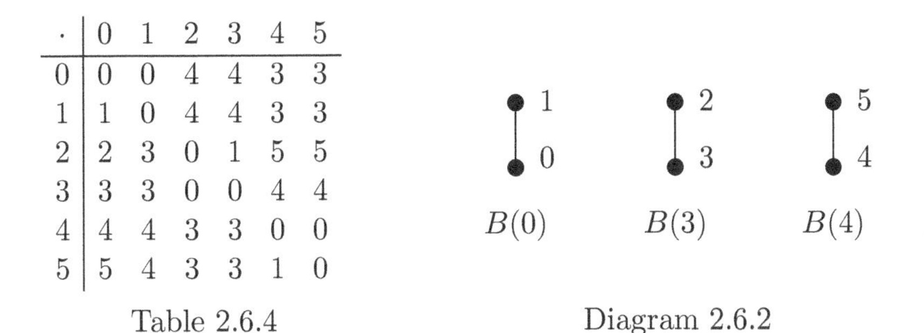

$\cdot$	0	1	2	3	4	5
0	0	0	4	4	3	3
1	1	0	4	4	3	3
2	2	3	0	1	5	5
3	3	3	0	0	4	4
4	4	4	3	3	0	0
5	5	4	3	3	1	0

Table 2.6.4

Diagram 2.6.2

The solid weak BCC-algebra with the operation $\cdot$ defined by Table 2.6.4 has three branches: $B(0) = \{0, 1\}$, $B(3) = \{2, 3\}$ and $B(4) = \{4, 5\}$. □

We start the study of solid weak BCC-algebras with the following instrumental but straightforward observations:

Lemma 2.6.6. *In a solid weak BCC-algebra $\mathfrak{X}$ for elements x, y belonging to the same branch we have*

$$xy \leq z \Longleftrightarrow xz \leq y$$

for any $z \in X$.

Proof. Indeed, if elements x, y belong to the same branch and $z \in X$, then, by (1.6), $xy \cdot z = 0$ is equivalent to $xz \cdot y = 0$. □

Corollary 2.6.7. *If in a solid weak BCC-algebra $\mathfrak{X}$ elements xy and zy (or xy and xz) belong to the same branch, then*

$$xy \cdot xz \leq zy. \tag{2.21}$$

Proof. In this case we have $(xy \cdot xz) \cdot zy = (xy \cdot zy) \cdot xz = 0$, by (BCC1). □

Corollary 2.6.8. *In a solid weak BCC-algebra $\mathfrak{X}$ for elements x, y, z belonging to the same branch* (2.21) *has been satisfied.*

Proof. By Theorem 2.4.20. □

Lemma 2.6.9. *In a solid weak BCC-algebra $\mathfrak{X}$ we have*

$$x \cdot xy \leq y$$

for all x, y belonging the same branch $B(a), a \in I(X)$.

Proof. Indeed, according to the definition, for x, y from the same branch we have

$$(x \cdot xy)y = xy \cdot xy = 0,$$

which proves $x \cdot xy \leq y$. □

Corollary 2.6.10. *In a solid weak BCC-algebra $\mathfrak{X}$ if $x, y \in B(a)$, $a \in I(X)$, then $x \cdot xy$, $y \cdot yx \in B(a)$.*

Proof. From Theorem 2.4.18, since $x, y \in B(a), a \in I(X)$, we have $x \cdot xy \in B(a \cdot aa) = B(a)$. □

Corollary 2.6.11. *In a solid weak BCC-algebra $\mathfrak{X}$ for all $a \in I(X)$ and $x \in B(a)$ we have $x \cdot xa = a$.*

Theorem 2.6.12. *In solid weak BCC-algebras the map φ is an endomorphism.*

Proof. By using (BCC2) and (1.6) several times, we get

$$\begin{aligned}\varphi(x)\varphi(y) &= 0x \cdot 0y \\ &= ((xy \cdot xy)x) \cdot 0y \\ &= ((xy \cdot x) \cdot xy) \cdot 0y \\ &= ((xx \cdot y) \cdot xy) \cdot 0y \\ &= (0y \cdot xy) \cdot 0y \\ &= (0y \cdot 0y) \cdot xy \\ &= 0 \cdot xy \\ &= \varphi(xy)\end{aligned}$$

for all $x, y \in X$. □

The next theorem is Corollary 2.5.5 in the version for solid weak BCC-algebras.

Theorem 2.6.13. *$\mathfrak{I}(\mathfrak{X})$ is the greatest group-like BCI-subalgebra of each solid weak BCC-algebra $\mathfrak{X}$.*

Proof. Comparing Theorem 2.2.6 and Corollary 2.5.4, we see that $\mathfrak{I}(\mathfrak{X})$ is the greatest group-like subalgebra of each weak BCC-algebra $\mathfrak{X}$. Thus, by Theorem 2.5.7, there exists a group $\langle X; *, 0\rangle$ such that $ab = a * b^{-1}$ for $a, b \in I(X)$. Since $\mathfrak{X}$ is solid, φ is its endomorphism. Hence $0a \cdot 0b = 0 \cdot ab$ for $a, b \in I(X)$, i.e., $a^{-1} * b = (a * b^{-1})^{-1} = b * a^{-1}$ in the corresponding group. The penultimate equality is possible only in a commutative group, but in this case $ab \cdot c = ac \cdot b$, which means that $\mathfrak{I}(\mathfrak{X})$ is a BCI-algebra. □

Example 2.6.14. A group-like weak BCC-algebra induced by the symmetric group S_3 shows that in nonsolid weak BCC-algebras, $\mathfrak{I}(\mathfrak{X})$ is not a BCI-algebra. □

Lemma 2.6.15. *In any solid weak BCC-algebra $\mathfrak{X}$*

$$ax = ab$$

for all $a, b \in I(X)$ and $x \in B(b)$.

Proof. Let $a, b \in I(X)$. Then for any $x \in B(b)$ we have $b \leq x$, which, by (1.5), implies $ax \leq ab$. But $\mathfrak{I}(\mathfrak{X})$ is a subalgebra of $\mathfrak{X}$ (Theorem 2.2.12), which implies $ab \in I(X)$. This means that ab is an initial element of X. Thus $ax = ab$. □

Lemma 2.6.16. *If in a solid weak BCC-algebra $\mathfrak{X}$ $ax = ab$ holds for some $a, x \in X$ and $b \in I(X)$, then $x \in B(b)$.*

Proof. If $ax = ab$ holds for some $a, x \in X$ and $b \in I(X)$, then, according to (BCC1), we have

$$0 = (ab \cdot xb) \cdot ax = (ab \cdot ax) \cdot xb = 0 \cdot xb.$$

We could use the condition (1.6) because ax and ab belong to the same branch. Thus $0 \leq xb$.

This, by Theorem 2.4.20, means that x and b are in the same branch and $x \in B(b)$. □

Corollary 2.6.17. *Elements x, y of a solid weak BCC-algebra $\mathfrak{X}$ are in the same branch if and only if $ax = ay$ for some $a \in I(X)$.*

Proof. If elements x, y belong to the branch $B(b)$, where $b \in I(X)$, then from Lemma 2.6.15 it follows $ax = ab = ay$ for all $a \in I(X)$.

Conversely, if $ax = ay$ for some $a \in I(X)$, then

$$0 = (ax \cdot yx) \cdot ay = (ax \cdot ay) \cdot yx = 0 \cdot yx.$$

Thus $yx \in B(0)$. Theorem 2.4.20 completes the proof. □

Now we will look at the operation of exponentiation in solid weak BCC-algebras and present some related results.

Lemma 2.6.18. *In a solid weak BCC-algebra $\mathfrak{X}$, the condition*

$$0(xy)^k = 0(ab)^k \tag{2.22}$$

holds for each natural k, $x \in B(a), y \in B(b)$ and $a, b \in I(X)$.

Proof. Since the elements xy and ab belong to the same branch, we may use the condition (1.6).

We have $ab \leq xy$ implies $0 \cdot xy = 0 \cdot ab$, by Theorem 2.4.18 and Lemma 2.2.2(7). Hence

$$\begin{aligned} 0(xy)^k &= (0 \cdot xy)(xy)^{k-1} = (0 \cdot ab)(xy)^{k-1} = (0 \cdot xy^{k-1})(ab) \\ &= \ldots = (0 \cdot xy)(ab)^{k-1} = (0 \cdot ab)(ab)^{k-1} = 0(ab)^k. \end{aligned}$$

□

Theorem 2.6.19. *In any solid weak BCC-algebra $\mathfrak{X}$ the following identity:*

$$0x^k \cdot 0y^k = 0(xy)^k \tag{2.23}$$

is satisfied for any $x, y \in X$ and each natural k.

Proof. Let $x \in B(a)$. Then, by Lemma 2.2.2(7), $a \leq x$ implies $0x = 0a$. Let us suppose that $0x^k = 0a^k$ for some natural k. Then also

$$0a^k \cdot x \leq 0a^k \cdot a,$$

by (1.5). Consequently,

$$0x^{k+1} = 0x^k \cdot x = 0a^k \cdot x \leq 0a^k \cdot a = 0a^{k+1},$$

which means that $0x^{k+1} = 0a^{k+1}$ because $0a^{k+1} \in I(X)$. So,

$$0x^k = 0a^k \tag{2.24}$$

is valid for all $x \in B(a)$ and each natural k.
Similarly,

$$0y^k = 0b^k \tag{2.25}$$

for all $y \in B(b)$ and each natural k.

Combining (2.23) with (2.24), (2.25) and (2.22) we see that a weak BCC-algebra $\mathfrak{X}$ satisfies the identity (2.23) if and only if

$$0a^k \cdot 0b^k = 0(ab)^k$$

holds for $a, b \in I(X)$.

But in view of Theorem 2.6.13 and Theorem 2.5.7 in the group $\langle I(X); *, 0\rangle$ the last equation $\mathfrak{X}$ can be written in the form

$$a^{-k} * b^k = (a * b^{-1})^{-k}.$$

Since $\langle I(X); *, 0\rangle$ is a commutative group (by Theorem 2.5.8), this equation is valid for all $a, b \in I(X)$. Hence (2.23) is satisfied for all $x, y \in X$ and all natural k.

The proof is completed. □

Corollary 2.6.20. *The map $\varphi_k(x) = 0x^k$ is an endomorphism of each solid weak BCC-algebra $\mathfrak{X}$.*

Definition 2.6.21. A weak BCC-algebra for which φ_k is an endomorphism is called *k-strong*. In the case $k = 1$, we say that it is *strong*.

A solid weak BCC-algebra is strong for every k. The converse statement is not true.

Example 2.6.22. A weak BCC-algebra $\mathfrak{X}$ with the operation $\cdot$ defined by Table 1.1.2 is not solid, but it is strong for every k.

For the proof, observe that in this weak BCC-algebra $0x = 0$ for $x \in B(0)$, $0x = 2$ for $x \in B(2)$. It is easy to see that this algebra is 1-strong. We have also $0x^2 = 0$ for all $x \in \{0,1,2,3\}$ and, of course, it is 2-strong.

Since in this algebra $0x^k = 0$ for even k, and $0x^k = 0x$ for odd k, it is strong for every k. □

Example 2.6.23. Direct computations show that a group-like weak BCC-algebra induced by the symmetric group S_3 is k strong for $k = 5$ and $k = 6$ but not for $k \in \{1,2,3,4,7,8\}$. □

Theorem 2.6.24. *A weak BCC-algebra $\mathfrak{X}$ is strong if and only if $\mathfrak{I}(\mathfrak{X})$ is a BCI-algebra.*

Proof. Indeed, if $\mathfrak{X}$ is strong, then $0a \cdot 0b = 0 \cdot ab$ holds for all $a, b \in I(X)$. Thus, in the group $\langle I(X); *, 0\rangle$ we have $a^{-1} * b = (a * b^{-1})^{-1} = b * a^{-1}$, which means that the group $\langle I(X); *, 0\rangle$ is commutative. Hence, by Theorem 2.5.8 $\langle I(X); \cdot, 0\rangle$ is a BCI-algebra.

On the other side, according to Lemma 2.2.2(7), for any $x \in B(a)$, $y \in B(b)$ we have $0x = 0a$ and $0y = 0b$. So, if $\mathfrak{I}(\mathfrak{X})$ is a BCI-algebra, then for any $a, b, c \in I(X)$ we have $ab \cdot c = ac \cdot b$.
Consequently,

$$\begin{aligned}
0x \cdot 0y &= 0a \cdot 0b \\
&= ((ab \cdot ab)a) \cdot 0b \\
&= ((ab \cdot a) \cdot ab) \cdot 0b \\
&= ((aa \cdot b) \cdot ab) \cdot 0b \\
&= (0b \cdot ab) \cdot 0b \\
&= (0b \cdot 0b) \cdot ab \\
&= 0 \cdot ab \\
&= 0 \cdot xy,
\end{aligned}$$

because $xy \in B(ab)$, by Theorem 2.4.18.
This completes the proof. □

Corollary 2.6.25. *A strong weak BCC-algebra is k-strong for every k.*

Proof. In a strong weak BCC-algebra $\mathfrak{X}$ the group $\langle I(X); *, 0\rangle$ is commutative and $0z^k = 0c^k$ for every $z \in B(c)$. Thus, by (2.22)

$$0x^k \cdot 0y^k = 0a^k \cdot 0b^k = a^{-k} * b^k = (a * b^{-1})^{-k} = 0(ab)^k = 0(xy)^k$$

for all $x \in B(a)$ and $y \in B(b)$. □

Example 2.6.23 shows that the converse statement is not true, i.e., there are weak BCC-algebras which are strong for some k but not for $k = 1$.

Corollary 2.6.26. *A weak BCC-algebra $\mathfrak{X}$ in which $\mathfrak{I}(\mathfrak{X})$ is a BCI-algebra is strong for every k.*

Corollary 2.6.27. *In any strong weak BCC-algebra $\mathfrak{X}$ we have*

$$0 \cdot 0x^k = 0 \cdot (0x)^k \tag{2.26}$$

for every $x \in X$ and every natural k.

Proof. Putting $x = 0$ and $y = x$ in (2.23), we get (2.26). □

Lemma 2.6.28. *In a solid weak BCC-algebra $\mathfrak{X}$ for elements belonging to the same branch, the following identity is satisfied:*

$$x(x \cdot xy) = xy. \tag{2.27}$$

Proof. Let $x, y \in B(a)$ for some $a \in I(X)$. Then $x \cdot xy \leq y$, by Lemma 2.6.9. This implies $x \cdot xy \in B(a)$ because comparable elements belong to the same branch. Moreover, by (1.5), we also have

$$xy \leq x(x \cdot xy). \tag{2.28}$$

Since $\mathfrak{X}$ is solid,

$$x(x \cdot xy) \cdot xy = (x \cdot xy)(x \cdot xy) = 0.$$

Thus

$$x(x \cdot xy) \leq xy. \tag{2.29}$$

Combining the inequalities (2.28) with (2.29), we get (2.27).
This completes the proof. □

We present some generalizations of the above result.

Lemma 2.6.29. *In a solid weak BCC-algebra* $\mathfrak{X}$

$$x(x \cdot xy)^2 = xy^2$$

for x, y *belonging to the same branch.*

Proof. Indeed, using Lemma 2.6.28, we obtain

$$x(x \cdot xy)^2 = x(x \cdot xy) \cdot (x \cdot xy) = xy \cdot (x \cdot xy) = x(x \cdot xy) \cdot y = xy \cdot y = xy^2.$$

□

Theorem 2.6.30. *In a super solid weak BCC-algebra* $\mathfrak{X}$

$$x(x \cdot xy)^n = xy^n$$

for all natural n *and* x, y *belonging to the same branch.*

Proof. For $n = 1$ this theorem coincides with Lemma 2.6.28, for $n = 2$ with Lemma 2.6.29.

For $n \geq 3$, by Lemma 2.6.28, we have

$$\begin{aligned} x(x \cdot xy)^n &= x(x \cdot xy) \cdot (x \cdot xy)^{n-1} \\ &= xy \cdot (x \cdot xy)^{n-1} \\ &= (xy \cdot (x \cdot xy)) \cdot (x \cdot xy)^{n-2} \\ &= (x(x \cdot xy))y \cdot (x \cdot xy)^{n-2} \\ &= (xy \cdot y) \cdot (x \cdot xy)^{n-2} \\ &= ((xy \cdot y) \cdot (x \cdot xy)) \cdot (x \cdot xy)^{n-3}. \end{aligned}$$

Since, by the assumption, x, y belong to the same branch $B(a)$, by Corollary 2.6.10, also $x \cdot xy \in B(a)$. Thus

$$\begin{aligned} ((xy \cdot y) \cdot (x \cdot xy)) \cdot (x \cdot xy)^{n-3} &= (xy \cdot (x \cdot xy))y \cdot (x \cdot xy)^{n-3} \\ &= (x(x \cdot xy) \cdot y)y \cdot (x \cdot xy)^{n-3} \\ &= (xy \cdot y)y \cdot (x \cdot xy)^{n-3} \\ &= xy^3 \cdot (x \cdot xy)^{n-3} \\ &\cdots\cdots\cdots\cdots \\ &= xy^{n-1} \cdot (x \cdot xy) \\ &= \ldots = xy^n. \end{aligned}$$

This completes the proof. □

Theorem 2.6.31. *Any solid weak BCC-algebra $\mathfrak{X}$ can be extended to a solid weak BCC-algebra containing one more element.*

Proof. Let $\mathfrak{X} = \langle X; \cdot, 0\rangle$ be a solid weak BCC-algebra and let $\theta \notin X$. Then the structure $\mathfrak{X}' = \langle X \cup \{\theta\}; \star, 0\rangle$ with the operation $\star$ defined in Lemma 1.2.11 is a solid weak BCC-algebra.

For proof, observe that the above construction saves the number of branches. Indeed, $\theta \in B(0)$ since $0 < \theta < y$ for every $y \in B(0)$. So, $\langle X; \cdot, 0\rangle$ and $\langle X'; \star, 0\rangle$ have the same initial elements and the same branches determined by nonzero initial elements. The branch $B(0)$ has in $\langle X'; \star, 0\rangle$ one element more than in $\langle X; \cdot, 0\rangle$.

From the assumption and from Definition 2.6.1, it follows that we have to prove that the condition (1.6) is satisfied only in the cases, when x or y is equal to θ.

Case 1^0. $x = \theta$. Then $y \in B(0)$.
The subcases, when $z \in B(0)$ are obvious.

Let us suppose that $y \neq 0$ and $z \in B(a)$ for $a \in I(X)$. Then we have $\theta y \cdot z = 0z$.

On the other hand, $\theta z \cdot y = 0z \cdot y$. But $0z, 0z \cdot y \in X$, $\mathfrak{X}$ is solid and $0z \in I(X)$. So, we can apply the Lemma 2.6.15 and we have $0z \cdot y = 0z \cdot 0 = 0z$. Hence in this case (1.6) is satisfied.

If $y = 0$ and $z \in B(a)$ for $a \in I(X)$, then $\theta y \cdot z = \theta z = 0z$. On the other hand, $\theta z \cdot y = 0z \cdot y$. We have the same situation as before.

Case 2^0. $y = \theta$.
We have $x\theta \cdot z = xz$. On the other hand, $xz \cdot \theta = xz$. Also in this case condition (1.6) is satisfied.

The proof is completed. □

Corollary 2.6.32. *For every natural $n \geq 6$ there are at least two non-isomorphic proper solid weak BCC-algebras containing n elements.*

Proof. For $n = 6$ we have two such weak BCC-algebras (Example 2.6.5). Applying the above construction to these two algebras, we obtain two algebras with 7 elements, and so on. In this way, for any $n \geq 7$ we can construct at least two new weak BCC-algebras containing n elements. □

Theorem 2.6.33. *Any BCK-algebra $\mathfrak{X}$ can be embedded into a solid weak BCC-algebra as its $B(0)$ branch.*

Proof. Let $\langle X; \cdot, 0\rangle$ be a BCK-algebra and let $\theta \notin X$ be a fixed element. Then, as it is not difficult to see, $\langle X \cup \{\theta\}; \star, 0\rangle$ with the operation defined by Lemma 1.2.1 is a solid weak BCC-algebra containing $\langle X; \cdot, 0\rangle$ as its subalgebra.

This weak BCC-algebra contains two branches: $B(0)$ and $B(\theta)$. The first branch coincides with the initial BCK-algebra $\langle X; \cdot, 0\rangle$, the second contains only one element. □

Theorem 2.6.34. *Any BCK-algebra $\mathfrak{X}$ can be embedded into a solid weak BCC-algebra $\mathfrak{Y}$ without trivial branches.*

Proof. Let $\langle X; \cdot, 0\rangle$ be a BCK-algebra and $\langle Y; *, 0\rangle$ a solid weak BCC-algebra without trivial branches such that $X \cap Y = \{0\}$. On $X \cup Y$ we define a common operation $\star$ by putting

$$x \star y = \begin{cases} xy & \text{if} \quad x, y \in X, \\ x * y & \text{if} \quad x, y \in Y, \\ 0 * y & \text{if} \quad x \in X, \ y \in Y \setminus \{0\}, \\ x & \text{if} \quad x \in Y, \ y \in X. \end{cases}$$

Then long but simple calculations show that $\langle X \cup Y; \star, 0\rangle$ is a solid weak BCC-algebra. The operation $\star$ coincides on X with the operation of $\langle X; \cdot, 0\rangle$. On Y it coincides with the operation of $\langle Y; *, 0\rangle$. Hence the natural order $\leq_{X\cup Y}$ of $\langle X \cup Y; \star, 0\rangle$ coincides on X with the natural order $\leq_X$ of $\langle X; \cdot, 0\rangle$, and on Y with the natural order $\leq_Y$ of $\langle Y; *, 0\rangle$. Since for $x \in X$ and $y \in B(0) \subset Y$ we have $x \star y = 0 * y = 0$, i.e., $x \leq_{X\cup Y} y$, each element of X is smaller than each nonzero element of the branch $B(0)$ of a weak BCC-algebra $\langle Y; *, 0\rangle$. □

Corollary 2.6.35. *Any BCC-algebra can be embedded into a weak BCC-algebra without trivial branches.*

Proof. We can use the same construction. Obtained weak BCC-algebra will be solid only in the case when the initial BCC-algebra will be a BCK-algebra. □

The idea of the above construction is based on gluing graphs presented in the following example.

Example 2.6.36. Consider a BCK-algebra $\langle\{0,1,2,3\};\ \cdot\ ,0\rangle$ with the operation $\cdot$ defined by Table 2.6.5:

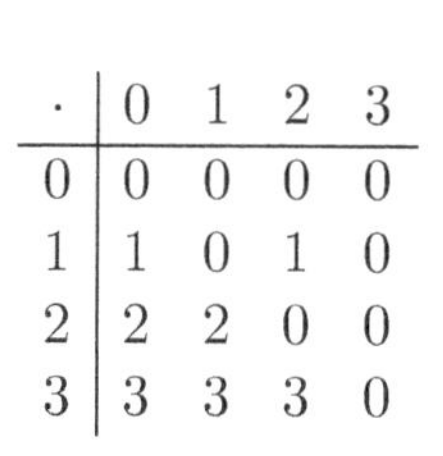

$\cdot$	0	1	2	3
0	0	0	0	0
1	1	0	1	0
2	2	2	0	0
3	3	3	3	0

Table 2.6.5

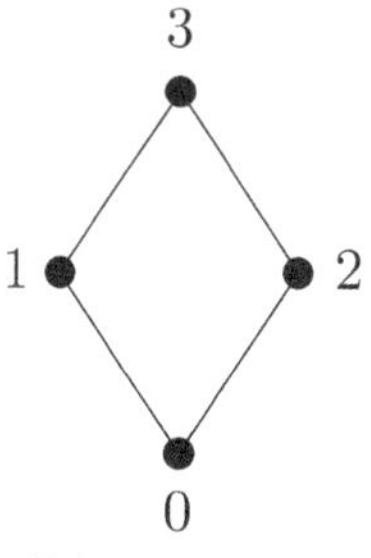

Diagram 2.6.3

and a solid weak BCC-algebra $\langle\{0,a,b,c,d\};\ *\ ,0\rangle$ with the operation $*$ defined by Table 2.6.6.

$*$	0	a	b	c	d
0	0	0	b	b	b
a	a	0	b	b	b
b	b	b	0	0	0
c	c	b	a	0	a
d	d	b	a	a	0

Table 2.6.6

Diagram 2.6.4

The above construction gives a solid weak BCC-algebra with the following operation $\star$:

$\star$	0	1	2	3	a	b	c	d
0	0	0	0	0	0	b	b	b
1	1	0	1	0	0	b	b	b
2	2	2	0	0	0	b	b	b
3	3	3	3	0	0	b	b	b
a	a	a	a	a	0	b	b	b
b	b	b	b	b	b	0	0	0
c	c	c	c	c	b	a	0	a
d	d	d	d	d	b	a	a	0

Table 2.6.7.

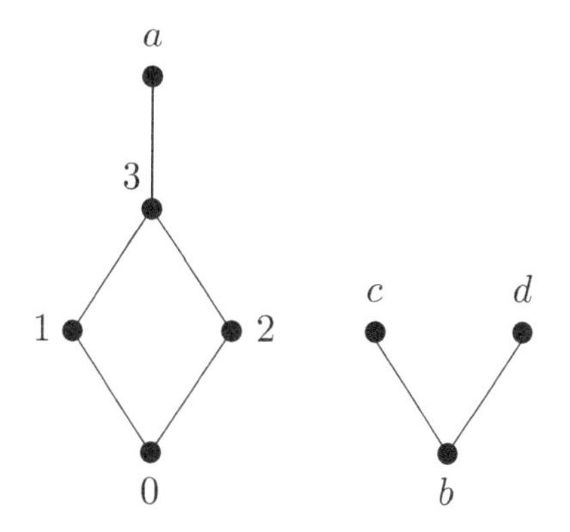

Diagram 2.6.5.

□

2.7 Nilpotent Weak BCC-Algebras

A special role in weak BCC-algebras is played by elements having a finite order, i.e., elements for which there exists some natural k such that $0x^k = 0$ (cf. [131]). We characterize sets of such elements and prove that the properties of such elements can be described by the properties of initial elements of branches containing these elements.

Definition 2.7.1. An element x of a weak BCC-algebra $\mathfrak{X}$ is called *nilpotent*, if there exists some positive integer k such that $0x^k = 0$. The smallest k with this property is called the *nilpotency index* of x and is denoted by $n(x)$. A weak BCC-algebra in which all elements are nilpotent is called *nilpotent.*

By $N_k(X)$ we denote the set of all nilpotent elements $x \in X$ such that $n(x) = k$. $N(X)$ denotes the set of all nilpotent elements of $\mathfrak{X}$. It is clear that $N_1(X) = B(0)$.

Example 2.7.2. In weak BCC-algebras defined by Tables 1.1.2 and 1.1.3, we have $n(0) = n(1) = 1$, $n(2) = n(3) = 2$. In a weak BCC-algebra defined by Table 2.2.1, there are no elements with $n(x) = 2$ but there are three elements with $n(x) = 3$ and three with $n(x) = 1$. □

Theorem 2.7.3. *Elements belonging to the same branch have the same nilpotency index.*

Proof. Let $x \in B(a)$. Then $a \leq x$, which, by Lemma 2.2.2(7), implies $0a = 0x$. This, together with $a \leq x$, by (1.5) and (1.4), gives

$$0x^2 \leq 0a \cdot x \leq 0a^2.$$

Hence $a \leq x$ implies $0x^2 \leq 0a^2$.

In the same manner

$$0x^k \leq 0a^k \text{ implies } 0x^{k+1} \leq 0a^{k+1},$$

which, by induction, proves

$$0x^m \leq 0a^m$$

for every $x \in B(a)$ and any natural m.
Thus

$$0a^m = 0 \text{ implies } 0x^m = 0. \tag{2.30}$$

On the other hand, from $0x^m = 0$ we obtain $0 \leq 0a^m$. This implies $0 = 0a^m$ since $0, 0a^m \in I(X)$ and elements of $I(X)$ are incomparable, i.e.,

$$0x^m = 0 \text{ implies } 0a^m = 0. \tag{2.31}$$

Finally, (2.30) and (2.31) give

$$0x^m = 0 \text{ if and only if } 0a^m = 0,$$

which means $n(x) = n(a)$ for every $x \in B(a)$. □

Corollary 2.7.4. *A weak BCC-algebra $\mathfrak{X}$ is nilpotent if and only if its subalgebra $\mathfrak{I}(\mathfrak{X})$ is nilpotent.*

Corollary 2.7.5. $x \in B(a) \cap N_k(X) \Longrightarrow B(a) \subset N_k(X)$.

The above results show that the study of nilpotency of given weak BCC-algebras can be reduced to the study of nilpotency of its initial elements.

Theorem 2.7.6. *Let $\mathfrak{X}$ be a weak BCC-algebra. If $\mathfrak{I}(\mathfrak{X})$ is a BCI-algebra, then $N_k(X)$ is a subalgebra and a BCC-ideal of $\mathfrak{X}$ for every k.*

Proof. Obviously $0 \in N_k(X)$ for every k. Let $x, y \in N_k(X)$.

Then $0x^k = 0y^k = 0$ and $0x^k = 0a^k = 0$, $0y^k = 0b^k = 0$ for some $a, b \in I(X)$.

Since $I(X)$ is a BCI-algebra, by Theorem 2.6.19, we have

$$0 = 0a^k \cdot 0b^k = 0(ab)^k.$$

Hence $ab \in N_k(X)$. Consequently,

$$xy \in B(a)B(b) = B(ab) \subset N_k(X).$$

So, $N_k(X)$ is a subalgebra of $\mathfrak{X}$.

Let now $x \in B(a)$, $y \in B(b)$, $z \in B(c)$. If y, $xy \cdot z \in N_k(X)$, then also b, $ab \cdot c \in N_k(X)$. Thus $0b^k = 0$ and

$$0(ac)^k = 0a^k \cdot 0c^k = (0a^k \cdot 0b^k) \cdot 0c^k = 0(ab \cdot c)^k = 0,$$

which implies $ac \in N_k(X)$. This, together with Theorem 2.4.18 and Corollary 2.7.5, implies $xz \in B(ac) \subset N_k(X)$.
Therefore, $N_k(X)$ is a BCC-ideal of $\mathfrak{X}$. Clearly, it is a BCK-ideal, too. □

Corollary 2.7.7. *$\mathfrak{N}_k(\mathfrak{X})$ is a subalgebra of each solid weak BCC-algebra $\mathfrak{X}$.*

Theorem 2.7.8. *$\mathfrak{N}(\mathfrak{X})$ is a subalgebra of each weak BCC-algebra $\mathfrak{X}$, in which $\mathfrak{I}(\mathfrak{X})$ is a BCI-algebra.*

Proof. Since $N(X) = \bigcup_{k \in N} N_k(X)$ and $0 \in N_k(X)$ for every k, the set $N(X)$ is nonempty. Let $x \in B(a)$, $y \in B(b)$.

If $x, y \in N(X)$ and $n(x) = m$, $n(y) = n$, then $0x^m = 0y^n = 0$. From this, by Theorem 2.7.3, we obtain $0a^m = 0b^n = 0$, which in the group $\langle I(X); *, 0\rangle$ can be written in the form $a^{-m} = b^{-n} = 0$. But $\mathfrak{I}(\mathfrak{X})$ is a BCI-algebra, hence $\langle I(X); *, 0\rangle$ is a commutative group. Thus

$$0 = (a^{-m})^n * (b^n)^m = a^{-mn} * b^{mn} = (a * b^{-1})^{-mn} = 0(ab)^{mn},$$

by Theorem 2.5.7. Hence $ab \in N(X)$. This implies $xy \in B(ab) \subset N(X)$. Therefore, $\mathfrak{N}(\mathfrak{X})$ is a subalgebra of X. □

Corollary 2.7.9. *$\mathfrak{N}(\mathfrak{X})$ is a subalgebra of each solid weak BCC-algebra.*

Corollary 2.7.10. *Any solid weak BCC-algebra $\mathfrak{X}$ with finite $I(X)$ is nilpotent.*

Proof. Indeed, $\mathfrak{I}(\mathfrak{X})$ is a greatest group-like BCI-algebra contained in any solid weak BCC-algebra (Theorem 2.6.13). Hence the group $\langle I(X); *, 0\rangle$ is commutative. If it is finite, then each of its elements has finite order k. Thus $0a^k = 0 * a^{-k} = 0$ for every $a \in I(X)$. Consequently $B(a) \subset N_k(X) \subset N(X)$ for every $a \in I(X)$. Therefore, $X = N(X)$. □

Corollary 2.7.11. *A solid weak BCC-algebra $\mathfrak{X}$ is nilpotent if and only if each element of the group $\langle I(X); *, 0\rangle$ has finite order.*

Corollary 2.7.12. *In a solid weak BCC-algebra $\mathfrak{X}$ the nilpotency of each $x \in N(X)$ is a divisor of $Card(I(X))$.*

3

Subclasses of BCC-Algebras

From the first chapter we know that the classes of BCC-algebras and weak BCC-algebras do not form a variety. However, there are subclasses of them that do. In this chapter, we cover the most known of them chiefly when some classical identities are satisfied only by elements belonging to the same branch (cf. for example [58]).

We will start with the so-called commutative solid weak BCC-algebras and show their connections with lattices. Further, we study various types of implicative weak BCC-algebras and so-called weak BCC-algebras with condition (S). In addition, we will look at and present the properties of subclasses such as fuzzy, hyper and other (weak) BCC-algebras.

3.1 Commutative Solid Weak BCC-Algebras

In any weak BCC-algebra $\mathfrak{X}$ we can define a binary operation $\wedge$ by putting

$$x \wedge y = y \cdot yx$$

for all $x, y \in X$.
A weak BCC-algebra satisfying the identity

$$x \cdot xy = y \cdot yx, \tag{3.1}$$

i.e.,

$$y \wedge x = x \wedge y,$$

is called *commutative.*

Theorem 3.1.1. *A weak BCC-algebra $\mathfrak{X}$ satisfying condition* (3.1) *is a commutative BCK-algebra.*

Proof. In any weak BCC-algebra $\mathfrak{X}$ satisfying (3.1) we have

$$0 \cdot 0x = x \cdot x0 = 0,$$

i.e., $\varphi^2(x) = 0$ for every $x \in X$.

Thus $\varphi(x) = \varphi^3(x) = 0$, by Lemma 2.2.2(5). Hence in this algebra $0 \leq x$ for every $x \in X$. This means that $\mathfrak{X}$ is a BCC-algebra with the condition (3.1).

But in any BCC-algebra $\mathfrak{X}$ we have $0 \leq yx$ for all $x, y \in X$, which by (1.5) implies $y \cdot yx \leq y$. This means that in $\mathfrak{X}$ the following inequality is satisfied:

$$x \cdot xy = y \cdot yx \leq y.$$

Consequently, it satisfies the identity $(x \cdot xy)y = 0$ and, by (1.2), it is a BCK-algebra. □

It is well known that the class of all commutative BCK-algebras forms a variety. Then, based on this fact and Theorem 3.1.1, we can formulate the following corollary:

Corollary 3.1.2. *The class of all commutative weak BCC-algebras forms a variety. It is a subvariety of that of the class of all commutative BCK-algebras.*

However, there are weak BCC-algebras, in which identity (3.1) is satisfied by elements belonging to the same branch.

Definition 3.1.3. A weak BCC-algebra $\mathfrak{X}$, in which (3.1) is satisfied by elements belonging to the same branch, is called *branchwise commutative.*

From Theorem 3.1.1 it follows that the assumption in the definition that the elements belong to the same branch is necessary. We show it in the following example:

Example 3.1.4. A weak BCC-algebra $\langle\{0, 1, 2, 3, 4\}; \cdot, 0\rangle$ with the operation $\cdot$ defined by Table 3.1.1 has two branches: $B(0) = \{0, 1, 2\}$ and $B(3) = \{3, 4\}$. In this algebra, the condition (3.1) is satisfied only in the case when x, y are in the same branch.

Indeed, for the elements of B(0) we have $0 = 0 \cdot 0x = x \cdot (x \cdot 0) = 0$ and

$$1 = 1 \cdot 0 = 1 \cdot (1 \cdot 2) = 2 \cdot (2 \cdot 1) = 2 \cdot 1 = 1.$$

For the elements of B(3) we have

$$3 = 3 \cdot 0 = 3 \cdot (3 \cdot 4) = 4 \cdot (4 \cdot 3) = 4 \cdot 1 = 3.$$

But for elements belonging to different branches we have

$$3 = 1 \cdot 3 = 1 \cdot (1 \cdot 3) \neq 3 \cdot (3 \cdot 1) = 3 \cdot 3 = 0.$$

·	0	1	2	3	4
0	0	0	0	3	3
1	1	0	0	3	3
2	2	1	0	4	3
3	3	3	3	0	0
4	4	3	3	1	0

Table 3.1.1 □

Theorem 3.1.5. *For a solid weak BCC-algebra $\mathfrak{X}$, the following conditions are equivalent:*

(1) $\mathfrak{X}$ *is branchwise commutative,*

(2) $xy = x(y \cdot yx)$ *for* $x, y \in B(a), a \in I(X)$,

(3) $x = y \cdot yx$ *for* $x \leq y$,

(4) $x \cdot xy = y \cdot y(x \cdot xy)$ *for* $x, y \in B(a), a \in I(X)$.

Proof. (1) $\Longrightarrow$ (2) Let $x, y \in B(a)$ for some $a \in I(X)$. Then, by Corollary 2.6.10, elements $y \cdot yx$ and $x \cdot xy$ are in $B(a)$. Thus

$$x(y \cdot yx) \cdot xy = (x \cdot xy)(y \cdot yx) = 0.$$

Hence

$$x(y \cdot yx) \leq xy. \tag{3.2}$$

From (3.1) we have

$$x(x(x \cdot xy)) = (x \cdot xy)((x \cdot xy)x). \tag{3.3}$$

Now we have

$$\begin{aligned} xy \cdot (x(y \cdot yx)) &= xy \cdot (x(x \cdot xy)) && \text{by (3.1)} \\ &= (x(x(x \cdot xy)))y && \text{by (1.6)} \\ &= (x \cdot xy)((x \cdot xy)x) \cdot y && \text{by (3.3)} \\ &= (x \cdot xy)(xx \cdot xy) \cdot y && \text{by (1.6)} \end{aligned}$$

$$= (x \cdot xy)(0 \cdot xy) \cdot y$$

$$= (x \cdot xy)0 \cdot y \qquad \text{from Theorem 2.4.20}$$

$$= (x \cdot xy)y$$

$$= xy \cdot xy$$

$$= 0.$$

Thus

$$xy \leq x(y \cdot yx). \tag{3.4}$$

Combining (3.2) with (3.4) we get (2).

(2) $\Longrightarrow$ (3) For $x \leq y$, by (2), we obtain $0 = x(y \cdot yx)$.

On the other hand, for all $x, y \in B(a), a \in I(X)$ we have

$$(y \cdot yx)x = yx \cdot yx = 0.$$

Thus

$$x(y \cdot yx) = (y \cdot yx)x = 0,$$

which, by (BCC5), implies (3).

(3) $\Longrightarrow$ (4) Let $x, y \in B(a)$ for some $a \in I(X)$.
Then $(x \cdot xy)y = 0$, i.e., $x \cdot xy \leq y$. This, by (3), implies (4).

(4) $\Longrightarrow$ (1) If $x, y \in B(a)$, then also $y \cdot yx \in B(a)$. Thus

$$(x \cdot xy)(y \cdot yx) = (y(y(x \cdot xy)))(y \cdot yx) = (y(y \cdot yx))(y(x \cdot xy)).$$

From this, applying Lemma 2.6.28, we obtain

$$(x \cdot xy)(y \cdot yx) = yx \cdot (y(x \cdot xy)).$$

But,

$$yx \cdot (y(x \cdot xy)) = (y(y(x \cdot xy)))x = (x \cdot xy)x = xx \cdot xy = 0,$$

because $xy \in B(0)$. Hence $(x \cdot xy)(y \cdot yx) = 0$.

In a similar way, we show $(y \cdot yx)(x \cdot xy) = 0$, which, together with (BCC5), shows that a solid weak BCC-algebra satisfying (4) is branchwise commutative. □

In the investigations of branchwise commutative solid weak BCC-algebras a special role is played by the set

$$A(b) = \{x \in X \mid x \leq b\}$$

called *initial part* of a weak BCC-algebra $\mathfrak{X}$ determined by $b \in X$.

Theorem 3.1.6. *A solid weak BCC-algebra $\mathfrak{X}$ is branchwise commutative if and only if*

(1) *each branch of X is a semilattice with respect to the operation $\wedge$ defined by $x \wedge y = y \cdot yx$ or equivalently,*

(2) *$A(x) \cap A(y) = A(x \wedge y)$ for all x, y belonging to the same branch.*

Proof. Let $\mathfrak{X}$ be a branchwise commutative solid weak BCC-algebra. Then, by Lemma 2.6.9, for $x, y \in B(a)$, $a \in I(X)$, we have

$$y \wedge x = x \wedge y = x \cdot xy \leq y.$$

Hence $x \wedge y \in B(a)$. Moreover, for any $p \leq q$, we have

$$p = p0 = p \cdot pq = q \wedge p = p \wedge q.$$

Thus

$$p \leq q \Longrightarrow p = p \wedge q. \tag{3.5}$$

We prove that $p \wedge q$ is the greatest lower bound for any $p, q \in B(a)$. Let us consider an arbitrary element $z \in X$ such that $z \leq p$ and $z \leq q$. Then, obviously, $z \in B(a)$ and

$$z(p \wedge q) = z(q \cdot qp) = (z \wedge q)(q \cdot qp),$$

by (3.5). Thus

$$z(p \wedge q) = (q \cdot qz)(q \cdot qp) = q(q \cdot qp) \cdot qz$$

because $q \cdot qp = p \wedge q \in B(a)$ for $p, q \in B(a)$. The last expression, by Lemma 2.6.28, is equal to $qp \cdot qz$. Hence $z(p \wedge q) = qp \cdot qz = 0$, because $qp \leq qz$ for $z \leq p$. So, $z \leq p \wedge q$. This means that $p \wedge q$ is the greatest lower bound of p and q. Hence the greatest lower bound of each $p, q \in B(a)$ is in B(a), so $B(a)$ is a semilattice with respect to $\wedge$. This proves (1).

Now, if (1) is satisfied and $z \in A(x) \cap B(y)$ for some $x, y \in B(a)$, $a \in I(X)$, then $z \leq x$ and $z \leq y$. Thus $z \in B(a)$ and $z \leq x \wedge y$, since $x \wedge y \in B(a)$ is the greatest lower bound of x and y. Hence $z \in A(x \wedge y)$. Consequently, $A(x) \cap A(y) \subseteq A(x \wedge y)$.

On the other hand, for any $z \in A(x \wedge y)$, by Lemma 2.6.9, we have

$$z \leq x \wedge y = y \cdot yx \leq x.$$

Hence $z \in A(x)$. Since

$$(x \wedge y)y = (y \cdot yx)y = yy \cdot yx = 0 \cdot yx = 0,$$

we have $x \wedge y \leq y$, which implies $z \in A(y)$. So, $z \in A(x) \cap A(y)$. Thus $A(x \wedge y) \subseteq A(x) \cap A(y)$ and hence $A(x \wedge y) = A(x) \cap A(y)$. This proves (2).

Finally, if (2) is satisfied, then

$$A(x \wedge y) = A(x) \cap A(y) = A(y) \cap A(x) = A(y \wedge x).$$

Hence $x \wedge y \in A(y \wedge x)$ and $y \wedge x \in A(x \wedge y)$. Therefore, $y \wedge x \leq x \wedge y$ and $x \wedge y \leq y \wedge x$, which implies $x \wedge y = y \wedge x$. So, X is branchwise commutative. □

In the proof of the next theorem, we will need the following well-known result of the theory of BCK-algebras (cf. [117]).

Lemma 3.1.7. *If p is the greatest element of a commutative BCK-algebra $\mathfrak{X}$, then $(X; \leq)$ is a distributive lattice with respect to the operations $x \wedge y = y \cdot yx$ and $x \vee y = p(px \wedge py)$.*

Theorem 3.1.8. *In a solid branchwise commutative weak BCC-algebra $\mathfrak{X}$, for every $p \in X$, the set $A(p)$ is a distributive lattice with respect to the operations $x \wedge y = y \cdot yx$ and $x \vee_p y = p(px \wedge py)$.*

Proof. 1^0. We prove that $(A^p; \leq)$, where

$$A^p = \{px \mid x \in A(p)\},$$

is a distributive lattice.

First, we show that $\mathfrak{A}^p$ is a subalgebra of $\mathfrak{B}(0)$. It is clear that $0 = pp \in A^p$ and $A(p) \subseteq B(a)$ for some $a \in I(X)$. Thus $A^p \subset B(0)$, by Theorem 2.4.20. Obviously $a \leq x$ for every $x \in A(p)$. Consequently, $px \leq pa$. Hence pa is the greatest element of A^p.

Let px, py be arbitrary elements of A^p. Then obviously $yx \in B(0)$ and $z = p \cdot yx \in A^p$ because $zp = (p \cdot yx)p = 0$. Moreover, by (BCC1$'$)

$$zx = (p \cdot yx)x = px \cdot yx \leq py.$$

Since

$$(px \cdot pz) \cdot zx = (px \cdot zx) \cdot pz = 0,$$

we have also $px \cdot pz \leq zx \leq py$. Therefore,

$$0 = (px \cdot pz) \cdot py = (px \cdot py) \cdot pz,$$

i.e.,

$$px \cdot py \leq pz. \tag{3.6}$$

On the other hand, since a weak BCC-algebra $\mathfrak{X}$ is branchwise commutative, for every $y \in A(p)$, according to Theorem 3.1.5(3), we have $p \cdot py = y$. Hence $px \cdot py = (p \cdot py)x = yx$. But

$$pz \cdot yx = p(p \cdot yx) \cdot yx = (p \cdot yx)(p \cdot yx) = 0.$$

Thus $pz \leq yx = px \cdot py$, which together with (3.6) gives

$$px \cdot py = pz.$$

Hence $\mathfrak{A}^p$ is a subalgebra of $\mathfrak{B}(0)$.

Obviously, $\mathfrak{B}(0)$ as a BCC-algebra contained in $\mathfrak{X}$ is commutative, and consequently it is a commutative BCK-algebra. Thus $\mathfrak{A}^p$ is a commutative BCK-algebra, too. By Lemma 3.1.7, $(A^p; \leq)$ is a distributive lattice.

2^0. Now we show that $(A(p); \leq)$ is a distributive lattice. It is clear that p is the greatest element of $A(p)$.

Let $x, y \in A(p)$. Then $px, py \in A^p$ and from the fact that $(A^p; \leq)$ is a lattice, it follows that there exists the last upper bound $pz \in A^p$, i.e.,

$$px \vee_p py = pz. \tag{3.7}$$

Observe that for $x, y \in A(p)$ we have

$$py \leq px \Longleftrightarrow x \leq y. \tag{3.8}$$

Indeed, in view of (1.5), $x \leq y$ implies $py \leq px$.

Similarly, $py \leq px$ implies $p \cdot px \leq p \cdot py$. But $\mathfrak{X}$ is branchwise commutative, hence, by Theorem 3.1.5(3), for every $v \in A(p)$ we have $p \cdot pv = v$. Therefore,

$$x = p \cdot px \leq p \cdot py = y.$$

From (3.7) and (3.8) it follows that z is the greatest lower bound for

x and y. Hence $x \wedge y = z$. Moreover, we have $p(x \wedge y) = pz$ and $px \vee_p py = pz$, which implies

$$p(x \wedge y) = px \vee_p py. \tag{3.9}$$

Analogously, we can prove that for all $x, y \in A(p)$ there exists $x \vee_p y$ and

$$p(x \vee_p y) = px \wedge py. \tag{3.10}$$

Therefore, $(A(p); \leq)$ is a lattice.

Since (3.9) and (3.10) are satisfied in A^p and $(A^p; \leq)$ is a distributive lattice, we have

$$px \vee_p (py \wedge pz) = (px \vee_p py) \wedge (px \vee_p pz)$$

for all $px, py, pz \in A^p$.

This, in view of (3.9) and (3.10), gives

$$p(x \wedge (y \vee_p z)) = p((x \wedge y) \vee_p (x \wedge z)).$$

Consequently,

$$p \cdot p(x \wedge (y \vee_p z)) = p \cdot p((x \wedge y) \vee_p (x \wedge z))$$

and

$$x \wedge (y \vee_p z) = (x \wedge y) \vee_p (x \wedge z),$$

because $x \wedge (y \vee_p z)$, $(x \wedge y) \vee_p (x \wedge z) \in A(p)$. This means that $(A(p); \leq)$ is a distributive lattice.

In this lattice $x \vee_p y = p(p(x \vee_p y)) = p(px \wedge py)$.

This completes the proof. □

Definition 3.1.9. A weak BCC-algebra $\mathfrak{X}$ is called *restricted* if every its branch has the greatest element.

The greatest element of the branch $B(a)$ will be denoted by 1_a. By N_a we will denote the unary operation $N_a : X \to X$ defined by $N_a x = 1_a x$.

Lemma 3.1.10. *The main properties of the operation N_a in restricted solid weak BCC-algebra $\mathfrak{X}$ are as follows:*

(1) $N_a 1_a = 0$ *and* $N_a 0 = 1_a$,

(2) $N_a N_a x \leq x$,

(3) $(N_a x)y = (N_a y)x$,

(4) $x \leq y \Longrightarrow N_a y \leq N_a x$,

(5) $N_a x N_a y \leq yx$,

(6) $N_a N_a N_a x = N_a x$,

where $x, y \in B(a)$, $a \in X$.

Proof. (1) is obvious.

(2) follows from (1.3).

(3) follows from (1.6). Indeed,

$$(N_a x)y = (1_a x)y = (1_a y)x = (N_a y)x.$$

To prove (4), let $x \leq y$. Then, from (1.5) it follows $1_a y \leq 1_a x$, which is equivalent to $N_a y \leq N_a x$.

To prove (5), we have

$$N_a x N_a y = 1_a x 1_a y = (1_a \cdot 1_a y)x \leq yx,$$

by (1.6), (2) and (1.4).

The condition (6) follows from (2.27). □

Definition 3.1.11. A restricted weak BCC-algebra $\mathfrak{X}$ is called *involutory* if $N_a N_a x = x$ holds for every $x \in B(a)$, $a \in I(X)$. An element x satisfying this condition is called an *involution*.

Simple examples of involutions, in restricted solid weak BCC-algebras, are $a \in I(X)$ and 1_a.

In an involutory weak BCC-algebra, the map $N_a : B(a) \to B(0)$ is one-to-one. Thus, in an involutory weak BCC-algebra $\mathfrak{X}$ with finite $B(0)$, all branches are finite.

Theorem 3.1.12. *Any branchwise commutative restricted solid weak BCC-algebra* $\mathfrak{X}$ *is involutory.*

Proof. Indeed,

$$N_a N_a x = 1_a(1_a x) = x(x 1_a) = x0 = x$$

for every $x \in B(a)$, $a \in I(X)$. □

Lemma 3.1.13. *In an involutory solid weak BCC-algebra* $\mathfrak{X}$ *the condition*

$$xy = N_a y N_a x$$

is valid for all $a \in I(X)$ *and* $x, y \in B(a)$.

Proof. In fact, $xy = (N_a N_a x)y = (1_a \cdot 1_a x)y = 1_a y \cdot 1_a x = N_a y N_a x$. □

Theorem 3.1.14. *A solid weak BCC-algebra* $\mathfrak{X}$ *is involutory if and only if*

$$xN_a y = yN_a x$$

holds for all $a \in I(X)$ *and* $x, y \in B(a)$.

Proof. According to Theorem 2.4.20, $N_a x, N_a y \in B(0)$ for $x, y \in B(a)$. Thus $yN_a x, xN_a y \in B(a)$, and consequently

$$xN_a y \cdot yN_a x = (x \cdot 1_a y)(y \cdot 1_a x) = x(y \cdot 1_a x) \cdot 1_a y = (1_a \cdot 1_a x)(y \cdot 1_a x) \cdot 1_a y.$$

Since, by (BCC1$'$),

$$(1_a \cdot 1_a x)(y \cdot 1_a x) \leq 1_a y,$$

it follows

$$(1_a \cdot 1_a x)(y \cdot 1_a x) \cdot 1_a y = 0.$$

Hence $xN_a y \cdot yN_a x = 0$.
Analogously, we show $yN_a x \cdot xN_a y = 0$, which, by (BCC5), implies $xN_a y = yN_a x$.

On the other hand, if $xN_a y = yN_a x$ holds for all $a \in I(X)$ and $x, y \in B(a)$, then for $y = 1_a$ and arbitrary $x \in B(a)$ we have

$$x = x \cdot 1_a 1_a = xN_a 1_a = 1_a N_a x = N_a N_a x,$$

which means that this weak BCC-algebra is involutory.
The proof is completed. □

Theorem 3.1.15. *For an involutory solid weak BCC-algebra* $\mathfrak{X}$ *the following statements are equivalent:*

(1) *each branch of* X *is a lower semilattice,*

(2) *each branch of* X *is a lattice.*

Moreover, if $(B(a); \leq)$ *is a lattice, then*

$$x \wedge y = N_a(N_a x \vee_a N_a y) \text{ and } x \vee_a y = N_a(N_a x \wedge N_a y).$$

Proof. (1) $\Longrightarrow$ (2) Since each branch $B(a)$ of X is a lower semilattice, then $N_ax \wedge N_ay$ exists for all $x, y \in B(a)$, i.e., for all $N_ax, N_ay \in B(0)$. Hence

$$N_ax \wedge N_ay \leq N_ax,$$

which gives

$$zN_ax \leq z(N_ax \wedge N_ay)$$

for each $z \in B(a)$. Similarly,

$$zN_ay \leq z(N_ax \wedge N_ay).$$

This means that $z(N_ax \wedge N_ay)$ is an upper bound of zN_ax and zN_ay.

Let us assume that $u \in B(a)$ for some $a \in I(X)$ is another upper bound for zN_ax and zN_ay. From $zN_ax \leq u$ and $zN_ay \leq u$ we have $zu \leq z(zN_ax)$ and $zu \leq z(zN_ay)$. But $z(zN_av) \leq N_av$ for $v, z \in B(a)$ because $N_av \in B(0)$ and $zN_av \in B(a)$. Hence $zu \leq N_ax$ and $zu \leq N_ay$, which implies $zu \leq N_ax \wedge N_ay$. Thus

$$z(N_ax \wedge N_ay) \leq z(zu) \leq u$$

and $z(N_ax \wedge N_ay)$ is the least upper bound of zN_ax and zN_ay. Therefore, for every $x, y, z \in B(a)$, there exists the least upper bound of zN_ax and zN_ay, i.e.,

$$zN_ax \vee_a zN_ay.$$

In particular, for every $x, y \in B(a)$, there exists

$$1_aN_ax \vee_a 1_aN_ay = N_aN_ax \vee_a N_aN_ay = x \vee_a y.$$

This shows that $(B(a); \leq)$ is an upper semilattice. Consequently, $B(a)$ is a lattice.

(2) $\Longrightarrow$ (1) This case is obvious.

Since $(B(a); \leq)$ is a lattice for every $a \in I(X)$, using the same argumentation as in the second part of the proof of Theorem 3.1.8 we can show that

$$N_a(x \wedge y) = N_ax \vee_a N_ay$$

for $x, y \in B(a)$.

Thus,

$$x \wedge y = N_aN_a(x \wedge y) = N_a(N_ax \vee_a N_ay).$$

Analogously, $N_a(x \vee_a y) = N_a x \wedge N_a y$ implies

$$x \vee_a y = N_a N_a(x \vee_a y) = N_a(N_a x \wedge N_a y).$$

This completes the proof. $\square$

Now we will consider n-fold branchwise commutative solid weak BCC-algebras, which are a generalization of branchwise commutative solid weak BCC-algebras.

Definition 3.1.16. A weak BCC-algebra $\mathfrak{X}$ is called *n-fold branchwise commutative* (shortly: *n-b commutative*) if there exists a natural number n such that

$$xy = x(y \cdot yx^n) \tag{3.11}$$

holds for $x, y \in B(a),\ a \in I(X)$.

Theorem 3.1.5(2) shows that for $n = 1$ we are dealing with a branchwise commutative weak BCC-algebra.

Theorem 3.1.17. *For a solid weak BCC-algebra $\mathfrak{X}$ the following conditions are equivalent:*

(1) *$\mathfrak{X}$ is n-b commutative,*

(2) $x \cdot xy \leq y \cdot yx^n$ *for* $x, y \in B(a),\ a \in I(X)$,

(3) $x \leq y \Longrightarrow x \leq y \cdot yx^n$.

Proof. (1) $\Longrightarrow$ (2) Let $x, y \in B(a)$ for some $a \in B(a),\ a \in I(X)$. Then $xy \in B(0)$ and consequently $x(y \cdot yx^n) \in B(0)$. This, by Theorem 2.4.20, shows that x and $y \cdot yx^n$ are in the same branch. Hence $y \cdot yx^n \in B(a)$ and

$$(x \cdot xy)(y \cdot yx^n) = x(y \cdot yx^n) \cdot xy = xy \cdot xy = 0.$$

Therefore, $x \cdot xy \leq y \cdot yx^n$.

(2) $\Longrightarrow$ (3) Because x, y are belonging to the same branch, we can use Lemma 2.6.28. We get

$$x(x \cdot xy) = xy = 0,$$

by assumption. So, we have

$$x \leq x \cdot xy \leq y \cdot yx^n,$$

i.e.,

$$x \leq y \cdot yx^n.$$

(3) $\Longrightarrow$ (1) Since $0 \leq xy$ for $x, y \in B(a)$, $a \in I(X)$, we have

$$x \cdot xy \leq x. \tag{3.12}$$

Consequently, $yx \leq y(x \cdot xy)$. This implies

$$yx \cdot x \leq y(x \cdot xy) \cdot x = yx \cdot (x \cdot xy)$$

and

$$yx \cdot (x \cdot xy) \leq y(x \cdot xy) \cdot (x \cdot xy).$$

Therefore,

$$yx^2 = yx \cdot x \leq y(x \cdot xy) \cdot (x \cdot xy) = y(x \cdot xy)^2,$$

i.e.,

$$yx^2 \leq y(x \cdot xy)^2.$$

From this inequality we obtain

$$yx^2 \cdot x \leq y(x \cdot xy)^2 \cdot x. \tag{3.13}$$

By (3.12), we get

$$y(x \cdot xy)^2 \cdot x \leq y(x \cdot xy)^2 \cdot (x \cdot xy),$$

which, together with (3.13), gives

$$yx^3 \leq y(x \cdot xy)^3.$$

Repeating the above procedure, we can see that

$$yx^n \leq y(x \cdot xy)^n$$

holds for every natural n. Hence

$$x(y \cdot yx^n) \leq x(y \cdot y(x \cdot xy)^n). \tag{3.14}$$

Obviously, $x \cdot xy \leq y$ for x, y belonging to the same branch $B(a)$, $a \in I(X)$. Applying (3) to the last inequality we obtain

$$x \cdot xy \leq y \cdot y(x \cdot xy)^n.$$

Hence, by (1.5) and Lemma 2.6.28, we conclude

$$x(y \cdot y(x \cdot xy)^n) \leq x(x \cdot xy) = xy.$$

Consequently,

$$x(y \cdot y(x \cdot xy)^n) \leq xy. \tag{3.15}$$

Combining (3.14) and (3.15), we get

$$x(y \cdot yx^n) \leq xy. \tag{3.16}$$

Thus $x(y \cdot yx^n) \in B(0)$. This, by Theorem 2.4.20, means that $y \cdot yx^n \in B(a)$. But in this case $yx^n \in B(0)$.

Indeed, if $yx^n \in B(b)$ for some $b \in I(X)$, then $y \cdot yx^n \in B(a)B(b) = B(ab)$, by Corollary 2.4.17.

So, $y \cdot yx^n \in B(a) \cap B(ab)$. Thus $B(a) = B(ab)$, i.e., $a = ab$. Hence $0 = ab \cdot a = aa \cdot b = 0b$. Therefore, $b \in B(0)$ and $b = 0$, because $b \in I(X)$.

This means that

$$(y \cdot yx^n)y = 0 \cdot yx^n = 0.$$

Consequently, $y \cdot yx^n \leq y$ which, by (1.5), implies $xy \leq x(y \cdot yx^n)$.

Comparing the last inequality with (3.16) we obtain $xy = x(y \cdot yx^n)$. This completes the proof. □

In 1981 in [20] W.H. Cornish formulated the condition

(J) $\quad x \cdot x(y \cdot yx) = y \cdot y(x \cdot xy)$.

He proved that the class of all BCK-algebras satisfying the condition (J) forms a variety.

An analogous result holds for BCC-algebras, although, in general, BCC-algebras with the condition (J) are not BCK-algebras. For example, in the proper BCC-algebras defined by Tables 1.2.10, 1.2.11 and 1.2.12, the condition (J) holds. In other BCC-algebras of order 4, this identity is not satisfied.

It can be easily seen that condition (J) holds in all BCC-algebras $\langle X'; \star, 0\rangle$ constructed in Lemma 1.2.10, Corollary 1.2.23 and Theorem 1.2.26 if it holds in the initial BCC-algebra $\langle X; \cdot, 0\rangle$.

Moreover, if it holds in all BCC-algebras $\langle X_\xi, \cdot_\xi, 0_\xi\rangle$, $\xi \in \Lambda$, then it holds also in a BCC-algebra constructed in Example 1.2.16.

Theorem 3.1.18. *The class of all BCC-algebras satisfying condition (J) is a variety which has (BCC1), (BCC3) and (J) as its equational base. These identities are independent.*

Proof. Let us suppose that $\mathfrak{X} = \langle X; \cdot, 0\rangle$ satisfies (BCC1), (BCC3) and the condition (J). First, we prove that this algebra satisfies (BCC5). Indeed, if $xy = yx = 0$, then (J) and (BCC3) yields $x = y$. Hence (BCC5) holds.
Thus, by Corollary 1.1.24, $\mathfrak{X}$ is a weak BCC-algebra.

By putting in (J) $y = 0$, we get

$$x \cdot x(0 \cdot 0x) = 0 \cdot 0(x \cdot x0) \Longleftrightarrow x \cdot x(0 \cdot 0x) = 0,$$

i.e.,

$$x \leq x(0 \cdot 0x).$$

By combining it with Lemma 2.2.2(2) we obtain

$$0 \cdot 0x \leq x(0 \cdot 0x).$$

Replacing x by $0x$ and using the identity (5) from Lemma 2.2.2 we obtain

$$0x = 0(0 \cdot 0x) \leq 0x \cdot 0(0 \cdot 0x) = 0x \cdot 0x = 0,$$

which gives $0x = 0$. This proves (BCC4).
Hence $\mathfrak{X}$ is a BCC-algebra.

To prove the independence of these axioms, let us consider the algebra $\langle\{0, 1, 2\}; \cdot, 0\rangle$ with the operation $\cdot$ defined by the following table

$\cdot$	0	1	2
0	0	0	0
1	1	2	0
2	2	1	0

Table 3.1.2

It can be easily seen that (BCC3) is satisfied. By easy computing, we can check that the algebra satisfies (J), too. However, since

$$((1 \cdot 1) \cdot (2 \cdot 1)) \cdot (1 \cdot 2) = (2 \cdot 1) \cdot 0 = 1 \cdot 0 = 1$$

it does not satisfy (BCC1), i.e., (BCC1) is independent.

Let us now note that an algebra $\langle X; \cdot, 0\rangle$ with $xy = 0$ for all $x, y \in X$ satisfies (BCC1) and (J) but it does not satisfy (BCC3). Hence (BCC3) is independent.

From the fact that there are BCC-algebras that do not satisfy (J), the condition (J) is independent, too.

The proof is completed. □

3.2 Quasi-Commutative Solid Weak BCC-Algebras

In this section, we investigate the quasi-commutative solid weak BCC-algebras studied by many mathematicians and show many interesting results (cf. for example [102]). We also provide an interesting, specific proving method.

In a weak BCC-algebra $\mathfrak{X}$ for nonnegative integers m, n, we define a polynomial $Q_{m,n}(x, y)$ by putting:

$$Q_{m,n}(x, y) = (x \cdot xy)(xy)^m \cdot (yx)^n.$$

Definition 3.2.1. A weak BCC-algebra $\mathfrak{X}$ is called *quasi-commutative of type* $(m, n; i, j)$ if there exist two pairs of nonnegative integers i, j and m, n such that

$$Q_{m,n}(x, y) = Q_{i,j}(y, x)$$

or equivalently

$$(x \cdot xy)(xy)^m \cdot (yx)^n = (y \cdot yx)(yx)^i \cdot (xy)^j$$

holds for all $x, y \in \mathfrak{X}$. If the above identity holds only for all x, y belonging to the same branch $B(a)$, $a \in I(X)$, then we say that this weak BCC-algebra is *branchwise quasi-commutative* (or *b-quasi-commutative* for short).

Exchanging x and y in $Q_{m,n}(x, y) = Q_{i,j}(y, x)$ we see that *a weak BCC-algebra $\mathfrak{X}$ is quasi-commutative of type $(i, j; m, n)$ if and only if it is quasi-commutative of type $(m, n; i, j)$.*

Example 3.2.2. We give two simple examples of b-quasi-commutative weak BCC-algebras:

(1) A group-like weak BCC-algebra $\mathfrak{X}$ is b-quasi-commutative of any type since each its branch has only one element.

(2) A weak BCC-algebra $\mathfrak{X}$ is branchwise commutative (commutative) if and only if it is b-quasi-commutative (quasi-commutative) of type $(0, 0; 0, 0)$. □

Theorem 3.2.3. *A b-quasi-commutative solid weak BCC-algebra $\mathfrak{X}$ of type $(0, k; 0, 0)$ is branchwise commutative.*

Proof. Let $\mathfrak{X}$ be a weak BCC-algebra satisfying the assumption. Then

$$Q_{0,k}(x,y) = (x \cdot xy)(yx)^k = y \cdot yx = Q_{0,0}(y,x)$$

for $x, y \in B(a),\ a \in I(X)$.

For $k = 0$, it is obviously branchwise commutative.

Let $k > 0$. Then $yx \in B(0)$. Hence $0 \leq yx$. This, by (1.5), implies $(x \cdot xy)(yx) \leq x \cdot xy$. Thus

$$y \cdot yx = (x \cdot xy)(yx)^k \leq (x \cdot xy)(yx)^{k-1} \leq \ldots \leq (x \cdot xy)(yx) \leq x \cdot xy,$$

i.e., $y \cdot yx \leq x \cdot xy$.

Interchanging x and y we get $x \cdot xy = y \cdot yx$. □

Theorem 3.2.4. *In a solid weak BCC-algebra $\mathfrak{X}$, the following inequalities:*

$$(1)\quad Q_{n-1,n}(x,y) \geq Q_{n,n}(x,y) \geq Q_{n,n+1}(x,y) \geq Q_{n+1,n+1}(x,y),$$
$$(2)\quad Q_{n-1,n}(x,y) \geq Q_{n,n}(y,x) \geq Q_{n,n+1}(x,y) \geq Q_{n+1,n+1}(y,x)$$

are valid for all natural n and $x, y \in B(a),\ a \in I(X)$.

Proof. (1) Observe that $x \cdot xy \in B(a)$ and $(x \cdot xy)(xy)^k \in B(a)$ for every k and $x, y \in B(a)$. The first is a consequence of Corollary 2.6.10, the second follows from the fact that $0 \leq xy$ implies $a \cdot xy \leq a$, i.e., $a \cdot xy = a$, because $a \in I(X)$. Therefore $a = a \cdot (xy)^k \leq (x \cdot xy)(xy)^k$. Thus using (BCC1$'$) and (1.4) we obtain

$$\begin{aligned}
Q_{n,n}(x,y) \cdot Q_{n-1,n}(x,y) = \\
&= ((x \cdot xy)(xy)^n \cdot (yx)^n)((x \cdot xy)(xy)^{n-1} \cdot (yx)^n) \\
&\leq ((x \cdot xy)(xy)^n \cdot (yx)^{n-1})((x \cdot xy)(xy)^{n-1} \cdot (yx)^{n-1}) \\
&\leq ((x \cdot xy)(xy)^n \cdot (yx)^{n-2})((x \cdot xy)(xy)^{n-1} \cdot (yx)^{n-2}) \\
&\leq \ldots \leq (x \cdot xy)(xy)^n \cdot (x \cdot xy)(xy)^{n-1} \\
&\leq \ldots \leq (x \cdot xy)(xy) \cdot (x \cdot xy) \\
&= (x \cdot xy)(x \cdot xy) \cdot xy \\
&= 0 \cdot xy \\
&= 0.
\end{aligned}$$

Thus

$$Q_{n,n}(x,y) \cdot Q_{n-1,n}(x,y) = 0,$$

which proves

$$Q_{n,n}(x,y) \leq Q_{n-1,n}(x,y).$$

Similarly,

$$\begin{aligned}
Q_{n,n+1}(x,y) \cdot Q_{n,n}(x,y) = \\
&= ((x \cdot xy)(xy)^n \cdot (yx)^{n+1})((x \cdot xy)(xy)^n \cdot (yx)^n) \\
&\leq ((x \cdot xy)(xy)^n \cdot (yx)^n)((x \cdot xy)(xy)^n \cdot (yx)^{n-1}) \\
&\leq \ldots \leq ((x \cdot xy)(xy)^n \cdot yx) \cdot (x \cdot xy)(xy)^n \\
&= ((x \cdot xy)(xy)^n \cdot (x \cdot xy)(xy)^n) \cdot yx \\
&= 0 \cdot yx \\
&= 0.
\end{aligned}$$

Hence

$$Q_{n,n+1}(x,y) \leq Q_{n,n}(x,y).$$

The last inequality of (1) is a consequence of the first. So, we have proven the inequality (1).

(2) If $x, y \in B(a)$, $a \in I(X)$, then $xy, yx \in B(0)$ and hence $x \cdot xy, y \cdot yx \in B(a)$, by Theorem 2.4.20 and Corollary 2.6.10. From this, analogously as in the proof of (1), we can deduce that, for every natural n, elements $(x \cdot xy)(xy)^{n-1}$ and $(x \cdot xy)(xy)^{n-1} \cdot (yx)^n$ are in $B(a)$. Therefore, using this fact, we obtain the inequality

$$Q_{n,n}(y,x) \leq Q_{n-1,n}(x,y).$$

Indeed,

$$\begin{aligned}
Q_{n,n}(y,x) \cdot Q_{n-1,n}(x,y) = \\
&= ((y \cdot yx)(yx)^n \cdot (xy)^n)((x \cdot xy)(xy)^{n-1} \cdot (yx)^n) \\
&= ((y \cdot yx)(yx)^n \cdot ((x \cdot xy)(xy)^{n-1} \cdot (yx)^n))(xy)^n \\
&\leq ((y \cdot yx) \cdot (x \cdot xy)(xy)^{n-1})(xy)^n \\
&= (y \cdot yx)(xy)^n \cdot (x \cdot xy)(xy)^{n-1} \\
&\leq (y \cdot yx)(xy) \cdot (x \cdot xy)
\end{aligned}$$

$$\leq (y \cdot yx)x$$
$$= 0.$$

Similarly,

$$
\begin{aligned}
Q_{n,n+1}(x,y) \cdot Q_{n,n}(y,x) = \\
&= ((x \cdot xy)(xy)^n \cdot (yx)^{n+1})((y \cdot yx)(yx)^n \cdot (xy)^n) \\
&= ((x \cdot xy)(xy)^n \cdot ((y \cdot yx)(yx)^n \cdot (xy)^n))(yx)^{n+1} \\
&\leq ((x \cdot xy) \cdot (y \cdot yx)(yx)^n)(yx)^{n+1} \\
&= (x \cdot xy)(yx)^{n+1} \cdot (y \cdot yx)(yx)^n \\
&\leq (x \cdot xy)(yx) \cdot (y \cdot yx) \\
&\leq (x \cdot xy)y \\
&= 0.
\end{aligned}
$$

This proves that

$$Q_{n,n+1}(x,y) \leq Q_{n,n}(y,x).$$

The last inequality of (2) is a consequence of the first. □

Theorem 3.2.5. *Every solid weak BCC-algebra $\mathfrak{X}$, which is decomposed into a finite number of finite branches is b-quasi-commutative of some type of the form $(m,m;m,m+1)$.*

Proof. For any $a \in I(X)$ the branch $B(a)$ is finite. Hence for each pair of elements $x, y \in B(a)$, the sequence (2) from Theorem 3.2.4 is finite. This means that for all $x, y \in B(a)$, $a \in I(X)$ there exists natural $n' = n(x,y)$ such that $Q_{n,n}(x,y) = Q_{n,n+1}(y,x)$ for all $n \geq n'$. Since $I(X)$ is finite, for every

$$m \geq \max\{n(x,y) \mid x, y \in B(a),\, a \in I(X)\}$$

and x, y belonging to the same branch we have $Q_{m,m}(x,y) = Q_{m,m+1}(y,x)$, which shows that $\mathfrak{X}$ is quasi-commutative of type $(m,m;m,m+1)$. □

Corollary 3.2.6. *Any finite solid weak BCC-algebra is b-quasi-commutative of some type $(m,m;m,m+1)$.*

Theorem 3.2.7. *If a proper weak BCC-algebra $\mathfrak{X}$ is quasi-commutative of type $(i,j;m,n)$, then $i - j + m - n + 1 \neq \pm 1$.*

Proof. Since, by the assumption, a weak BCC-algebra $\mathfrak{X}$ is proper, it has at least two branches, i.e., there exists $a \in I(X)$ such that $a \neq 0$. For this a we have $Q_{i,j}(0,a) \cdot Q_{m,n}(a,0) = 0$, because $\mathfrak{X}$ is quasi-commutative of type $(i,j;m,n)$.

By Theorem 2.2.12, $\mathfrak{I}(\mathfrak{X})$ is a subalgebra of $\mathfrak{X}$. By Theorem 2.2.6 and Corollary 2.5.4, it is a group-like subalgebra. Hence, by Theorem 2.5.7, there exists a group $\langle I(X); *, 0\rangle$ such that $xy = x * y^{-1}$ for $x, y \in I(X)$. Thus

$$\begin{aligned}
0 &= Q_{i,j}(0,a) \cdot Q_{m,n}(a,0) \\
&= ((0 \cdot 0a)(0a)^i \cdot (a0)^j)((a \cdot a0)(a0)^m \cdot (0a)^n) \\
&= (a(0a)^i \cdot a^j)(0a^m \cdot (0a)^n) \\
&= (a^{1+i} * a^{-j}) * (a^{-m} * a^n)^{-1} \\
&= a^{1+i-j+m-n}.
\end{aligned}$$

For $i - j + m - n + 1 = \pm 1$, from the above we obtain $a^{\pm 1} = 0$, which implies $a = 0$. But this is contrary to our assumption on a. Therefore, it must be $i - j + m - n + 1 \neq \pm 1$. □

Theorem 3.2.8. *For $i - j + m - n + 1 \neq \pm 1$ there exists a group-like quasi-commutative weak BCC-algebra $\mathfrak{X}$ of type $(i,j;m,n)$.*

Proof. Let $k = |i - j + m - n + 1|$. By Theorem 3.2.7, we have $k \neq 1$. Let us consider a group-like weak BCC-algebra $\langle X; \cdot, 0\rangle$ induced by a commutative group $\langle X; *, 0\rangle$. Then, as it is not difficult to verify,

$$Q_{i,j}(x,y) \cdot Q_{m,n}(y,x) = (x^{-1} * y)^{i-j+m-n+1} = (x^{-1} * y)^{\pm k}.$$

This means that for $k = 0$, each group-like weak BCC-algebra, induced by a commutative group, is quasi-commutative of type $(i,j;m,n)$. For $k > 1$ such weak BCC-algebra will be induced by a cyclic group of order k. □

Theorem 3.2.9. *An algebra $\langle X; \cdot, 0\rangle$ of type $(2,0)$ is a quasi-commutative weak BCC-algebra of type $(i,j;m,n)$ if and only if it satisfies the following three identities:*

(1) $(xy \cdot zy) \cdot xz = 0$,

(2) $x0 = x$,

(3) $Q_{i,j}(x, y) = Q_{m,n}(y, x)$.

Proof. The necessity is obvious. To show the sufficiency, by Theorem 1.1.24, we only need to verify the axiom (BCC5) because (BCC1) coincides with (1) and (BCC3) coincides with (2).

If $xy = yx = 0$, then

$$Q_{i,j}(x, y) = (x \cdot xy)(xy)^i \cdot (yx)^j = x$$

and

$$Q_{m,n}(y, x) = (y \cdot yx)(yx)^m \cdot (xy)^n = y.$$

This, by (3), implies $x = y$ and completes the proof. □

Corollary 3.2.10. *The class of quasi-commutative weak BCC-algebras of a fixed type forms a variety.*

In the next theorem we show that the class of quasi-commutative weak BCC-algebras of a fixed type can be defined by only two equalities.

Theorem 3.2.11. *An algebra $\langle X; \cdot, 0\rangle$ of type $(2, 0)$ is a quasi-commutative weak BCC-algebra of type $(i, j; m, n)$ if and only if it satisfies the following identities:*

(α) $u((xy \cdot zy) \cdot xz) = u$,

(β) $Q_{i,j}(x, y) = Q_{m,n}(y, x) \cdot 0$

for any $x, y, z \in X$.

Proof. The necessity is obvious. To prove sufficiency, we will show that any algebra $\langle X; \cdot, 0\rangle$ satisfying the conditions (α), (β) also satisfies the conditions (1), (2), (3) from the previous theorem.

Let $\theta = (00 \cdot 00) \cdot 00$. Then, by ($\alpha$), we have

$$\theta\theta = \theta((00 \cdot 00) \cdot 00) = \theta.$$

Using (α) once again, for every $u \in X$ we obtain

$$u((\theta\theta \cdot \theta\theta) \cdot \theta\theta) = u,$$

which, in view of $\theta\theta = \theta$, gives $u\theta = u$.

By putting $y = z = \theta$ in (α) and applying just proven identity

$u\theta = u$, we get $u \cdot xx = u$ for all $x, u \in X$.
This means that

$$u(xx)^k = u \tag{3.17}$$

for any natural k. In particular $0(00)^k = 0$. Hence

$$Q_{i,j}(0,0) = (0 \cdot 00)(00)^i \cdot (00)^j = 0(00)^j = 0.$$

Similarly, $Q_{m,n}(0,0) = 0$. This, by (β), implies $00 = 0$. Consequently, $u0 = u \cdot 00 = u$ for every $u \in X$. So, the condition (2) from Theorem 3.2.9 is satisfied. Combining (2) and (β) we obtain the condition (3).

Observe that (3.17) for $u = xx$ implies $xx \cdot (xx)^k = xx$. From (3.17) we also obtain $0(xx)^k = 0$ for every k. Hence

$$\begin{aligned}
Q_{i,j}(xx,0) &= (xx \cdot (xx \cdot 0))(xx \cdot 0)^i \cdot (0 \cdot xx)^j \\
&= (xx \cdot xx)(xx)^i \cdot 0^j \\
&= xx \cdot (xx)^{i+1} \\
&= xx
\end{aligned}$$

and

$$\begin{aligned}
Q_{m,n}(0,xx) &= (0 \cdot (0 \cdot xx))(0 \cdot xx)^m \cdot (xx \cdot 0)^n \\
&= (00 \cdot 0^m) \cdot (xx)^n \\
&= 0(xx)^n \\
&= 0,
\end{aligned}$$

which together with the just proven (3) gives $xx = 0$ for every $x \in X$.
By putting $(xy \cdot zy) \cdot xz = u$ in (α) we have

$$u = u((xy \cdot zy) \cdot xz) = uu = 0.$$

This means that $(xy \cdot zy) \cdot xz = 0$, so any algebra $\langle X; \cdot, 0 \rangle$ satisfying (α), (β) satisfies also (3), and consequently it is a quasi-commutative weak BCC-algebra of type $(i, j; m, n)$. □

Theorem 3.2.12. *If a solid weak BCC-algebra $\mathfrak{X}$ is quasi-commutative of type $(i, j; m, n)$, then its branch $B(0)$ is a quasi-commutative BCK-algebra of one of the following three types: $(i, i; i, i)$, $(j, j; j, j)$ and $(n, j; j, n)$.*

The proof of this theorem is based on the following lemma.

Lemma 3.2.13. *In a quasi-commutative solid weak BCC-algebra $\mathfrak{X}$ of type $(i,j;m,n)$ we have*

(1) $xy^{i+1} = xy^{n+1}$,

(2) $xy^{j+1} = xy^{m+1}$

for $x, y \in B(0)$.

Proof. According to Theorem 1.1.27, $\mathfrak{B}(0)$ is the greatest BCC-algebra contained in $\mathfrak{X}$. Since $\mathfrak{X}$ is solid, for all $x, y, z \in B(0)$ we have $xy \cdot z = xz \cdot y$. Thus, $\mathfrak{B}(0)$ is a BCK-algebra.

First, observe that

$$x(x \cdot xy)^k = xy^k \tag{3.18}$$

for $x, y \in B(0)$ and any natural k.
Indeed, for $k = 1$ it is valid by Lemma 2.6.28.
If it is valid for some k, then for $k+1$ we have

$$\begin{aligned} x(x \cdot xy)^{k+1} &= \\ &= x(x \cdot xy)^k \cdot (x \cdot xy) \\ &= xy^k \cdot (x \cdot xy) && \text{by the assumption on } k \\ &= (xy^{k-1} \cdot (x \cdot xy))y \\ &= (xy^{k-2} \cdot (x \cdot xy))y^2 \\ &= \ldots = (x(x \cdot xy))y^k \\ &= xy \cdot y^k && \text{by Lemma 2.6.28} \\ &= xy^{k+1}. \end{aligned}$$

Therefore, (3.18) is valid for every natural k.
Hence

$$\begin{aligned} Q_{i,j}(x, xy) &= (x(x \cdot xy))(x \cdot xy)^i \cdot (xy \cdot x)^j \\ &= x(x \cdot xy)^{i+1} \cdot 0^j && \text{because } xy \cdot x = 0 \\ &= x(x \cdot xy)^{i+1}. \end{aligned}$$

Likewise,

$$\begin{aligned} Q_{m,n}(xy,x) &= (xy \cdot (xy \cdot x))(xy \cdot x)^m \cdot (x \cdot xy)^n \\ &= xy \cdot (x \cdot xy)^n \\ &= x(x \cdot xy)^n \cdot y \\ &= xy^{n+1}. \end{aligned}$$

Further, since $\mathfrak{X}$ is quasi-commutative of type $(i, j; m, n)$, we have

$$Q_{i,j}(x, xy) = Q_{m,n}(xy, x).$$

Thus, $xy^{i+1} = xy^{n+1}$. This proves the first identity.

The second equation follows from the fact that any quasi-commutative weak BCC-algebra $\mathfrak{X}$ of type $(i, j; m, n)$ is also quasi-commutative of type $(m, n; i, j)$. □

Proof of Theorem 3.2.12. Let a solid weak BCC-algebra $\mathfrak{X}$ be quasi-commutative of type $(i, j; m, n)$. Then, in particular,

$$(x \cdot xy)(xy)^i \cdot (yx)^j = (y \cdot yx)(yx)^m \cdot (xy)^n$$

for $x, y \in B(0)$.

Since $yx \in B(0)$, the second identity of Lemma 3.2.13 shows that

$$(y \cdot yx)(yx)^m = y(yx)^{m+1} = y(yx)^{j+1} = (y \cdot yx)(yx)^j.$$

Thus

$$(x \cdot xy)(xy)^i \cdot (yx)^j = (y \cdot yx)(yx)^j \cdot (xy)^n$$

for all $x, y \in B(0)$.

Hence $Q_{i,j}(x, y) = Q_{j,n}(y, x)$ for $x, y \in B(0)$.

So, $\mathfrak{B}(0)$ is quasi-commutative of type $(i, j; j, n)$. Obviously, it is also quasi-commutative of type $(j, n; i, j)$.

Repeating the above procedure we can show that $\mathfrak{B}(0)$ is quasi-commutative of type $(j, n; n, j)$.

This implies that it is quasi-commutative of type $(n, j; j, n)$.

For $j = n$ it is quasi-commutative of type $(j, j; j, j)$.

Thus in a quasi-commutative solid weak BCC-algebra of type

$(i, j; m, j)$, the subalgebra $\mathfrak{B}(0)$ is quasi-commutative of type $(j, j; j, j)$.

Finally, let us consider the case $i = j$, i.e., the quasi-commutativity of type $(i, i; m, n)$. From the first part of this proof it follows that in this case $\mathfrak{B}(0)$ is quasi-commutative of type $(i, i; i, n)$. Thus for $x, y \in B(0)$ and $i = j$ we have

$$(x \cdot xy)(xy)^i \cdot (yx)^i = (y \cdot yx)(yx)^i \cdot (xy)^n.$$

Since

$$(y \cdot yx)(yx)^i \cdot (xy)^n \leq (y \cdot yx)(yx)^i \cdot (xy)^i$$

for $i \leq n$ and $x, y \in B(0)$, the above implies

$$(x \cdot xy)(xy)^i \cdot (yx)^i \leq (y \cdot yx)(yx)^i \cdot (xy)^i.$$

Exchanging x and y we obtain

$$(y \cdot yx)(yx)^i \cdot (xy)^i \leq (x \cdot xy)(xy)^i \cdot (xy)^i,$$

which together with the previous inequality gives

$$(x \cdot xy)(xy)^i \cdot (yx)^i = (y \cdot yx)(yx)^i \cdot (xy)^i.$$

Therefore, in this case $\mathfrak{B}(0)$ is quasi-commutative of type $(i, i; i, i)$.

Corollary 3.2.14. *Suppose that $\mathfrak{X}$ is a quasi-commutative BCK-algebra of type $(i, j; m, n)$. Then its type of quasi-commutativity can be reduced to one of the following types: $(i, i; i, i)$, $(j, j; j, j)$ and $(n, j; j, n)$.*

3.3 Implicative Solid Weak BCC-Algebras

Implicative and positive implicative BCC-algebras originate from the systems of positive implicational calculus and weak positive implicational calculus in the implicational functor in logical systems.

In this section, we will also deal with some generalized implicative and positive implicative solid weak BCC-algebras called φ-implicative (cf. [58]).

Definition 3.3.1. A weak BCC-algebra $\mathfrak{X}$ is called *implicative* if

$$x \cdot yx = x \tag{3.19}$$

holds for all $x, y \in X$.

Putting $x = 0$ in the above equality, we get $0 \cdot y0 = 0$. This means that $0y = 0$ for any $y \in X$. Such weak BCC-algebra is not proper, it is a BCC-algebra and it has only one branch.

Definition 3.3.2. A weak BCC-algebra $\mathfrak{X}$ is called *branchwise implicative* if (3.19) holds for all x, y belonging to the same branch of X.

Example 3.3.3. The weak BCC-algebra $\langle\{0,1,2,3,4,5\}; \cdot, 0\rangle$ with the operation $\cdot$ defined by the following table:

$\cdot$	0	1	2	3	4	5
0	0	0	4	4	2	2
1	1	0	4	4	2	2
2	2	2	0	0	4	4
3	3	2	1	0	4	4
4	4	4	2	2	0	0
5	5	4	3	3	1	0

Table 3.3.1

is proper because $(5 \cdot 3) \cdot 2 \neq (5 \cdot 2) \cdot 3$. It has three branches: $B(0)$, $B(2)$, $B(4)$. Direct computation shows that it is branchwise commutative and branchwise implicative. Since $1 \cdot (1 \cdot 2) \neq 2 \cdot (2 \cdot 1)$ and $1 \cdot (2 \cdot 1) = 4$, it is neither commutative nor implicative. □

We begin with a few results describing branchwise implicativity in solid weak BCC-algebras.

Theorem 3.3.4. *Any solid branchwise implicative weak BCC-algebra $\mathfrak{X}$ is branchwise commutative.*

Proof. Let $x, y \in B(a)$ for some $a \in I(X)$. Then $x \cdot xy \leq y$, by Lemma 2.6.9. Hence x, y, $x \cdot xy \in B(a)$. Thus

$$(x \cdot xy) \cdot y(x \cdot xy) = x \cdot xy \tag{3.20}$$

since $\mathfrak{X}$ is branchwise implicative. Moreover, from $x \cdot xy \leq y$ and (1.4), it follows

$$(x \cdot xy) \cdot y(x \cdot xy) \leq y \cdot y(x \cdot xy).$$

Hence, by (3.20) and Lemma 2.6.9, we obtain

$$x \cdot xy \leq y \cdot y(x \cdot xy) \leq x \cdot xy.$$

So, $y \cdot y(x \cdot xy) = x \cdot xy$ for all $x, y \in B(a)$.

Theorem 3.1.5(4) completes the proof. □

Theorem 3.3.5. *In a solid branchwise implicative weak BCC-algebra* $\mathfrak{X}$ *the equation*

$$xy \cdot 0y = (xy \cdot 0y)y \cdot 0y$$

is satisfied for all $x, y \in B(a)$, $a \in I(X)$.

Proof. Let $\mathfrak{X}$ be branchwise implicative and solid. Then, according to BCC1, for all $x, y \in B(a)$, we have

$$(xy \cdot 0y)x = 0. \tag{3.21}$$

So, $xy \cdot 0y \in B(a)$ and xy, $x(xy \cdot 0y) \in B(0)$. Therefore,

$$\begin{aligned}
(xy \cdot 0y) \cdot x(xy \cdot 0y) &= (xy \cdot x(xy \cdot 0y)) \cdot 0y \\
&= (x \cdot x(xy \cdot 0y))y \cdot 0y \\
&= ((xy \cdot 0y) \cdot (xy \cdot 0y)x)y \cdot 0y && \text{by Theorem 3.3.4} \\
&= ((xy \cdot 0y)0)y \cdot 0y && \text{by (3.21)} \\
&= (xy \cdot 0y)y \cdot 0y.
\end{aligned}$$

Hence

$$(xy \cdot 0y) \cdot x(xy \cdot 0y) = (xy \cdot 0y)y \cdot 0y.$$

Since $xy \cdot 0y$ and x are in the same branch, the implicativity shows that

$$(xy \cdot 0y) \cdot x(xy \cdot 0y) = xy \cdot 0y,$$

which, together with the previous equation, gives

$$xy \cdot 0y = (xy \cdot 0y)y \cdot 0y.$$

□

Lemma 3.3.6. *In a solid weak BCC-algebra $\mathfrak{X}$ for x, y belonging to the same branch, the following inequality holds:*

$$(xy \cdot 0y)y \cdot 0y \leq ((xy \cdot y) \cdot 0y) \cdot 0y.$$

Proof. Indeed, if $x, y \in B(a)$, $a \in I(X)$, then, as in the previous proof, we can see that also $xy \cdot 0y \in B(a)$. Hence

$$\begin{aligned} ((xy \cdot 0y)y \cdot 0y)(((xy \cdot y) \cdot 0y) \cdot 0y) \leq \\ &\leq (xy \cdot 0y)y \cdot ((xy \cdot y) \cdot 0y) && \text{by (BCC1}^{'}\text{)} \\ &= ((xy \cdot 0y)((xy \cdot y) \cdot 0y))y \\ &\leq (xy \cdot (xy \cdot y))y && \text{by (BCC1}^{'}\text{)} \\ &\leq (x \cdot xy)y = xy \cdot xy = 0, \end{aligned}$$

i.e.,

$$((xy \cdot 0y)y \cdot 0y)(((xy \cdot y) \cdot 0y) \cdot 0y) = 0.$$

This implies

$$((xy \cdot 0y)y) \cdot 0y \leq ((xy \cdot y) \cdot 0y) \cdot 0y.$$

The proof is completed. □

Theorem 3.3.7. *If $I(X)$ is a BCK-ideal of a branchwise commutative solid weak BCC-algebra $\mathfrak{X}$ and*

$$xy \cdot 0y = ((xy \cdot y) \cdot 0y) \cdot 0y \tag{3.22}$$

is valid for all x, y belonging to the same branch $B(a)$, $a \in I(X)$, then $\mathfrak{X}$ is branchwise implicative.

Proof. Let $x, y \in B(a)$ for some $a \in I(X)$. Then

$$\begin{aligned} x(x \cdot yx) \cdot 0x &= (x(x \cdot yx) \cdot 0) \cdot 0x \\ &= (x(x \cdot yx) \cdot (x \cdot yx)(x \cdot yx)) \cdot 0x \\ &= (x(x \cdot yx) \cdot (x(x \cdot yx) \cdot yx)) \cdot 0x && x,\ x \cdot yx \in B(a) \\ &= (yx \cdot (yx \cdot x(x \cdot yx))) \cdot 0x, \end{aligned}$$

because yx, $x(x \cdot yx) \in B(0)$ and $\mathfrak{X}$ is branchwise commutative.
So, we have

$$x(x \cdot yx) \cdot 0x = (yx \cdot (yx \cdot x(x \cdot yx))) \cdot 0x. \tag{3.23}$$

Therefore,

$$
\begin{aligned}
(yx \cdot (yx \cdot x(x \cdot yx))) \cdot 0x &= (yx \cdot 0x)(yx \cdot x(x \cdot yx)) \\
&= (((yx \cdot x) \cdot 0x) \cdot 0x)(yx \cdot x(x \cdot yx)),
\end{aligned}
$$

by (3.22). And we get

$$
(yx \cdot (yx \cdot x(x \cdot yx))) \cdot 0x = (((yx \cdot x) \cdot 0x) \cdot 0x)(yx \cdot x(x \cdot yx)). \quad (3.24)
$$

Since, by (BCC1$'$), $(yx \cdot x) \cdot 0x \leq yx \cdot 0 = yx$ and $yx \in B(0)$ implies $(yx \cdot x) \cdot 0x \in B(0)$, from (3.23) and (3.24) we obtain

$$
\begin{aligned}
x(x \cdot yx) \cdot 0x &= (((yx \cdot x) \cdot 0x) \cdot 0x)(yx \cdot x(x \cdot yx)) \\
&= (((yx \cdot x) \cdot 0x)(yx \cdot x(x \cdot yx))) \cdot 0x \\
&= (((yx \cdot x)(yx \cdot x(x \cdot yx))) \cdot 0x) \cdot 0x,
\end{aligned}
$$

because, as it is not difficult to see, $yx \cdot x, 0x \in B(0a)$.

Now, from the fact that $yx \cdot x(x \cdot yx)$ and $x(x \cdot yx)$ are in $B(0)$ and $\mathfrak{X}$ is branchwise commutative, we have

$$
\begin{aligned}
x(x \cdot yx) \cdot 0x &= (((yx \cdot x)(yx \cdot x(x \cdot yx))) \cdot 0x) \cdot 0x \\
&= (((yx \cdot (yx \cdot x(x \cdot yx)))x) \cdot 0x) \cdot 0x \\
&= (((x(x \cdot yx) \cdot (x(x \cdot yx) \cdot yx))x) \cdot 0x) \cdot 0x.
\end{aligned}
$$

From this, in view of $x \cdot yx \in B(a)$, we get

$$
\begin{aligned}
x(x \cdot yx) \cdot 0x &= (((x(x \cdot yx) \cdot ((x \cdot yx)(x \cdot yx)))x) \cdot 0x) \cdot 0x \\
&= (((x(x \cdot yx) \cdot 0)x) \cdot 0x) \cdot 0x \\
&= ((x(x \cdot yx))x \cdot 0x) \cdot 0x \\
&= ((xx \cdot (x \cdot yx)) \cdot 0x) \cdot 0x \\
&= ((0 \cdot (x \cdot yx)) \cdot 0x) \cdot 0x \\
&= ((0x \cdot (0 \cdot yx)) \cdot 0x) \cdot 0x \qquad \text{by Theorem 2.6.12} \\
&= ((0x \cdot 0) \cdot 0x) \cdot 0x \\
&= (0x \cdot 0x) \cdot 0x \\
&= 0 \cdot 0x \in I(X),
\end{aligned}
$$

by Theorem 2.2.6. Hence $x(x{\cdot}yx){\cdot}0x \in I(X)$. Also $0x \in I(X)$. Since, by the assumption, $I(X)$ is a BCK-ideal of $\mathfrak{X}$, we obtain $x(x \cdot yx) \in I(X)$. But $x(x \cdot yx) \in B(0)$, so $x(x \cdot yx) \in I(X) \cap B(0)$. Thus $x(x \cdot yx) = 0$. This means that $x \leq x \cdot yx \leq x$. Consequently, $x \cdot yx = x$. Therefore $\mathfrak{X}$ is branchwise implicative.

The proof is completed. □

The example presented below shows that in the last theorem the assumption on $I(X)$ is essential.

Example 3.3.8. Let us take a weak BCC-algebra $\mathfrak{X}$ from the Example 2.6.3. Because $\langle\{0,1,3,4\};\ \cdot\,,0\rangle$ is a BCI-algebra (cf. the algebra I_{4-2-1} on p. 337 in [146]), to show that $\mathfrak{X}$ is a weak BCC-algebra it is sufficient to check the axiom (BCC1) only in the case when at least one of the elements x, y, z is equal to 2.

Such defined weak BCC-algebra is proper since $(2 \cdot 3) \cdot 4 \neq (2 \cdot 4) \cdot 3$. It is also branchwise commutative and satisfies (3.22) but it is not branchwise implicative, because $1 \cdot (2 \cdot 1) = 0 \neq 1$.

We have $1 \cdot 3,\ 3 \in I(X)$ but $1 \notin I(X)$. Hence $I(X)$ is not a BCK-ideal of $\mathfrak{X}$. □

Positive implicative BCK-algebras originate from the systems of positive implicational calculus in some logical systems. They were introduced and initially researched by K. Iseki in 1975 and together with S. Tanaka in 1978. There are many interesting results for these algebras by many authors. Also positive implicative BCC-algebras, as a generalization of positive implicative BCK-algebras, form a usable subclass. In this part of the book, we show results concerned with the positive implicative BCC-algebras and their role among other subclasses of BCC-algebras.

Definition 3.3.9. A BCC-algebra is called *positive implicative* if it satisfies the identity

$$xy \cdot y = xy. \tag{3.25}$$

Let us consider a solid weak BCC-algebra $\mathfrak{X}$ with the condition (3.25). If $x, y \in B(a)$, $a \in I(X)$, then $xy \cdot y \in B(aa \cdot a) = B(0a)$, and $xy \in B(0)$. Hence $0a = 0$, which implies $a = 0$, because $\mathfrak{I}(\mathfrak{X})$ is a group-like subalgebra. This means that such algebra is not proper and it is consisting of only one branch $B(0)$. So, $\mathfrak{X}$ is a BCC-algebra.

Example 3.3.10. The proper BCC-algebra $\mathfrak{X}$ with the operation $\cdot$ defined by Table 1.2.11 is an example of a BCC-algebra, which is positive implicative but it is neither commutative, because

$$1 = 1 \cdot 0 = 1 \cdot (1 \cdot 2) = 2 \cdot (2 \cdot 1) = 2 \cdot 2 = 0$$

nor implicative, because

$$1 \cdot (2 \cdot 1) = 1 \cdot 2 = 0 \neq 1.$$

But it is proper because $(2 \cdot (2 \cdot 3)) \cdot 3 \neq 0$. □

In the theorem below, we give the independent base of the class of positive implicative BCC-algebras

Theorem 3.3.11. *The class of all positive implicative BCC-algebras has* $(BCC1)$, $(BCC3)$, $(BCC5)$ *and* (3.25) *as its independent base.*

Proof. By putting $y = z = 0$ in (BCC1), we obtain $xx = 0$, i.e., (BCC2) is satisfied. Now, putting $x = y$ in (3.25) we obtain (BCC4). Hence an algebra satisfying (BCC1), (BCC3), (BCC5) and (3.25) is a positive implicative BCC-algebra.

Now we will prove the independence of the base of a positive implicative BCC-algebra.

The algebra with the operation $\cdot$ defined by Table 1.1.5 satisfies all these axioms with the exception of (BCC5). The algebra $\langle X; \cdot, 0\rangle$ with the operation $\cdot$ defined as $xy = 0$ satisfies all axioms with the exception of (BCC3). Similarly, the algebra $\langle X; \cdot, 0\rangle$ with the operation $\cdot$ defined as $xy = x$ shows that the axiom (BCC1) is independent. Since there are BCC-algebras which are not positive implicative, the axiom (3.25) is independent, too.

This completes the proof. □

Let us characterize the positive implicative BCC-algebras.

Theorem 3.3.12. *If in a weak BCC-algebra* $\mathfrak{X}$ *each proper branch is linear, then* X *is branchwise implicative if it is branchwise commutative and satisfies the identity*

$$xy \cdot 0y = ((xy)y \cdot 0y) \cdot 0y. \tag{3.26}$$

Proof. First, we consider the case when $x, y \in B(0)$. Then $yx \in B(0)$ because $\mathfrak{B}(0)$ is a subalgebra of $\mathfrak{X}$. So, (3.26) has the form

$$xy = xy \cdot y, \tag{3.27}$$

i.e., $\mathfrak{B}(0)$ is a positive implicative BCC-algebra.

Moreover, from $0 \leq yx$ and (1.5) we obtain

$$x \cdot yx \leq x. \tag{3.28}$$

Since $\mathfrak{X}$ is branchwise commutative, therefore,

$$x(x \cdot yx) = yx \cdot (yx \cdot x),$$

whence, according to (3.27), we obtain $yx \cdot (yx \cdot x) = 0$. So,

$$x \leq x \cdot yx,$$

which together with (3.28) proves $x \cdot yx = x$ for all $x, y \in B(0)$.

Now we consider the case when $x, y \in B(a)$, where $a \in I(X)$ and $a \neq 0$. Then $xy \in B(0)$ and, for $y = xy$, the condition (3.26) has the form

$$(x \cdot xy)(0 \cdot xy) = ((x \cdot xy)(xy \cdot (0 \cdot xy)))(0 \cdot xy)$$

whence, according to the above assumption on x and y, we obtain

$$x \cdot xy = (x \cdot xy) \cdot xy. \tag{3.29}$$

Since, by assumption, the branch $B(a)$ is linear, x and y are comparable, i.e., $xy = 0$ or $yx = 0$. In the second case, the branchwise commutativity implies $y = y0 = y \cdot yx = x \cdot xy$, which together with (3.29) gives $y = y \cdot xy$.

In the case $xy = 0$ the proof is analogous, but in (3.29) we must exchange x and y.

This completes the proof. □

From the first part of the above proof, the following corollary follows:

Corollary 3.3.13. *Any commutative and positive implicative BCC-algebra is an implicative BCK-algebra.*

Let us note that the conditions mentioned in Theorem 3.3.12 cannot be omitted.

Example 3.3.14. A weak group-like BCC-algebra satisfies (3.26), but it is not branchwise commutative. □

Example 3.3.15. A weak BCC-algebra $\langle \{0,1,2,3\};\ \cdot, 0\rangle$ with the operation $\cdot$ defined by the following table:

$\cdot$	0	1	2	3
0	0	0	0	3
1	1	0	0	3
2	2	1	0	3
3	3	3	3	0

Table 3.3.2

is branchwise commutative, but (3.26) is not satisfied in this algebra. Indeed, we have

$$(2\cdot 1)\cdot(0\cdot 1) = 1\cdot 0 = 1.$$

On the other side,

$$(((2\cdot 1)\cdot 1)\cdot(0\cdot 1))\cdot(0\cdot 1) = ((1\cdot 1)\cdot 0)\cdot 0 = 0.$$

Since

$$1\cdot(2\cdot 1) = 1\cdot 1 = 0 \neq 1,$$

this algebra is not branchwise implicative. □

Theorem 3.3.16. *A solid branchwise implicative weak BCC-algebra $\mathfrak{X}$ is branchwise positive implicative if and only if it is a commutative BCK-algebra.*

Proof. By Theorem 3.3.4, a solid branchwise implicative weak BCC-algebra $\mathfrak{X}$ is branchwise commutative. If it is branchwise positive implicative, then for elements from the same branch, say $B(a)$ for $a \in I(X)$, we have $xy = xy\cdot y$. The left side of the equation belongs to $B(aa) = B(0)$, while the right side belongs to $B(aa\cdot a) = B(0a)$. So, we get $B(0) = B(0a)$. This means that this weak BCC-algebra coincides with the branch $B(0)$. Hence, it is a commutative BCK-algebra.

Conversely, if a commutative BCK-algebra is implicative, then

$$(xy\cdot y)\cdot xy = (xy\cdot xy)y = 0y = 0$$

and

$$xy\cdot(xy\cdot y) = y(y\cdot xy) = yy = 0,$$

by commutativity and implicativity. Hence

$$(xy \cdot y) \cdot xy = xy \cdot (xy \cdot y) = 0,$$

which implies $xy \cdot y = xy$.

So, $\mathfrak{X}$ is positive implicative. □

As a consequence of the above results, we obtain a well-known characterization of implicative BCK-algebras presented below.

Theorem 3.3.17. *A BCK-algebra is implicative if and only if it is both commutative and positive implicative.*

Proof. Indeed, in an implicative BCK-algebra $\mathfrak{X}$ we have

$$xy = xy \cdot (y \cdot xy) = xy \cdot y$$

for all $x, y \in X$. Hence, $\mathfrak{X}$ is positive implicative.

By Theorem 3.3.4, it also is commutative.

On the other hand, in any commutative and positive implicative BCK-algebra $\mathfrak{X}$, we have

$$x(x \cdot yx) = yx \cdot (yx \cdot x) = yx \cdot yx = 0$$

and

$$(x \cdot yx)x = xx \cdot yx = 0 \cdot yx = 0$$

for all $x, y \in X$.

This implies $x \cdot yx = x$. So, $\mathfrak{X}$ is implicative. □

Since the class of BCI-algebras is included in the class of weak BCC-algebras, also a positive implicative BCI-algebra is a BCK-algebra. Therefore, a positive implicative weak BCC-algebra should be defined differently. One way was proposed by J. Meng and X. L. Xin [120]. They defined a positive implicative BCI-algebra as a BCI-algebra satisfying the identity

$$xy = (xy \cdot y) \cdot 0y.$$

Using this definition, one can prove that *a BCI-algebra is implicative if and only if it is both positive implicative and commutative.* Unfortunately, in the proof of this result, a critical role plays the identity (1.6). So, this proof cannot be transferred to weak BCC-algebras.

In connection with this, W.A. Dudek [42] introduced a new class of positive implicative weak BCC-algebras he called *φ-implicative.*

Definition 3.3.18. A weak BCC-algebra $\mathfrak{X}$ is called *φ-implicative* if it satisfies the identity

$$xy = xy \cdot y(0 \cdot 0y), \tag{3.30}$$

i.e.,

$$xy = xy \cdot y\varphi^2(y).$$

If (3.30) is satisfied only by elements belonging to the same branch, then we say that such weak BCC-algebra is *branchwise φ-implicative.*

It is clear that in the case of BCC-algebras, the conditions (3.30) and (3.25) are equivalent. Thus a BCC-algebra is φ-implicative if and only if it is positive implicative. For BCI-algebras and weak BCC-algebras, it is not the case. A group-like weak BCC-algebra is a simple example of a φ-implicative weak BCC-algebra, which is not positive implicative.

Example 3.3.19. Let us consider the BCI-algebra $\langle\{0, 1, 2, 3, 4\}; \cdot, 0\rangle$ with the operation $\cdot$ defined by Table 3.3.3 (cf. the proper BCI-algebra I_{5-2-3}, p. 339 in [146]).

This BCI-algebra has three branches: $B(0)$, $B(2)$ and $B(4)$. Since $3 \cdot 2 \neq ((3 \cdot 2) \cdot 2) \cdot \varphi(2)$, it is not (branchwise) positive implicative, but it is φ-implicative.

Indeed, the case $x \leq y$ is obvious because $0 \leq y\varphi^2(y)$ implies $y\varphi^2(y) \in B(0)$. The case $y \in I(X)$ is similar. The remaining five cases can be checked by standard simple calculations.

$\cdot$	0	1	2	3	4
0	0	0	4	4	2
1	1	0	4	4	2
2	2	2	0	0	4
3	3	2	1	0	4
4	4	4	2	2	0

Table 3.3.3

□

Theorem 3.3.20. *A solid weak BCC-algebra $\mathfrak{X}$ is branchwise implicative if and only if it is branchwise φ-implicative and branchwise commutative.*

Proof. Let $\mathfrak{X}$ be a branchwise implicative solid weak BCC-algebra. Then, by Theorem 3.3.4, it is branchwise commutative.

Next, for $x, y \in B(a), a \in I(X)$ we obtain

$$\begin{aligned}(xy \cdot y\varphi^2(y)) \cdot xy &= (xy \cdot xy) \cdot y\varphi^2(y)\\ &= 0 \cdot y\varphi^2(y)\\ &= 0y \cdot 0\varphi^2(y)\\ &= \varphi(y)\varphi^3(y)\\ &= \varphi(y)\varphi(y)\\ &= 0,\end{aligned}$$

because φ is an endomorphism (Theorem 2.6.12) and by Lemma 2.2.2(5). Thus

$$xy \cdot y\varphi^2(y) \leq xy. \tag{3.31}$$

Moreover, from the fact that a weak BCC-algebra $\mathfrak{X}$ is branchwise commutative and $xy,\ y\varphi^2(y) \in B(0)$, we obtain

$$\begin{aligned}xy \cdot (xy \cdot y\varphi^2(y)) &= y\varphi^2(y) \cdot (y\varphi^2(y) \cdot xy)\\ &= y\varphi^2(y) \cdot ((y \cdot xy)\varphi^2(y)) \qquad \text{since } \varphi^2(y) \in B(a)\\ &= y\varphi^2(y) \cdot y\varphi^2(y) \qquad \text{from (3.19)}\\ &= 0.\end{aligned}$$

Hence

$$xy \leq xy \cdot y\varphi^2(y). \tag{3.32}$$

Comparing (3.31) and (3.32), we get

$$xy = xy \cdot y\varphi^2(y),$$

so this weak BCC-algebra is φ-implicative.

Conversely, let a solid weak BCC-algebra $\mathfrak{X}$ be branchwise φ-implicative and branchwise commutative. Then $x \cdot yx \in B(a)$ for any $x, y \in B(a)$, $a \in I(X)$. Hence

$$x(x \cdot yx) \cdot yx = (x \cdot yx)(x \cdot yx) = 0.$$

Consequently,

$$x(x \cdot yx) = x(x \cdot yx) \cdot 0 = x(x \cdot yx) \cdot (x(x \cdot yx) \cdot yx).$$

But yx and $x(x \cdot yx)$ are in $B(0)$ and $\mathfrak{X}$ is branchwise commutative, so we also have

$$x(x \cdot yx) \cdot (x(x \cdot yx) \cdot yx) = yx \cdot (yx \cdot x(x \cdot yx)).$$

Thus

$$x(x \cdot yx) = yx \cdot (yx \cdot x(x \cdot yx)).$$

Since, by Theorem 2.4.18, elements yx, $yx \cdot x\varphi^2(x)$ and $yx \cdot x(x \cdot yx)$ are in $B(0)$, from the above, in view of φ-implicativity of $\mathfrak{X}$ and Corollary 2.6.7, we obtain

$$\begin{aligned} x(x \cdot yx) &= (yx \cdot x\varphi^2(x))(yx \cdot x(x \cdot yx)) \\ &\leq x(x \cdot yx) \cdot x\varphi^2(x) \\ &\leq \varphi^2(x)(x \cdot yx), \end{aligned}$$

because $x(x \cdot yx)$, $x\varphi^2(x) \in B(0)$. So, we have

$$x(x \cdot yx) \leq \varphi^2(x)(x \cdot yx). \tag{3.33}$$

On the other hand, by Lemma 2.2.2(2) and by (1.4), we have

$$\varphi^2(x)(x \cdot yx) \leq x(x \cdot yx). \tag{3.34}$$

Comparing (3.33) and (3.34) we get

$$x(x \cdot yx) = \varphi^2(x)(x \cdot yx). \tag{3.35}$$

Moreover, from $\varphi^2(x) \in B(a)$, we get $a \leq \varphi^2(x)$, which, by Theorem 2.2.8(3) and Lemma 2.2.2(5), implies $a = \varphi^2(a) = \varphi^4(x) = \varphi^2(x)$. Thus (3.35) because $x \cdot yx \in B(a)$, implies

$$x(x \cdot yx) = \varphi^2(x)(x \cdot yx) = a(x \cdot yx) = 0,$$

Hence $x \leq x \cdot yx$.

On the other hand,

$$(x \cdot yx)x = xx \cdot yx = 0 \cdot yx = 0,$$

which together with the previous inequality gives

$$x \cdot yx = x.$$

We have proven that the weak BCC-algebra $\mathfrak{X}$ is branchwise implicative.

This completes the proof. □

Definition 3.3.21. A weak BCC-algebra $\mathfrak{X}$ is called *weakly positive implicative* if it satisfies the identity

$$xy \cdot z = (xz \cdot z) \cdot yz. \tag{3.36}$$

If (3.36) is satisfied only by elements belonging to the same branch $B(a)$, $a \in I(X)$, then we say that the weak BCC-algebra $\mathfrak{X}$ is *branchwise weakly positive implicative.*

Example 3.3.22. Routine and easy calculations show that a (improper) weak BCC-algebra $\langle\{0,1,2\};\ \cdot, 0\rangle$ with the operation $\cdot$ defined by the following table:

$\cdot$	0	1	2
0	0	0	2
1	1	0	2
2	2	2	0

Table 3.3.4

is weakly positive implicative. □

Lemma 3.3.23. *A solid weakly positive implicative weak BCC-algebra $\mathfrak{X}$ satisfies the identity*

$$xy = (xy \cdot y) \cdot 0y. \tag{3.37}$$

Proof. By putting $y = 0$ in (3.36) we obtain the identity $xz = (xz \cdot z) \cdot 0z$, which is equivalent to (3.37). □

Theorem 3.3.24. *A solid weakly positive implicative weak BCC-algebra $\mathfrak{X}$ is branchwise φ-implicative.*

Proof. Let $\mathfrak{X}$ be a solid weakly positive implicative weak BCC-algebra. Then for $x, y \in B(a)$, $a \in I(X)$, by Theorem 2.6.12 and Lemma 2.2.2(5), we have

$$\begin{aligned}
(xy \cdot (y(0 \cdot 0y))) \cdot xy &= (xy \cdot y\varphi^2(y)) \cdot xy \\
&= (xy \cdot xy) \cdot y\varphi^2(y) \\
&= 0 \cdot y\varphi^2(y) \\
&= \varphi(y)\varphi^3(y) \\
&= \varphi(y)\varphi(y) \\
&= 0.
\end{aligned}$$

Hence

$$xy \cdot y(0 \cdot 0y) \leq xy. \tag{3.38}$$

On the other hand,

$$\begin{aligned} xy \cdot (xy \cdot y(0 \cdot 0y)) &= xy \cdot (xy \cdot y\varphi^2(y)) \\ &= ((xy \cdot y) \cdot 0y)(xy \cdot y\varphi^2(y)) && \text{by } (3.37) \\ &= (xy \cdot y)\varphi(y) \cdot (xy \cdot y\varphi^2(y)) \\ &= (xy \cdot y)(xy \cdot y\varphi^2(y)) \cdot \varphi(y) \\ &= (xy \cdot (xy \cdot y\varphi^2(y)))y \cdot \varphi(y), \end{aligned}$$

i.e.,

$$xy \cdot (xy \cdot y(0 \cdot 0y)) = (xy \cdot (xy \cdot y\varphi^2(y)))y \cdot \varphi(y), \tag{3.39}$$

because, according to Theorem 2.4.18, we have $xy \cdot y$, $\varphi(y) \in B(0a)$ and xy, $xy \cdot y\varphi^2(y) \in B(0)$.

Since, in this case, also $y\varphi^2(y) \in B(0)$, therefore,

$$(xy \cdot (xy \cdot y\varphi^2(y))) \cdot y\varphi^2(y) = (xy \cdot y\varphi^2(y))(xy \cdot y\varphi^2(y)) = 0.$$

Thus

$$xy \cdot (xy \cdot y\varphi^2(y)) \leq y\varphi^2(y),$$

which, by (1.4), implies

$$(xy \cdot (xy \cdot y\varphi^2(y)))y \cdot \varphi(y) \leq (y\varphi^2(y))y \cdot \varphi(y). \tag{3.40}$$

Hence

$$\begin{aligned} xy \cdot (xy \cdot y(0 \cdot 0y)) &= (xy \cdot (xy \cdot y\varphi^2(y)))y \cdot \varphi(y) && \text{by } (3.39) \\ &\leq (y\varphi^2(y))y \cdot \varphi(y) && \text{by } (3.40) \\ &= (yy \cdot \varphi^2(y)) \cdot \varphi(y) \\ &= 0\varphi^2(y) \cdot \varphi(y) \\ &= \varphi^3(y)\varphi(y) \\ &= \varphi(y)\varphi(y) && \text{from Lemma 2.2.2(5)} \\ &= 0. \end{aligned}$$

This proves

$$xy \leq xy \cdot y(0 \cdot 0y). \tag{3.41}$$

Combining (3.38) and (3.41) we get

$$xy = xy \cdot y(0 \cdot 0y).$$

So, $\mathfrak{X}$ is a solid branchwise φ-implicative weak BCC-algebra.
The proof is completed. □

The converse of Theorem 3.3.24 is not true. It is easy to see that the weak solid BCC-algebra with the operation $\cdot$ defined in Table 2.6.4 is branchwise φ-implicative, but it is not weakly positive implicative since $2 \cdot 5 = 5$ and $((2 \cdot 5) \cdot 5) \cdot (0 \cdot 5) = 4$.

3.4 Weak BCC-Algebras with Condition (S)

BCK-algebras with condition (S) were introduced by K. Iséki [75] and next generalized to BCI-algebras. Later, several authors extensively studied such algebras from different points of view. Today, BCK-algebras with condition (S) are an important class of BCK-algebras.

Below we extend this concept to the case of weak BCC-algebras and prove some properties of these algebras (cf. [58]).

We start with the consideration of the set $A(x, y)$ defined for two given elements x and y of a weak BCC-algebra $\mathfrak{X}$ in the following way:

$$A(x, y) = \{p \in X \,|\, px \leq y\} = \{p \in X \,|\, px \cdot y = 0\}.$$

In the following lemma, we give some simple properties of this set.

Lemma 3.4.1. *Let $\mathfrak{X}$ be a weak BCC-algebra. Then for $x, y, z, u \in X$ the following conditions are true:*

(1) $A(0, x) = A(x, 0)$,

(2) $0 \in A(x, y) \Longleftrightarrow 0 \in A(y, x)$,

(3) $x \in A(x, y) \Longleftrightarrow y \in B(0)$,

(4) $x \in B(0) \Longrightarrow y \in A(x, y)$,

(5) $A(x, y) \subset A(u, y)$ *for* $x \leq u$,

(6) $A(x, y) \subset A(x, z)$ *for* $y \leq z$,

(7) $u \leq z,\ z \in A(x,y) \Longrightarrow u \in A(x,y)$,

(8) $A(x,y) = A(y,x)$ *if* $\mathfrak{X}$ *is a BCI-algebra.*

Proof. (1) We have

$$A(0,x) = \{p \in X \,|\, 0 = p0 \cdot x = px = px \cdot 0\} = A(x,0).$$

(2) Let $0 \in A(x,y)$. Then $0x \leq y$. This, by (1.5), implies $0y \leq 0 \cdot 0x$, which, by Lemma 2.2.2(2), gives $0y \leq 0 \cdot 0x \leq x$. Hence $0y \leq x$ and $0 \in A(y,x)$.

(3) If $x \in A(x,y)$, then $0 = xx \cdot y = 0y$. Hence, of course, $y \in B(0)$. The converse is obvious.

(4) Let $x \in B(0)$. Then, by (BCC1$'$), we have

$$yx = yx \cdot 0 = yx \cdot 0x \leq y0 = y,$$

i.e., $y \in A(x,y)$.

(5) Let $p \in A(x,y)$ and let $x \leq u$. Then, by (1.5), $pu \leq px$ and, by (1.4), $pu \cdot y \leq px \cdot y = 0$. Hence $pu \cdot y = 0$ and $p \in A(u,y)$. Therefore, $A(x,y) \subset A(u,y)$.

(6) Let $p \in A(x,y)$. Then $px \leq y$ and $y \leq z$ implies $px \leq z$ because the relation $\leq$ is a partial order on X. Hence $A(x,y) \subset A(x,z)$.

(7) Let $z \in A(x,y)$. Then $u \leq z$, by (1.4), implies $ux \leq zx \leq y$, which, by transitivity of $\leq$, gives $ux \leq y$. Hence $u \in A(x,y)$.

(8) Here we have $A(x,y) = \{p \in X \,|\, px \cdot y = 0\}$, which by (1.6), is equal to $\{p \in X \,|\, py \cdot x = 0\} = A(y,x)$.

The proof is completed. □

Example 2.6.5 shows that in general $A(x,y) \neq A(y,x)$. Indeed, in a weak BCC-algebra defined in Table 2.6.4 $A(4,5) = \{2,3\}$ and $A(5,4) = \{3\}$. So, $A(4,5) \neq A(5,4)$.

Theorem 3.4.2. *Let* $\mathfrak{X}$ *be a solid weak BCC-algebra.*
If $x \in B(a), y \in B(b)$ *for* $a,b \in I(X)$, *then* $A(x,y)$ *is a nonempty subset of the branch* $B(a \cdot 0b)$.

Proof. Let $x \in B(a), y \in B(b)$, $a,b \in I(X)$. Then, by Lemma 2.2.2(7), $\varphi(y) = \varphi(b)$. The condition $\varphi^2(x) \in I(X) \cap B(a)$ is also true.

Indeed, $\varphi(x) = a \in I(X)$. But, from Theorem 2.2.12, $\mathfrak{I}(\mathfrak{X})$ is a subalgebra of $\mathfrak{X}$. Hence $\varphi^2(x) \in I(X)$.

Now, from $\varphi^2(x) = 0 \cdot 0x$, Theorem 2.4.18 and Theorem 2.2.8(3) it follows

$$\varphi^2(x) \in B(0 \cdot 0a) = B(\varphi^2(a)) = B(a).$$

Hence $\varphi^2(x) \in I(X) \cap B(a)$.

Moreover, since φ is an endomorphism (Theorem 2.6.12), we have

$$s = 0(0x \cdot y) = \varphi(\varphi(x)y) = \varphi^2(x)\varphi(y) = a\varphi(b) = a \cdot 0b.$$

Therefore,

$$sx \cdot y = (a \cdot 0b)x \cdot y = (ax \cdot 0b)y = (0 \cdot 0b)y = by = 0,$$

shows that $s \in A(x, y)$. Thus the set $A(x, y)$ is nonempty.

Let p be an arbitrary element of $A(x, y)$. Then px and y are in the same branch. Consequently,

$$s = 0(0x \cdot y) = 0 \cdot (pp \cdot x)y = 0 \cdot (px \cdot p)y = 0 \cdot (px \cdot y)p = 0 \cdot 0p = \varphi^2(p) \leq p,$$

by Lemma 2.2.2(2).

So, s is the least element of $A(x, y)$ and $A(x, y) \subset B(s)$. □

Definition 3.4.3. We say that a solid weak BCC-algebra $\mathfrak{X}$ is *with condition* (S) if each of its subsets $A(x, y)$ has the greatest element. The greatest element of $A(x, y)$ will be denoted by $x \circ y$.

Example 3.4.4. Let $\langle X; *, 0\rangle$ be a commutative group. Then $\langle X; \cdot, 0\rangle$ with the operation $xy = x * y^{-1}$ is a solid weak BCC-algebra in which each branch has only one element. Thus $A(x, y) \subset B(x \cdot 0y) = \{x \cdot 0y\}$. Consequently, $x \circ y = x \cdot 0y = x * y$. □

Example 3.4.5. Each finite solid weak BCC-algebra decomposed into linear branches is with condition (S) since each set $A(x, y)$ is a finite subset of some linear branch. □

Example 3.4.6. A solid weak BCC-algebra $\mathfrak{X}$ with the operation $\cdot$ defined by Table 2.6.6 (see Example 2.6.36) is not with condition (S) since $A(a, b) = \{b, c, d\} = B(b)$ has none greatest element. □

Since in a solid weak BCC-algebra $\mathfrak{X}$ with condition (S), for each $x, y \in X$, the set $A(x, y)$ has the greatest element $x \circ y$, so that $\circ$ it can be treated as a binary operation defined on X and $\langle X; \circ, 0\rangle$ can be considered an algebra of type $(2, 0)$.

Since in any case $A(x, 0) = A(0, x) = \{p \in X \mid p \leq x\}$, the groupoid $\langle X; \circ, 0\rangle$ has always the identity 0.

In the case of BCI-algebras with condition (S), $\langle X;\circ,0\rangle$ is a commutative semigroup (cf. [75], [76]). For weak BCC-algebras, it is not true.

For example, a solid weak BCC-algebra defined by Table 1.1.2 is with condition (S), but in this algebra $1\circ 2=2$ and $2\circ 1=3$. Therefore $1\circ 2\neq 2\circ 1$.

Similarly, $(2\circ 2)\circ 2\neq 2\circ(2\circ 2)$, because $(2\circ 2)\circ 2 = 1\circ 2=2$ and $2\circ(2\circ 2)=2\circ 1=3$. Therefore, $\circ$ is not associative.

For some BCI-algebras (described in [25] and [27]) $\langle X;\circ,0\rangle$ is a commutative group. A similar situation takes place in the case of weak BCC-algebras.

To prove this fact, we need the following lemma:

Lemma 3.4.7. *In a weak BCC-algebra $\mathfrak{X}$ with condition (S)*

$$x\le y\Longrightarrow x\circ z\le y\circ z$$

for all $x,y,z\in X$.

Proof. If $x\le y$, then also $(x\circ z)y\le(x\circ z)x$, by (1.5). But, according to the definition, $(x\circ z)x\le z$. Thus $(x\circ z)y\le z$, which implies $x\circ z\le y\circ z$, because $y\circ z$ is the greatest element satisfying the inequality $py\le z$. □

Theorem 3.4.8. *Let $\mathfrak{X}=\langle X;\cdot,0\rangle$ be a weak BCC-algebra with condition (S). Then $\langle X;\circ,0\rangle$ is a group if and only if $\langle X;\cdot,0\rangle$ is group-like.*

Proof. Let $\langle X;\circ,0\rangle$ be a group. We consider an arbitrary element $x\in B(0)$. Denote by x^{-1} the inverse element of x in a group $\langle X;\circ,0\rangle$. By Lemma 3.4.7, from $0\le x$ it follows $x^{-1}=0\circ x^{-1}\le x\circ x^{-1}=0$. Hence $x^{-1}=0$. Thus $B(0)=\{0\}$. This, by Theorem 2.5.2(6), shows that a weak BCC-algebra $\langle X;\cdot,0\rangle$ is group-like.

Conversely, if a weak BCC-algebra $\langle X;\cdot,0\rangle$ is group-like, then each of its branches has only one element. Hence $px\le y$ means $px=y$, i.e., $p*x^{-1}=y$ in the corresponding group $\langle X;*,0\rangle$ (see Theorem 2.5.7). Thus $p=y*x$ is uniquely determined by $x,y\in X$. Therefore, $x\circ y=y*x$.

So, $\langle X;\circ,0\rangle$ is a group. □

Corollary 3.4.9. *Let $\langle X;\cdot,0\rangle$ be a weak BCC-algebra with condition (S). Then $\langle X;\circ,0\rangle$ is a commutative group if and only if $\langle X;\cdot,0\rangle$ is a group-like BCI-algebra.*

Proof. Indeed, $\langle X; \circ, 0\rangle$ is commutative if and only if $\langle X; *, 0\rangle$ is commutative, that is, if and only if $xy \cdot z = x * y^{-1} * z^{-1} = x * z^{-1} * y^{-1} = xz \cdot y$. □

Lemma 3.4.10. *In a solid weak BCC-algebra* $\mathfrak{X}$ *with condition* (S) *we have*

$$xy \cdot z = x(y \circ z) \tag{3.42}$$

for x, y *belonging to the same branch and* $z \in B(0)$.

Proof. Let $x, y \in B(a), a \in I(X)$ and $z \in B(0)$. Then

$$x(xy \cdot z) \cdot y = xy \cdot (xy \cdot z) \leq z \text{ implies } x(xy \cdot z) \cdot y \leq z.$$

Thus $x(xy \cdot z) \in A(y, z)$. Hence $x(xy \cdot z) \leq y \circ z$ and, by (1.5),

$$x(y \circ z) \leq x(x(xy \cdot z)) \tag{3.43}$$

Further, since $x(xy \cdot z) \in B(a)$, we have

$$x(x(xy \cdot z)) \cdot (xy \cdot z) = x(xy \cdot z) \cdot x(xy \cdot z) = 0.$$

And, as a consequence,

$$x(x(xy \cdot z)) \leq xy \cdot z. \tag{3.44}$$

By combining (3.43) and (3.44), we get

$$x(y \circ z) \leq x(x(xy \cdot z)) \leq xy \cdot z,$$

which gives

$$x(y \circ z) \leq xy \cdot z. \tag{3.45}$$

On the other hand, $y \circ z \in A(y, z) \subset B(a)$, by Theorem 3.4.2. This, by Theorem 2.4.20, gives $x(y \circ z) \in B(0)$. Thus

$$xy \cdot x(y \circ z) \leq (y \circ z)y. \tag{3.46}$$

Indeed, we have

$$(xy \cdot x(y \circ z)) \cdot (y \circ z)y = (xy \cdot (y \circ z)y) \cdot x(y \circ z).$$

Then, using (BCC1$'$), we get

$$xy \cdot (y \circ z)y \leq x(y \circ z)$$

and

$$(xy \cdot (y \circ z)y) \cdot x(y \circ z) = 0.$$

So, the inequation (3.46) is satisfied.
Consequently,

$$xy \cdot x(y \circ z) \leq (y \circ z)y \leq z.$$

Therefore,

$$0 = (xy \cdot x(y \circ z))z = (xy \cdot z) \cdot x(y \circ z),$$

which implies

$$xy \cdot z \leq x(y \circ z). \tag{3.47}$$

Finally, (3.45) together with (3.47) give (3.42).

The proof is completed. □

Corollary 3.4.11. *In a solid weak BCC-algebra with condition* (S) *the branch* $B(0)$ *satisfies identity* (3.42).

Corollary 3.4.12. *Any BCI-algebra with condition* (S) *satisfies identity* (3.42).

Theorem 3.4.13. *A solid weak BCC-algebra* $\mathfrak{X}$ *with condition* (S) *is restricted if and only if some its branches are restricted.*

Proof. Let us assume that some branch, for example $B(a), a \in I(X)$, is restricted and 1_a is the greatest element of $B(a)$. Then $xb \in B(0)$ for every $x \in B(b)$ and an arbitrary $b \in I(X)$. Thus $xb \cdot 0a \in B(0 \cdot 0a) = B(a)$, by Theorem 2.4.18. Hence $xb{\cdot}0a \leq 1_a$, i.e., $xb \leq 0a{\circ}1_a$, according to the definition of $0a \circ 1_a$. Consequently,

$$xb \cdot (0a \circ 1_a) = 0 \quad \text{and} \quad (0a \circ 1_a) \cdot 0a \leq 1_a. \tag{3.48}$$

Hence $(0a \circ 1_a) \cdot 0a \in B(a)$, i.e., $a \leq (0a \circ 1_a) \cdot 0a$. Since the map φ is an endomorphism (Theorem 2.6.12), from the last inequality, applying Lemma 2.2.2(7), we obtain

$$0a = 0((0a \circ 1_a) \cdot 0a) = 0(0a \circ 1_a) \cdot (0 \cdot 0a) = 0(0a \circ 1_a) \cdot a.$$

Therefore, by (BCC1$'$), we have

$$0 = (0(0a \circ 1_a) \cdot a) \cdot 0a \leq 0(0a \circ 1_a) \cdot 0 = 0(0a \circ 1_a),$$

i.e.,

$$0 \leq 0(0a \circ 1_a).$$

This, by (1.5) and Lemma 2.2.2(2), gives

$$0 = 0 \cdot 0(0a \circ 1_a) = \varphi^2(0a \circ 1_a) \leq 0a \circ 1_a.$$

So, $0a \circ 1_a \in B(0)$.

Now let $m = b \circ (0a \circ 1_a)$. Then for every $x \in B(b)$, according to Lemma 3.4.10 and (3.48), we have

$$xm = x(b \circ (0a \circ 1_a)) = xb \cdot (0a \circ 1_a) = 0,$$

which implies $x \leq m$. Therefore, m is the greatest element of the branch $B(b)$.

The converse statement is obvious. □

3.5 Initial Segments

In this section, we will investigate the initial segments. They have connections to both BCK- and BCC-ideals (cf. [60]). They also have links to positive implicative BCC-algebras. We will cover them in detail in this section.

For any fixed elements $a \leq b$ of X, where $\mathfrak{X}$ is a BCC-algebra the set

$$[a, b] = \{x \in X \mid a \leq x \leq b\} = \{x \in X \mid ax = xb = 0\}$$

is called a *segment* of X.
The segment

$$[0, b] = \{x \in X \mid x \leq b\} = \{x \in X \mid xb = 0\}$$

is called an *initial segment.* It turns out to be the *left annihilator* of b.

The annihilators are defined and researched mainly for BCK-algebras, although, for example, in the paper [67], they were examined for weakly standard BCC-algebras.

Since $[0, b]$ has two elements only in the case when $b \in X$ is an atom of $\mathfrak{X}$, from the result obtained in [59] it follows that a BCC-algebra in which all initial segments have at most two elements has the trivial structure.

Example 3.5.1. An algebra $\mathfrak{X}$ with $X = \{0, 1, 2, 3, 4, 5\}$ and the operation $\cdot$ defined by Table 2.1.1 is a proper BCC-algebra. Its initial segments have the form $[0, 1] = \{0, 1\}$, $[0, 2] = \{0, 1, 2\}$, $[0, 3] = \{0, 1, 2, 3\}$, $[0, 4] = \{0, 1, 4\}$, $[0, 5] = \{0, 5\}$. All these segments are BCK-chains.

On the other hand, BCC-algebras with operations defined by Tables 1.2.7, 1.2.8 and 1.2.9 are BCC-chains with 3 as the greatest element. Since these BCC-algebras are proper, they are not BCK-chains. A BCC-algebra with the operation $\cdot$ defined by Table 1.2.13 is a minimal proper BCC-algebra which coincides with some of its initial segments. It is not a BCC-chain because elements 1 and 2 are incomparable. Algebras with operations defined by Tables 1.2.10, 1.2.11, 1.2.12 and 1.2.14 are minimal proper BCC-algebras, which are a set-theoretic union of two different BCK-chains. □

Theorem 3.5.2. *Every initial segment of a BCC-algebra is a BCC-subalgebra.*

Proof. Let $[0, c]$ be a initial segment of $\mathfrak{X}$. Obviously $0 \in [0, c]$. If for $x, y \in X$ $x, y \in [0, c]$, then $x \leq c$ and $y \leq c$, which, by (1.4) and Theorem 1.1.30(8), implies $xy \leq cy \leq c$. Thus $xy \in [0, c]$, which proves that $[0, c]$ is a BCC-subalgebra. □

Theorem 3.5.3. *The set theoretic union of any two initial segments of a given BCC-algebra is a BCC-subalgebra.*

Proof. The proof follows from Theorem 1.1.30(8). Indeed, as it can be easily seen, in the case, when $x, y \in [0, a] \cup [0, b]$, we have $xy \leq x \leq a$. Hence $xy \in [0, a] \cup [0, b]$. □

Theorem 3.5.4. *A BCC-algebra containing at least two initial segments $[0, x]$ and $[0, y]$ such that $[0, x] \cap [0, y] = \{0\}$ and $xy \neq x$ is proper.*

Proof. Let us assume a contrary that $\mathfrak{X}$ is a BCK-algebra in which $[0, x] \cap [0, y] = \{0\}$ for some $x, y \in X$ such that $x \neq y$. Then, by (1.3), $x \cdot xy \leq y$ and, by Theorem 1.1.30(8), $x \cdot xy \leq x$.

Thus $x \cdot xy \in [0, x] \cap [0, y] = \{0\}$. Hence $x \leq xy$. This, together with Theorem 1.1.30(8), gives $xy = x$, which is a contradiction. Thus $\mathfrak{X}$ cannot be a BCK-algebra. It is a proper BCC-algebra. □

The following corollary follows from Theorem 1.2.8:

Corollary 3.5.5. *Every BCC-chain containing at most three elements is a BCK-chain.*

In general, initial segments are not BCC-ideals.

Example 3.5.6. It is easy to verify that an algebra $\mathfrak{X}$ with the operation $\cdot$ defined by Table 1.2.7 is a BCC-chain. We have $2 \cdot 1 \in [0, 1]$ and $1 < 2$, but $2 \notin [0, 1]$. Thus $[0, 1]$ is not a BCK-ideal.
From Theorem 2.3.4 it follows that it is not also a BCC-ideal. □

Theorem 3.5.7. *An initial segment $[0, c]$ of a BCC-algebra $\mathfrak{X}$ is a BCC-ideal if and only if*

$$xc \cdot z \leq c \Longrightarrow xz \leq c \tag{3.49}$$

holds for all $x, z \in X$.

Proof. Assume that the above implication holds. Of course, $0 \in [0, c]$. If $xy \cdot z$ and $y \in [0, c]$, then $xy \cdot z \leq c$ and $y \leq c$.

Now, $y \leq c$ imply, by (1.4) and (1.5), $xc \cdot z \leq xy \cdot z$. Thus $xc \cdot z \leq c$, which, by the assumption, implies $xz \leq c$. Hence $xz \in [0, c]$, i.e., $[0, c]$ is a BCC-ideal.

The converse follows from the definition of BCC-ideal. □

Corollary 3.5.8. *If $[0, c]$ is a BCC-ideal of $\mathfrak{X}$, then for every $x \in X$*

$$xc \leq c \Longrightarrow x \leq c. \tag{3.50}$$

Proof. By putting $z = 0$ in (3.49), we get (3.50). □

Using the same method as in Theorem 3.5.7, we can prove the following corollary:

Corollary 3.5.9. *An initial segment $[0, c]$ of a BCC-algebra $\mathfrak{X}$ is a BCK-ideal if and only if* (3.50) *holds for every $x \in X$.*

Corollary 3.5.10. *If a nontrivial segment $[0, c]$ is a BCK-ideal or a BCC-ideal of $\mathfrak{X}$, then $xc \neq c$ for all nonzero $x \in X$.*

Proof. For $c \neq 0$, let $[0, c]$ be a BCK-ideal. If $xc = c$ for some $x \in X$, then $xc \in [0, c]$ and, as a consequence, $x \leq c$, which is a contradiction since in this case we obtain $0 = xc = c$. □

Corollary 3.5.11. *If the condition* $(xc \cdot z)c = xz \cdot c$ *holds for* $x, z \in X$, *then* $[0, c]$ *is a BCC-ideal of* $\mathfrak{X}$.

If a BCC-algebra $\mathfrak{X}$ satisfies the identity

$$(xy \cdot z)y = xz \cdot y \tag{3.51}$$

for all $x, y, z \in X$, then all initial segments are BCC-ideals.

For $z = 0$ the identity has the form of (3.25), which means that $\mathfrak{X}$ is positive implicative.

Moreover, for $z = xy$ in (3.51) we get (1.2), which means that $\mathfrak{X}$ is a BCK-algebra. In the theory of BCK-algebras, the conditions (3.51) and (3.25) are equivalent. Indeed, if (3.25) is satisfied, then

$$(xy \cdot z)y = (xy \cdot y)z = xy \cdot z = xz \cdot y$$

by, respectively, (1.6), (3.51) and again (1.6). In the theory of BCC-algebras, the both conditions are not equivalent. There are proper BCC-algebras, in which only (3.25) holds. For example, direct computations show that a BCC-algebra $\mathfrak{X}$ with the operation $\cdot$ defined by Table 1.2.13 is positive implicative.

But the condition (3.51) is not satisfied. Indeed,

$$((3 \cdot 2) \cdot 1) \cdot 2 = (1 \cdot 1) \cdot 2 = 0 \cdot 2 = 0.$$

On the other side,

$$(3 \cdot 1) \cdot 2 = 3 \cdot 2 = 1.$$

From the above remarks it follows:

Theorem 3.5.12. *A BCC-algebra* $\mathfrak{X}$ *satisfying* (3.51) *is a positive implicative BCK-algebra, in which all initial segments are* BCK*-ideals of* $\mathfrak{X}$.

Theorem 3.5.13. *A BCC-algebra* $\mathfrak{X}$ *in which all initial segments have at most two elements is a positive implicative BCK-algebra. Initial segments of such BCK-algebra are* BCK*-ideals of* $\mathfrak{X}$.

Proof. The proof follows from Lemma 2.3.15, Theorem 2.3.17 and Corollary 2.1.7. □

Theorem 3.5.14. *If* $[0, c]$ *is a two-elements BCC-ideal of a BCC-algebra* $\mathfrak{X}$, *then*

$$xc \cdot z = c \text{ implies } xz = c$$

for $x, z \in X$.

Proof. Indeed, by Theorem 3.5.7, from $xc \cdot z = c$ it follows $xz \leq c$. Let $xz = 0$. Then, by (1.4) and 1.1.30(8), we get

$$x \leq z \Longrightarrow xc \leq zc \leq z \Longrightarrow xc \cdot z \leq zz = 0,$$

i.e., $xc \cdot z = 0$, which is a contradiction.
Therefore, it must be, $xz = c$. □

Theorem 3.5.15. $[0, c]$ *is a BCC-ideal of a BCC-algebra* $\mathfrak{X}$ *if and only if the relation* Θ *defined by*

$$(x, y) \in \Theta \Longleftrightarrow xy \leq c \text{ and } yx \leq c$$

is a congruence.

Proof. From Theorem 2.4.2 it follows that Θ is a congruence relation. By Corollary 2.4.10, it is regular. And finally, Theorem 2.4.11 completes the proof.

Conversely, observe that if Θ is a congruence determined by the segment $[0, c]$, then $(x, 0) \in \Theta$ if and only if $x \in [0, c]$. This means that $[0, c] = C_0$ and from Lemma 2.4.4 it follows that $[0, c]$ is a *BCC*-ideal. □

As a generalization of Corollary 3.5.10 we have:

Corollary 3.5.16. *A two-element initial segment* $[0, c]$ *is a BCK-ideal if and only if* $xc \neq c$ *for every* $x \in X$.

Proof. If $xc \neq c$ and $xc \leq c$, then $xc = 0$. Thus $[0, c]$ is a *BCK*-ideal. The converse statements follows from Corollary 3.5.10. □

Corollary 3.5.17. *If* $xc \cdot c = xc$ *for every* $x \in X$, *then* $[0, c]$ *is a BCK-ideal.*

As a generalization of the second statement of Theorem 3.5.13 we have the following corollary:

Corollary 3.5.18. *Initial segments of a positive implicative BCC-algebra are BCK-ideals.*

In general, initial segments of a positive implicative BCC-algebra are not BCC-ideals.

As an example we may consider a subalgebra $\mathfrak{S} = \langle\{0,1,2,5\};\cdot,0\rangle$ of $\mathfrak{X}$ from Example 3.5.1. This subalgebra is positive implicative (it is isomorphic to a BCC-algebra $\mathfrak{X}$ with the operation $\cdot$ defined in Table 1.2.11), but $[0,5]$ is not a BCC-ideal since $(2\cdot 5)\cdot 1 = 0 \in [0,5]$ and $2\cdot 1 = 2 \notin [0,5]$.

Theorem 3.5.19. *A finite BCC-chain of a positive implicative BCC-algebra is a BCK-chain with the trivial structure.*

Proof. We will use the induction method. For BCC-chains containing at most two elements, our statement is obviously true.

If $[0,c]$ has $n+1>2$ elements, then there exists $y\in[0,c]$ such that $[y,c]$ has only two elements. Thus $[0,y]$ has n elements and, by the assumption, has the trivial structure. Since, by Corollary 3.5.18, it is also a BCK-ideal, for every $x\in[0,y]$, from $cx\in[0,y]$ it follows $c\in[0,y]$, which is impossible because $y<c$. Thus $cx\notin[0,y]$, i.e., $y<cx\leq c$. Hence $cx=c$ for every $x<c$.

This completes the proof. □

Corollary 3.5.20. *A finite linearly ordered BCC-algebra is positive implicative if and only if it has the trivial structure.*

In this part of the section we give several methods of construction of BCC-algebras with given BCC-chains. Some general methods of constructions of proper BCC-algebras can be found in Section 1.2.

First, we observe that:

Theorem 3.5.21. *Any finite BCK-chain may be extended to a proper BCC-chain.*

The proof is based on the observation that any two-elements BCK-chains may be extended to three-elements BCK-chain with the trivial structure. Any three-elements BCK-chain may be extended to a proper BCC-chain by the following construction, which is a special case of the construction used in Theorem 1.2.24.

Corollary 3.5.22. *Let $\langle X;\cdot,0\rangle$ be a finite BCK-chain containing at least three elements and let c be its maximal element.*

Then $X' = X \cup \{a\}$, *where* $a \notin X$, *with the operation* $\cdot$ *defined as follows:*

$$xy = \begin{cases} xy & \text{if } x, y \in X, \\ 0 & \text{if } x \in X', \ y = a, \\ a & \text{if } x = a, \ y = 0, \\ c & \text{if } x = a, \ y \in X \end{cases}$$

is a proper BCC-chain.

Proof. The obtained structure is a BCC-chain with a as the maximal element. It is proper since $(a \cdot ay)y = ac \cdot y = cy \neq 0$ for any $0 < y < c$. □

As a simple consequence of Corollary 1.2.23 we obtain:

Corollary 3.5.23. *Let* $\langle X; \cdot, 0\rangle$ *be a finite proper BCC-chain. Then* $X' = X \cup \{a\}$, *where* $a \notin X$, *with the operation* $\cdot$ *defined as follows:*

$$xy = \begin{cases} xy & \text{if } x, y \in X, \\ 0 & \text{if } x \in X', \ y = a, \\ a & \text{if } x = a, \ y \in X \end{cases}$$

is a proper BCC-chain.

Proof. The obtained structure is a BCC-chain with the a as the maximal element. It is proper because the BCC-chain X is proper. □

From these two constructions it follows:

Corollary 3.5.24. *Any finite proper BCC-chain may be extended to at least two nonisomorphic proper BCC-chains of the same order.*

3.6 Fuzzy BCC-Subalgebras

L. A. Zadeh [148] introduced the notion of fuzzy sets. Until today, this concept has been used in many areas of mathematics: the theory of groups, topology, probability theory or functional analysis. In 1991, O.G. Xi [141] applied this concept to BCK-algebras. Fuzzy BCC-subalgebras were researched and described in a series of works by W.A. Dudek and Y.B. Jun (see Bibliography).

Let us start by defining the fuzzy BCC-subalgebras.

Definition 3.6.1. A function $\mu : X \longrightarrow [0,1]$ is called *a fuzzy set* in a BCC-algebra $\mathfrak{X}$.

Definition 3.6.2. A fuzzy set μ in a BCC-algebra $\mathfrak{X}$ is called *a fuzzy subalgebra* of $\mathfrak{X}$ if the following condition:

$$\mu(xy) \geq \min\{\mu(x), \mu(y)\}$$

is satisfied for all $x, y \in X$.

The set $\mu_t = \{x \in X \mid \mu(x) \geq t\}$ for fixed $t \in [0,1]$ is called a *level subset of* μ. If μ_t is a subalgebra, then it is called *a level subalgebra.*

A level subset of μ is under certain circumstances called an *upper level set* of μ and denoted as $U(\mu; t)$.

Respectively, by $L(\mu; t) = \{x \in X \mid \mu(x) \leq t\}$ will be denoted a *lower level set* of μ.

The image set of μ is denoted as $\mathrm{Im}(\mu)$ and defined by

$$\mu_{Im(\mu)}(y) = \begin{cases} \sup\limits_{z \in \mu^{-1}(y)} \mu(z) & \text{if } \mu^{-1}(y) = \{x \mid \mu(x) = y\} \text{ is nonempty,} \\ 0 & \text{otherwise.} \end{cases}$$

Example 3.6.3. Consider the BCC-algebra $\mathfrak{X} = \langle\{0,1,2,3,4\}; \cdot, 0\rangle$, with the operation $\cdot$ defined by Table 2.3.1. By routine calculations we know that a fuzzy set μ in X defined in the following way:

$$\mu(x) = \begin{cases} 0.4 & \text{if} \quad x = 4, \\ 0.8 & \text{if} \quad x \neq 4 \end{cases}$$

is a fuzzy subalgebra of $\mathfrak{X}$.

Then level subsets of μ look like this

$$\mu_t = \begin{cases} X & \text{if} \quad t \leq 0.4, \\ \{0,1,2,3\} & \text{if} \quad 0.4 < t \leq 0.8 \end{cases}$$

and finally $\mathrm{Im}(\mu) = \{0.4; 0.8\}$. □

Lemma 3.6.4. *If μ is a fuzzy subalgebra of a BCC-algebra $\mathfrak{X}$, then*

$$\mu(0) \geq \mu(x)$$

for any $x \in X$.

Proof. From (BCC2), for any $x \in X$ it follows

$$\mu(0) = \mu(xx) \geq \min\{\mu(x), \mu(x)\} = \mu(x).$$

The proof is completed. □

Theorem 3.6.5. *A fuzzy set μ of a BCC-algebra $\mathfrak{X}$ is a fuzzy subalgebra if and only if for every $t \in [0,1]$, μ_t is either empty or it is a subalgebra of $\mathfrak{X}$.*

Proof. Let μ be a fuzzy subalgebra of $\mathfrak{X}$ and let $\mu_t \neq \emptyset$. Then for any $x, y \in \mu_t$ we have

$$\mu(xy) \geq \min\{\mu(x), \mu(y)\} \geq t \Longrightarrow xy \in \mu_t.$$

Hence μ_t is a subalgebra of $\mathfrak{X}$.

Conversely, let μ_t be a subalgebra of $\mathfrak{X}$ and, for any $x, y \in X$, let us put $t = \min\{\mu(x), \mu(y)\}$. Then, of course, $xy \in \mu_t$. Hence

$$\mu(xy) \geq t, \text{ i.e., } \mu(xy) \geq \min\{\mu(x), \mu(y)\}.$$

Thus μ is a fuzzy subalgebra of $\mathfrak{X}$. □

Theorem 3.6.6. *Any subalgebra of a BCC-algebra $\mathfrak{X}$ can be realized as a level subalgebra of some fuzzy subalgebra of $\mathfrak{X}$.*

Proof. Let S be a subalgebra of $\mathfrak{X}$ and let μ be a fuzzy set in $\mathfrak{X}$ defined by

$$\mu(x) = \begin{cases} t & \text{if } \ x \in S, \\ 0 & \text{if } \ x \notin S, \end{cases}$$

where $t \in (0,1)$ is fixed. Of course, $\mu_t = S$.

Such defined μ is a fuzzy subalgebra of $\mathfrak{X}$.

Indeed, let $xy \in X$. We have three cases.

1^0. If $x, y \in S$, then also $xy \in S$. Hence $\mu(x) = \mu(y) = \mu(xy) = t$ and $\mu(xy) \geq \min\{\mu(x), \mu(y)\}$.

2^0. If $x, y \notin S$, then $\mu(x) = \mu(y) = 0$ and in the consequence $\mu(xy) \geq \min\{\mu(x), \mu(y)\} = 0$.

3^0. If at most one of x, y belongs to S, then at least one of $\mu(x)$ and $\mu(y)$ is equal to 0. Therefore, $\min\{\mu(x), \mu(y)\} = 0$ and $\mu(xy) \geq 0$.

The proof is completed. □

Note that in a finite BCC-subalgebra the number of its subalgebras is finite, whereas the number of level subsets may by (formally) infinite. In the next result we show that not all these level subsets are distinct.

Theorem 3.6.7. *Two level subalgebras μ_s, μ_t for $s < t$ of a fuzzy subalgebra $\mathfrak{X}$ are equal if and only if there is no $x \in X$ such that $s \leq \mu(x) < t$.*

Proof. Let for some $s < t$, $\mu_s = \mu_t$. If there exists $x \in X$ such that $s \leq \mu(x) < t$, then μ_t is a proper subset of μ_s, which contradicts the assumption.

Conversely, assume that there exists no $x \in X$ such that $s \leq \mu(x) < t$. If $x \in \mu_s$, then $\mu(x) \geq s$, and so $\mu(x) \geq t$ because $\mu(x)$ does not lie between s and t. Thus $x \in \mu_t$, which gives $\mu_s \subseteq \mu_t$.

The converse inclusion is obvious because $s < t$.
Therefore $\mu_s = \mu_t$. □

From the above theorem it follows that the set of all level subalgebras of a given fuzzy BCC-algebra $\mathfrak{X}$ is a chain. Indeed, we have $\mu(x) \leq \mu(0)$ for all $x \in X$, which implies μ_{t_0}, where $t_0 = \mu(0)$, is the smallest level subalgebra. Hence we have the chain

$$\mu_{t_0} \subset \mu_{t_1} \subset ... \subset \mu_{t_p} = X,$$

where $t_0 > t_1 > ... > t_p$.

Note that in Example 3.6.3 this chain has only two elements, namely $\mu_{t_0} = \{0, 1, 2, 3\}$ and $\mu_{t_1} = X$.

Corollary 3.6.8. *Let μ be a fuzzy subalgebra of $\mathfrak{X}$. If $\mathrm{Im}(\mu) = \{t_1, t_2, ..., t_n\}$, where $t_1 < t_2 < ... < t_n$, then the family of levels μ_{t_i} for $1 \leq i \leq n$, constitutes all level subalgebras of μ.*

Proof. Let μ_s, where $s \in [0, 1]$ and $s \notin \mathrm{Im}(\mu)$, be a some level subalgebra.

If $s < t_1$, then $\mu_{t_1} \subseteq \mu_s$. Since $\mu_{t_1} = X$, it follows that $\mu_s = X$ and $\mu_s = \mu_{t_1}$.

If $t_i < s < t_{i+1}$, then there is no $x \in X$ such that $s \leq \mu(x) < \mu_{t_{i+1}}$. Thus, by Theorem 3.6.7, $\mu_s = \mu_{t_{i+1}}$. This proves that for any $s \in [0, 1]$ with $s \leq \mu(0)$, the level subalgebra is one of $\{\mu_{t_i} : 1 \leq i \leq n\}$. □

It is interesting that two different fuzzy subalgebras of a BCC-algebra $\mathfrak{X}$ may have an identical family of level subalgebras. For example, putting in Example 3.6.3

$$\rho(x) = \begin{cases} 0.3 & \text{if} \quad x = 4, \\ 0.7 & \text{if} \quad x \neq 4, \end{cases}$$

we obtain a new fuzzy subalgebra of $\mathfrak{X}$ which has the same family of levels as μ.

We will specify it in the next results. We begin with the following lemma:

Lemma 3.6.9. *Let μ be a fuzzy subalgebra of a BCC-algebra $\mathfrak{X}$ with a finite image. If $\mu_s = \mu_t$ for $s, t \in \mathrm{Im}(\mu)$, then $s = t$.*

Proof. Let us assume, without loss of generality, that $s < t$.

Since $s \in \mathrm{Im}(\mu)$, there exists $x \in X$ such that $\mu(x) = s$. So, $x \in \mu_s$ and $x \notin \mu_t$, which is a contradiction. □

Theorem 3.6.10. *Let μ and ρ be two fuzzy subalgebras of a BCC-algebra $\mathfrak{X}$ with identical family of level subalgebras. If $\mathrm{Im}(\mu) = \{t_1, t_2, ..., t_n\}$ and $\mathrm{Im}(\rho) = \{s_1, s_2, ..., s_n\}$, where $t_1 > t_2 > ... > t_n$ and $s_1 > s_2 > ... > s_n$, then*

(1) $m = n$,

(2) $\mu_{t_i} = \rho_{s_i}$ for $i = 1, ..., n$,

(3) if $\mu(x) = t_i$, then $\rho(x) = s_i$ for $x \in X$ and $i = 1, ..., n$.

Proof. (1) By Corollary 3.6.8 fuzzy subalgebras μ and ρ of $\mathfrak{X}$ have (respectively) the only $\{\mu_{t_i}\}$ and $\{\rho_{s_i}\}$ as the families of level subalgebras. By assumption, these families are identical. So, $m = n$.

(2) It follows from Corollary 3.6.8 and Theorem 3.6.7.

(3) Let $x \in X$ be such that $\mu(x) = t_i$ and $\rho(x) = s_j$. From (2) and the assumption $\mu(x) = t_i$, it follows $x \in \rho_{s_i}$. Thus $\rho(x) \geq s_i$ and $s_j \geq s_i$, i.e., $\rho_{s_j} \subseteq \rho_{s_i}$ and, in the consequence, by (2), we have $\rho_{s_i} = \mu_{t_i} \subseteq \mu_{t_j} = \rho_{s_j}$. Thus $\rho_{s_i} = \rho_{s_j}$. But, from Lemma 3.6.9, we get $s_i = s_j$. Therefore, $\rho(x) = s_i$.

The proof is completed. □

Theorem 3.6.11. *Let μ and ρ be two fuzzy subalgebras of $\mathfrak{X}$ with identical family of level subalgebras.*
Then $\mu = \rho$ if and only if $\mathrm{Im}(\mu) = \mathrm{Im}(\rho)$.

Proof. We assume that $\mathrm{Im}(\mu) = \mathrm{Im}(\rho) = \{s_1, ...s_n\}$ and $s_1 > ... > s_n$. From Theorem 3.6.10 it follows that for any $x \in X$ there exists an element s_i such that $\mu(x) = s_i = \rho(x)$. Thus $\mu(x) = \rho(x)$ for all $x \in X$, which implies $\mu = \rho$. □

In the sequel of the section, we will investigate the t-norm and the s-norm. We will start with their definitions given in [1] and in [147].

Definition 3.6.12. By a *t-norm* T we mean a function

$$T : [0,1] \times [0,1] \longrightarrow [0,1]$$

satisfying the following conditions:

(T1) $T(x,1) = x$,
(T2) $T(x,y) \leq T(x,z)$ if $y \leq z$,
(T3) $T(x,y) = T(y,x)$,
(T4) $T(x,T(y,z)) = T(T(x,y)z)$

for all $x, y, z \in [0,1]$.

Definition 3.6.13. By a *s-norm* S we mean a binary operation on the interval $S : [0,1] \times [0,1] \longrightarrow [0,1]$ satisfying the following conditions:

(S1) $S(x,0) = x$,
(S2) $S(x,y) \leq S(x,z)$ if $y \leq z$,
(S3) $S(x,y) = S(y,x)$,
(S4) $S(x,S(y,z)) = S(S(x,y)z)$

for all $x, y, z \in [0,1]$.

Definition 3.6.14. A function $\mu : X \longrightarrow [0,1]$ is called a *T-fuzzy* (resp. *S-fuzzy*) BCC-subalgebra of $\mathfrak{X}$ if $\mu(xy) \geq T(\mu(x), \mu(y))$ (resp. $\mu(xy) \leq S(\mu(x), \mu(y))$) for all $x, y \in X$.

The full name of such subalgebra is *fuzzy BCC-subalgebra* of $\mathfrak{X}$ with respect to a t-norm T (resp. with respect to s-norm S).

The correlation between t-norm T and s-norm S is expressed as follows:

$$T(\alpha,\beta) \leq \min\{\alpha,\beta\} \leq \max\{\alpha,\beta\} \leq S(\alpha,\beta) \tag{3.52}$$

for all $\alpha, \beta \in [0,1]$.

Example 3.6.15. Let us consider a proper BCC-algebra $\mathfrak{X}$ with the operation $\cdot$ defined by Table 2.3.1. We define a fuzzy set μ in X by

$$\mu(x) = \begin{cases} 1 & \text{if } x \in \{0,1,2\}, \\ 0 & \text{otherwise} \end{cases}$$

Let $T_m : [0,1] \times [0,1] \longrightarrow [0,1]$ be a function defined by

$$T_m(\alpha, \beta) = \max\{\alpha + \beta - 1, 0\}$$

for all $\alpha, \beta \in [0, 1]$. T_m is a *t*-norm (cf. [147]).

Similarly, let S_m be an *s*-norm defined by

$$S_m(\alpha, \beta) = \min\{\alpha + \beta\} \text{ for all } \alpha, \beta \in [0, 1].$$

It is easy to check that μ satisfies the inequalities:

$$T_m(\mu(x), \mu(y)) \leq \mu(xy) \leq S_m(\mu(x), \mu(y))$$

for all $x, y \in X$. Hence

(1) μ is a T_m-fuzzy BCC-subalgebra of $\mathfrak{X}$,

(2) μ is a S_m-fuzzy BCC-subalgebra of $\mathfrak{X}$. □

The set of an idempotent elements of a BCC-algebra $\mathfrak{X}$, i.e., such that $xx = x$, is equal to $\{0\}$. But the set of all idempotent elements with respect to *s*-norm S (or S-idempotent), i.e., the set

$$E_S = \{\alpha \in [0, 1] \,|\, S(\alpha, \alpha) = \alpha\}$$

is closed with respect to the operation S. Moreover, from Definition 3.6.13 it follows that the structure $\langle E_S; S, 0\rangle$ is a commutative semigroup with 0 as a neutral element. For example, if $\mathrm{Im}(\mu) \subseteq E_S$, then μ is S-idempotent.

Definition 3.6.16. If an S-fuzzy BCC-subalgebra of $\mathfrak{X}$ is idempotent, then we say that μ is an *idempotent S-fuzzy BCC-subalgebra* of $\mathfrak{X}$.

Example 3.6.17. Let S be an *s*-norm defined by $S_0(\alpha, 0) = \alpha = S_0(0, \alpha)$ and $S_0(\alpha, \beta) = 1$ for $\alpha \neq 0$ and $\beta \neq 0$, where $\alpha, \beta \in [0, 1]$. Let $\mathfrak{X} = \langle\{0, 1, 2, 3\}; \cdot, 0\rangle$ be a BCC-algebra with the operation $\cdot$ defined by the following table:

$\cdot$	0	1	2	3
0	0	0	0	0
1	1	0	0	1
2	2	1	0	2
3	3	3	3	0

Table 3.6.1

We define a fuzzy set $\mu : X \longrightarrow [0, 1]$ by $\mu(0) = \alpha_0$, $\mu(1) = \mu(2) = \alpha_1$ and $\mu(3) = \alpha_2$, where $\alpha_0, \alpha_1, \alpha_2 \in [0, 1]$ with $\alpha_0 \leq \alpha_1 \leq \alpha_2$. Routine calculations give that μ is an S_0-fuzzy subalgebra of $\mathfrak{X}$, which is not idempotent. □

Since in Example 3.6.15 $\mathrm{Im}(\mu) \subseteq E_{S_m}$, μ is an idempotent fuzzy BCC-subalgebra of $\mathfrak{X}$.

Theorem 3.6.18. *Let* S_m *be the s-norm in* Example 3.6.15 *and let* $\mathfrak{A}$ *be a BCC-subalgebra of* $\mathfrak{X}$. *Then a fuzzy set* μ *in* X *defined in the following way:*

$$\mu(x) = \begin{cases} 1 & \text{if } x \in A, \\ 0 & \text{otherwise} \end{cases}$$

is an idempotent fuzzy BCC-subalgebra of $\mathfrak{X}$ *with respect to* S_m.

Proof. Let $x, y \in X$. If $x \notin A$ or $y \notin A$, then $\mu(x) = 1$ or $\mu(y) = 1$ and we get

$$S_m(\mu(x), \mu(y)) = 1 \geq \mu(xy).$$

Let us suppose that $x \in A$ and $y \in A$. Then, of course, $xy \in A$ and thus

$$\mu(xy) = 0 \leq S_m(\mu(x), \mu(y)).$$

So, $\mathrm{Im}(\mu) \subseteq E_{S_m}$, which completes the proof. □

Theorem 3.6.19. *If* μ *is an idempotent* S*-fuzzy BCC-subalgebra of* $\mathfrak{X}$, *then*

$$\mu(0) \leq \mu(x) \text{ and } \mu(xy) \leq \max\{\mu(x), \mu(y)\}$$

for all $x, y \in X$.

Proof. Indeed,

$$\mu(0) = \mu(xx) \leq S(\mu(x), \mu(x)) = \mu(x)$$

for all $x \in X$.
Similarly,

$$\begin{aligned} \max\{\mu(x), \mu(y)\} &= S(\max\{\mu(x), \mu(y)\}, \max\{\mu(x), \mu(y)\}) \\ &\geq S(\mu(x), \mu(y)) \\ &\geq \max\{\mu(x), \mu(y)\} \end{aligned}$$

for all $x, y \in X$. Hence

$$\mu(xy) \leq S(\mu(x), \mu(y)) = \max\{\mu(x), \mu(y)\}$$

for all $x, y \in X$.
The proof is completed. □

Theorem 3.6.20. *Let μ be a T-fuzzy BCC-algebra of a BCC-algebra $\mathfrak{X}$ and let $t \in [0,1]$. Then*

(1) *if $t = 1$, then μ_t is either empty or a $BCC-$ subalgebra of $\mathfrak{X}$,*

(2) *if $T = \min$, then μ_t is either empty or a $BCC-$ subalgebra of $\mathfrak{X}$.*

Moreover, $\mu(0) \geq \mu(x)$ for all $x \in X$.

Proof. (1) Let $t = 1$ and let $x, y \in \mu_t$. Then $\mu(x) \geq t = 1$ and $\mu(y) \geq t = 1$. From Definitions 3.6.12 and 3.6.14 it follows that

$$\mu(xy) \geq T(\mu(x), \mu(y)) \geq T(1,1) = 1,$$

so that $xy \in \mu_t$. Hence μ_t is a BCC-subalgebra of $\mathfrak{X}$.

(2) Let $T = \min$ and let $x, y \in \mu_t$. Then we have

$$\mu(xy) \geq T(\mu(x), \mu(y)) = \min(\mu(x), \mu(y)) \geq \min(t,t) = t$$

for all $t \in [0,1]$. So $xy \in \mu_t$.

Thus μ_t is a BCC-subalgebra of $\mathfrak{X}$.

Moreover, from (BCC2) we get

$$\mu(0) = \mu(xx) \geq T(\mu(x), \mu(x)) = \min(\mu(x), \mu(x)) = \mu(x).$$

This completes the proof. □

Corollary 3.6.21. *Let μ be a fuzzy BCC-subalgebra of $\mathfrak{X}$ with respect to an s-norm S and let $\alpha \in [0,1]$. Then*

(1) *if $S(\alpha, \alpha) = \alpha$, then the lower level set $L(\mu; \alpha)$ of μ is either empty or a BCC-subalgebra of $\mathfrak{X}$,*

(2) *if $S(\alpha, \beta) = \max\{\alpha, \beta\}$, then the lower level set $L(\mu; \alpha)$ of μ is either empty or a BCC-subalgebra of $\mathfrak{X}$.*
Moreover $\mu(0) \leq \mu(x)$, for all $x \in X$.

Corollary 3.6.22. *If μ in X is a fuzzy BCC-subalgebra of $\mathfrak{X}$ with respect to an idempotent s-norm S, then every nonempty lower set $L(\mu; \alpha)$ is a BCC-subalgebra of $\mathfrak{X}$.*

Theorem 3.6.23. *Let S be an s-norm and let μ be a fuzzy set in X. If every nonempty lower set $L(\mu; \alpha)$ is a BCC-subalgebra of $\mathfrak{X}$, then μ is an S-fuzzy BCC-subalgebra of $\mathfrak{X}$.*

Proof. Let us assume that $\mu(x_0y_0) > S(\mu(x_0), \mu(y_0))$ for some $x_0, y_0 \in X$. Then

$$\max\{\mu(x_0), \mu(y_0)\} \leq S(\mu(x_0), \mu(y_0)) < \alpha_0 < \mu(x_0y_0)$$

by taking

$$\alpha = \frac{1}{2}[\mu(x_0y_0) + S(\mu(x_0), \mu(y_0))].$$

It follows that $x_0, y_0 \in L(\mu; \alpha_0)$ and $x_0y_0 \notin L(\mu; \alpha_0)$.

This is a contradiction and hence μ satisfies the inequality $\mu(xy) \leq S(\mu(x), \mu(y))$ for all $x, y \in X$. □

Theorem 3.6.24. *Let μ be a T-fuzzy BCC-subalgebra of a BCC-algebra $\mathfrak{X}$. If there is a sequence $\{x_n\}$ in X such that $\lim_{n\to\infty} T(\mu(x_n), \mu(x_n)) = 1$, then $\mu(0) = 1$.*

Proof. Let $x \in X$. Then $\mu(0) = \mu(xx) \geq T(\mu(x), \mu(x))$. Therefore, $\mu(0) \geq T(\mu(x_n), \mu(x_n))$ for each $n \in \mathbb{N}$. Since

$$1 \geq \mu(0) \geq \lim_{n\to\infty} T(\mu(x_n), \mu(x_n)) = 1,$$

it follows that $\mu(0) = 1$, which completes the proof. □

For all $x \in X$ we define a mapping $\mu[\theta] : X \to [0, 1]$ by $\mu[\theta](x) = \mu(\theta(x))$, where μ is a fuzzy set in the base set X of a BCC-algebra $\mathfrak{X}$ and θ is a mapping $X \to X$.

Theorem 3.6.25. *If μ is a T-fuzzy BCC-subalgebra of a BCC-algebra $\mathfrak{X}$ and θ is an endomorphism of $\mathfrak{X}$, then $\mu[\theta]$ is a T-fuzzy BCC-subalgebra of $\mathfrak{X}$.*

Proof. For any $x, y \in X$, we have

$$\begin{aligned} \mu[\theta](xy) &= \mu(\theta(xy)) \\ &= \mu(\theta(x)\theta(y)) \\ &\geq T(\mu(\theta(x)), \mu(\theta(y))) \\ &= T(\mu[\theta](x), \mu[\theta](y)). \end{aligned}$$

Hence $\mu[\theta]$ is a fuzzy BCC-subalgebra of $\mathfrak{X}$. □

Definition 3.6.26. The *preimage* of v under f is the fuzzy set $\mu = v \circ f$, i.e., $\mu(x) = v(f(x))$ for all $x \in X$, where f is a mapping defined on a BCC-algebra $\mathfrak{X}$ and v is a fuzzy set in $f(X)$.

Theorem 3.6.27. *An epimorphic preimage of a T-fuzzy BCC-subalgebra is a T-fuzzy BCC-subalgebra.*

Proof. Let $f : X \longrightarrow X^{'}$ be an epimorphism of BCC-algebras, v a T-fuzzy BCC-subalgebra $\mathfrak{X}^{'}$ and let μ be the preimage of v under f. Then for any $x, y \in X$ we have

$$\begin{aligned}\mu(xy) &= v(f(xy)) \\ &= v(f(x)f(y)) \\ &\geq T(v(f(x)), v(f(y))) \\ &= T(\mu(x), \mu(y)).\end{aligned}$$

So, μ is a T-fuzzy BCC-subalgebra of $\mathfrak{X}^{'}$. □

Definition 3.6.28. Let μ be a fuzzy set in the set X of a BCC-algebra $\mathfrak{X}$ and let f be a mapping defined on X. The fuzzy set μ^f in $f(X)$ defined by

$$\mu^f(y) = \sup_{x \in f^{-1}(y)} \mu(x)$$

for all $y \in f(X)$ is called the *image* of μ under f.

A fuzzy set μ in X is said to have *sup property*, if for every subset $T \subseteq X$, there exists an element $t_0 \in T$ such that $\mu(t_0) = \sup_{t \in T} \mu(t)$.

Theorem 3.6.29. *An epimorphic image of a fuzzy BCC-subalgebra with sup property is a fuzzy BCC-algebra.*

Proof. Let $f : X \longrightarrow X^{'}$ be an epimorphism of BCC-algebras and let μ be a fuzzy BCC-subalgebra of $\mathfrak{X}$ with sub-property. Now let for $x^{'}, y^{'} \in X^{'}$ the elements $x_0 \in f^{-1}(x^{'})$ and $y_0 \in f^{-1}(y^{'})$ such that $\mu(x_0) = \sup_{t \in f^{-1}(x^{'})} \mu(t)$ and $\mu(y_0) = \sup_{t \in f^{-1}(y^{'})} \mu(t)$, respectively.
Then we get

$$\begin{aligned}\mu^f(x^{'}y^{'}) &= \sup_{z \in f^{-1}(x^{'}y^{'})} \mu(z) \\ &\geq \min\{\mu(x_0), \mu(y_0)\} \\ &= \min\{\sup_{t \in f^{-1}(x^{'})} \mu(t), \sup_{t \in f^{-1}(y^{'})} \mu(t)\} \\ &= \min\{\mu^f(x^{'}), \mu^f(y^{'})\}.\end{aligned}$$

Which means, that μ^f is a fuzzy BCC-subalgebra of $\mathfrak{X}'$. □

There is a stronger version of Theorem 3.6.29.

To present it, we will need the following definition of continuous t-norm (cf. [147]).

Definition 3.6.30. A t-norm T on [0,1] is called a *continuous t-norm* if T is a continuous function $T : [0,1] \times [0,1] \longrightarrow [0,1]$ with respect to the usual topology. Similarly, we define a *continuous s-norm.*

An example of a continuous t-norm is the function min.

Theorem 3.6.31. *Let T be a continuous t-norm and let f be an epimorphism on a BCC-algebra $\mathfrak{X}$. If μ is a T-fuzzy BCC-subalgebra of $\mathfrak{X}$, then μ^f is a T-fuzzy BCC-subalgebra of $f(\mathfrak{X})$.*

Proof. Let $A_1 = f^{-1}(y_1)$, $A_2 = f^{-1}(y_2)$ and $A_{12} = f^{-1}(y_1y_2)$, where $y_1, y_2 \in f(X)$. Let us consider the set

$$A_1A_2 = \{x \in X \,|\, x = a_1a_2 \text{ for some } a_1 \in A_1 \text{ and } a_2 \in A_2\}.$$

Let $x \in A_1A_2$, then $x = x_1x_2$ for some $x_1 \in A_1$ and $x_2 \in A_2$ and as a consequence

$$f(x) = f(x_1x_2) = f(x_1)f(x_2) = y_1y_2,$$

i.e., $x \in f^{-1}(y_1y_2) = A_{12}$, which implies $A_1A_2 \subseteq A_{12}$.
Hence

$$\begin{aligned}\mu^f(y_1y_2) &= \sup_{x \in f^{-1}(y_1y_2)} \mu(x) \\ &= \sup_{x \in A_{12}} \mu(x) \\ &\geq \sup_{x \in A_1A_2} \mu(x) \\ &\geq \sup_{x_1 \in A_1, x_2 \in A_2} \mu(x_1x_2) \\ &\geq \sup_{x_1 \in A_1, x_2 \in A_2} T(\mu(x_1), \mu(x_2)).\end{aligned}$$

From the assumption about T, for every $\xi > 0$ there exists a number $\psi > 0$ such that for $\sup\limits_{x_1 \in A_1} \mu(x_1) - x_1' \leq \psi$ and $\sup\limits_{x_2 \in A_2} \mu(x_2) - x_2' \leq \psi$, we have

$$T(\sup_{x_1 \in A_1} \mu(x_1), \sup_{x_2 \in A_2} \mu(x_2)) - T(x_1', x_2') \leq \xi.$$

Now let us choose $a_1 \in A_1$ and $a_2 \in A_2$ such that

$$\sup_{x_1 \in A_1} \mu(x_1) - \mu(a_1) \leq \psi \text{ and } \sup_{x_2 \in A_2} \mu(x_2) - \mu(a_2) \leq \psi.$$

Then

$$T(\sup_{x_1 \in A_1} \mu(x_1), \sup_{x_2 \in A_2} \mu(x_2)) - T(\mu(a_1), \mu(a_2)) \leq \xi.$$

Consequently,

$$\begin{aligned}\mu^f(y_1 y_2) &\geq \sup_{x_1 \in A_1, x_2 \in A_2} T(\mu(x_1), \mu(x_2)) \\ &\geq T(\sup_{x_1 \in A_1} \mu(x_1), \sup_{x_2 \in A_2} \mu(x_2)) \\ &= T(\mu^f(y_1), \mu^f(y_2)),\end{aligned}$$

which shows that μ^f is a T-fuzzy BCC-subalgebra of $f(\mathfrak{X})$. □

The next lemma comes from [1].

Lemma 3.6.32. *Let T be a t-norm. Then*

$$T(T(\alpha, \beta), T(\gamma, \delta)) = T(T(\alpha, \gamma), T(\beta, \delta))$$

for all $\alpha, \beta, \gamma, \delta \in [0, 1]$.

Theorem 3.6.33. *Let T be a t-norm and let $\mathfrak{X}$, with $X = X_1 \times X_2$, be the direct product BCC-algebra of BCC-algebras $\mathfrak{X}_1$ and $\mathfrak{X}_2$.*
If μ_1 (resp. μ_2) is a T-fuzzy BCC-subalgebra of $\mathfrak{X}_1$ (resp. $\mathfrak{X}_2$), then $\mu = \mu_1 \times \mu_2$ is a T-fuzzy BCC-algebra of $\mathfrak{X}$ defined by

$$\mu(x_1, x_2) = (\mu_1 \times \mu_2)(x_1, x_2) = T(\mu_1(x_1), \mu_2(x_2))$$

for all $(x_1, x_2) \in X_1 \times X_2$.

Proof. Let $x = (x_1, x_2)$ and $y = (y_1, y_2)$ be any elements of $X = X_1 \times X_2$. Then

$$\begin{aligned}\mu(xy) &= \mu((x_1, x_2)(y_1, y_2)) \\ &= \mu(x_1 y_1, x_2 y_2) \\ &= T(\mu_1(x_1 y_1), \mu_2(x_2 y_2)) \\ &\geq T(T(\mu_1(x_1), \mu_1(y_1)), T(\mu_2(x_2), \mu_2(y_2))) \\ &= T(T(\mu_1(x_1), \mu_2(x_2)), T(\mu_1(y_1), \mu_2(y_2))) \\ &= T(\mu(x_1, x_2), \mu(x_2, y_2)) \\ &= T(\mu(x), \mu(y)).\end{aligned}$$

Hence μ is a T-fuzzy BCC-subalgebra of $\mathfrak{X}$. □

The next results are a generalization of the direct product of two subalgebras to the product of n T-fuzzy BCC-subalgebras. The definition generalizes the domain of t-norm T to $\prod_{i=1}^{n}[0,1]$. The Definition 3.6.34 and Lemma 3.6.35 were first presented in [1].

Definition 3.6.34. The function $T_n : \prod_{i=1}^{n}[0,1] \longrightarrow [0,1]$ is defined by

$$T_n(\alpha_1, \alpha_2, \dots, \alpha_n) = T(\alpha_i, T_{n-1}(\alpha_1, \dots, \alpha_{i-1}, \alpha_{i+1}, \dots, \alpha_n))$$

for all $1 \leq i \leq n$, where $n \geq 2$, $T_2 = T$ and $T_1 = id$ (identity).

Lemma 3.6.35. *For a t-norm T and every $\alpha_i, \beta_i \in [0,1]$ where $1 \leq i \leq n$ and $n \geq 2$, we have*

$$\begin{aligned} & T_n(T(\alpha_1, \beta_1), T(\alpha_2, \beta_2), \dots, T(\alpha_n, \beta_n)) \\ & = T(T_n(\alpha_1, \alpha_2, \dots, \alpha_n), T_n(\beta_1, \beta_2, \dots, \beta_n)). \end{aligned}$$

Theorem 3.6.36. *Let T be a t-norm and let $\{\mathfrak{X}_i\}$ for $i \in \{1, ..., n\}$ be a finite collection of BCC-algebras and $\mathfrak{X}$ be the direct product BCC-algebra of $\{\mathfrak{X}_i\}$, where $X = \prod_{i=1}^{n} X_i$.*
Let μ_i be a T-fuzzy BCC-subalgebra of $\mathfrak{X}_i$, where $1 \leq i \leq n$. Then $\mu = \prod_{i=1}^{n} \mu_i$ defined by

$$\begin{aligned} \mu(x_1, x_2, \dots, x_n) &= (\prod_{i=1}^{n} \mu_i)(x_1, x_2, \dots, x_n) \\ &= T_n(\mu_1(x_1), \mu_2(x_2), \dots, \mu_n(x_n)) \end{aligned}$$

is a T-fuzzy BCC-subalgebra of the BCC-algebra $\mathfrak{X}$.

Proof. Let $x = (x_1, x_2, \dots, x_n)$ and $y = (y_1, y_2, \dots, y_n)$ be any elements of $X = \prod_{i=1}^{n} X_i$. Then

$$\begin{aligned}
\mu(xy) &= \\
&= \mu(x_1y_1, x_2y_2, \ldots, x_ny_n) \\
&= T_n(\mu_1(x_1y_1), \mu_2(x_2y_2), \ldots, \mu_n(x_ny_n)) \\
&\geq T_n(T(\mu_1(x_1), \mu_1(y_1)), T(\mu_2(x_2), \mu_2(y_2)), \ldots, T(\mu_n(x_n), \mu_n(y_n))) \\
&= T(T_n(\mu_1(x_1), \mu_2(x_2), \ldots, \mu_n(x_n)), T_n(\mu_1(y_1), \mu_2(y_2), \ldots, \mu_n(y_n))) \\
&= T(\mu(x_1, x_2, \ldots, x_n), \mu(y_1, y_2, \ldots, y_n)) \\
&= T(\mu(x), \mu(y)).
\end{aligned}$$

Hence μ is a T-fuzzy BCC-subalgebra of $\mathfrak{X}$.

The proof is completed. □

Definition 3.6.37. Let T be a t-norm and let μ and v be fuzzy sets in a BCC-algebra $\mathfrak{X}$. Then the T-product of μ and v, denoted by $[\mu \cdot v]_T$, is defined by

$$[\mu \cdot v]_T(x) = T(\mu(x), v(x))$$

for all $x \in X$.

Definition 3.6.38. We say that t-norm T^* *dominates* T, when

$$T^*(T(\alpha, \beta), T(\gamma, \delta)) \geq T(T^*(\alpha, \gamma), T^*(\beta, \delta))$$

for all $\alpha, \beta, \gamma, \delta \in [0, 1]$.

Theorem 3.6.39. *Let T be a t-norm and let μ and v be T-fuzzy BCC-subalgebras of a BCC-algebra $\mathfrak{X}$. If T^* is a t-norm which dominates T, then the T^*-product of μ and v, $[\mu \cdot v]_{T^*}$, is a T-fuzzy BCC-subalgebra of $\mathfrak{X}$.*

Proof. For any $x, y \in X$ we have

$$\begin{aligned}
[\mu \cdot v]_{T^*}(xy) &= T^*(\mu(xy), v(xy)) \\
&\geq T^*(T(\mu(x), \mu(y)), T(v(x), v(y))) \\
&\geq T(T^*(\mu(x), v(x)), T^*(\mu(y), v(y))) \\
&= T([\mu \cdot v]_{T^*}(x), [\mu \cdot v]_{T^*}(y)).
\end{aligned}$$

Hence $[\mu \cdot v]$ is a T-fuzzy BCC-subalgebra of $\mathfrak{X}$. □

Let $f : X \longrightarrow X^{'}$ be an epimorphism of BCC-algebras. Let T and T^* be t-norms such that T^* dominates T.

If μ and v are T-fuzzy BCC-subalgebras of $\mathfrak{X}^{'}$, then the T^*-product of μ and v, $[\mu \cdot v]_{T^*}$, is a T-fuzzy BCC-subalgebra of $\mathfrak{X}^{'}$.

Since every epimorphic preimage of a T-fuzzy BCC-subalgebra is a T-fuzzy BCC-subalgebra, the preimages $f^{-1}(\mu), f^{-1}(v)$ and $f^{-1}([\mu \cdot v]_{T^*})$ are T-fuzzy BCC-subalgebras of $\mathfrak{X}$.

The next theorem shows the relation between $f^{-1}([\mu \cdot v]_{T^*})$ and the T^*- product $[f^{-1}(\mu) \cdot f^{-1}(v)]_{T^*}$ of $f^{-1}(\mu)$ and $f^{-1}(v)$.

Theorem 3.6.40. *Let $f : X \longrightarrow X^{'}$ be an epimorphism of BCC-algebras. Let T and T^* be t-norms such that T^* dominates T. Let μ and v be T-fuzzy BCC-subalgebras of $\mathfrak{X}^{'}$. If $[\mu \cdot v]_{T^*}$ is the T^*-product of μ and v and $[f^{-1}(\mu) \cdot f^{-1}(v)]_{T^*}$ is the T^*-product of $f^{-1}(\mu)$ and $f^{-1}(v)$, then*

$$f^{-1}([\mu \cdot v]_{T^*}) = [f^{-1}(\mu) \cdot f^{-1}(v)]_{T^*}.$$

Proof. For any $x \in X$, we have

$$\begin{aligned}
[f^{-1}([\mu \cdot v]_{T^*})](x) &= [\mu \cdot v]_{T^*}(f(x)) \\
&= T^*(\mu(f(x)), v(f(x))) \\
&= T^*([f^{-1}(\mu)](x), [f^{-1}(v)](x)) \\
&= [f^{-1}(\mu) \cdot f^{-1}(v)]_{T^*}(x).
\end{aligned}$$

The proof is completed. □

In [49] the concept of zero invariant fuzzy BCC-algebras was introduced. We will now give their definition and show their properties.

Definition 3.6.41. A fuzzy set μ in X is said to be a *zero invariant (0-invariant)* if $\mu(0) = 0$.

Note that the fuzzy BCC-subalgebra of $\mathfrak{X}$ with respect to S_m in Theorem 3.6.18 is zero invariant.

In the next theorem, we answer the question: what condition must a fuzzy BCC-subalgebra of $\mathfrak{X}$ with respect to an s-norm satisfy in order to be zero invariant?

Theorem 3.6.42. *Let S be an s-norm and let μ be an S-fuzzy BCC-subalgebra of $\mathfrak{X}$. If there is a sequence $\{x_n\}$ in X such that*

$$\lim_{n\to\infty} \inf S(\mu(x_n), \mu(x_n)) = 0,$$

then μ is zero invariant.

Proof. For any $x \in X$, we have $\mu(0) = \mu(xx) \le S(\mu(x), \mu(x))$. Therefore, $\mu(0) \le S(\mu(x_n), \mu(x_n))$ for each $n \in N$. So, we have

$$0 \le \mu(0) \le \lim_{n\to\infty} \inf S(\mu(x_n), \mu(x_n)) = 0.$$

From the above, it follows that $\mu(0) = 0$. Hence μ is zero invariant. □

Now we define the mapping $\mu_\Theta : X \longrightarrow [0, 1]$ by $\mu_\Theta(x) = \mu(\Theta(x))$ for all $x \in X$, where μ is a fuzzy set in X and Θ is a mapping from X into itself.

Theorem 3.6.43. *Let S be an s-norm. If μ is an S-fuzzy BCC-subalgebra of $\mathfrak{X}$ and Θ is an endomorphism of $\mathfrak{X}$, then μ_Θ is an S-fuzzy BCC-subalgebra of $\mathfrak{X}$. Moreover, if μ is zero invariant, then so is μ_Θ.*

Proof. For any $x, y \in X$, we have

$$\begin{aligned}\mu_\Theta(xy) &= \mu(\Theta(xy)) \\ &= \mu(\Theta(x)\Theta(y)) \\ &\le S(\mu(\Theta(x)), \mu(\Theta(y))) \\ &= S(\mu_\Theta(x), \mu_\Theta(y)).\end{aligned}$$

Hence μ_Θ is an S-fuzzy BCC-subalgebra of $\mathfrak{X}$.

Let us assume that μ is zero invariant. Since $\Theta(0) = 0$, we get

$$\mu_\Theta(0) = \mu(\Theta(0)) = \mu(0) = 0.$$

This completes the proof. □

Let f be a mapping defined on X and let v be a fuzzy set in $f(X)$. Then from Definition 3.6.26 it follows that $f^{-1}(v)$ defined by $[f^{-1}](x) = v(f(x))$ for all $x \in X$, is the preimage of v under f.

Theorem 3.6.44. *Let S be an s-norm. An onto homomorphic preimage of a zero invariant S-fuzzy BCC-subalgebra is also a zero invariant S-fuzzy BCC-subalgebra.*

Proof. Let $f : X \longrightarrow Y$ be a homomorphism from a BCC-algebra $\mathfrak{X}$ onto a BCC-algebra $\mathfrak{Y}$ and let v be zero invariant S-fuzzy BCC-subalgebra of Y.

Then

$$\begin{aligned}[f^{-1}(v)](xy) &= v(f(xy)) = v(f(x)f(y)) \\ &\leq S(v(f(x)), v(f(y))) = S([f^{-1}(v)](x), [f^{-1}(v)](y))\end{aligned}$$

for all $x, y \in X$.

Hence $f^{-1}(v)$ S-fuzzy BCC-subalgebra of $\mathfrak{X}$.

Since

$$[f^{-1}(v)](0) = v(f(0)) = v(0) = 0,$$

$f^{-1}(v)$ is zero invariant. □

Definition 3.6.45. Let μ be a fuzzy set in X and let f be a mapping defined on X. The fuzzy set μ^f in $f(X)$ defined by

$$\mu^f(y) = \inf_{x \in f^{-1}(y)} \mu(x)$$

for all $y \in f(X)$ is called the *anti-image* of μ under f.

Theorem 3.6.46. *Let S be a continuous s-norm and let f be a homomorphism on $\mathfrak{X}$. If μ is a zero invariant S-fuzzy BCC-subalgebra of $\mathfrak{X}$, then the anti-image of μ under f is a zero invariant S-fuzzy BCC-subalgebra of $f(\mathfrak{X})$.*

Proof. Let $A_1 = f^{-1}(y_1), A_2 = f^{-1}(y_2)$ and $A_{12} = f^{-1}(y_1y_2)$, where $y_1, y_2 \in f(X)$. Let us consider the set

$$A_1A_2 = \{x \in X \mid x = a_1a_2 \text{ for some } a_1 \in A_1, a_2 \in A_2\}.$$

If $x \in A_1A_2$, then $x = x_1x_2$ for some $x_1 \in A_1, x_2 \in A_2$. So, we have

$$f(x) = f(x_1x_2) = f(x_1)f(x_2) = y_1y_2,$$

i.e.,

$$x \in f^{-1}(y_1y_2) = A_{12}.$$

Thus $A_1A_2 \subseteq A_{12}$.

It follows that

$$\begin{aligned}\mu^f(y_1y_2) &= \inf_{x\in f^{-1}(y_1y_2)} \mu(x)\\ &= \inf_{x\in A_{12}} \mu(x)\\ &\le \inf_{x\in A_1A_2} \mu(x)\\ &\le \inf_{x_1\in A_1, x_2\in A_2} \mu(x_1x_2)\\ &\le \inf_{x_1\in A_1, x_2\in A_2} S(\mu(x_1),\mu(x_2)).\end{aligned}$$

From assumption S is continuous. Therefore, if ϵ is any positive number, then there exists a number $\delta > 0$ such that

$$S(x_1^*, x_2^*) \le S\left(\inf_{x_1\in A_1} \mu(x_1), \inf_{x_2\in A_2} \mu(x_2)\right) + \epsilon,$$

whenever

$$x_1^* \le \inf_{x_1\in A_1} \mu(x_1) + \delta \text{ and } x_2^* \le \inf_{x_2\in A_2} \mu(x_2) + \delta.$$

Let us choose $a_1 \in A_1$ and $a_2 \in A_2$ such that

$$\mu(a_1) \le \inf_{x_1\in A_1} \mu(x_1) + \delta \text{ and } \mu(a_2) \le \inf_{x_2\in A_2} \mu(x_2) + \delta.$$

Then

$$S(\mu(a_1),\mu(a_2)) \le S\left(\inf_{x_1\in A_1} \mu(x_1), \inf_{x_2\in A_2} \mu(x_2)\right) + \epsilon.$$

As a consequence, we have

$$\begin{aligned}\mu^f(y_1y_2) &\le \inf_{x_1\in A_1, x_2\in A_2} S(\mu(x_1),\mu(x_2))\\ &\le S\left(\inf_{x_1\in A_1} \mu(x_1), \inf_{x_2\in A_2} \mu(x_2)\right)\\ &= S(\mu^f(y_1),\mu^f(y_2)).\end{aligned}$$

Moreover,

$$\mu^f(0) = \inf_{x\in f^{-1}(0)} \mu(x) \le \mu(0) = 0.$$

Hence $\mu^f(0) = 0$.

Thus μ^f is a zero variant fuzzy BCC-subalgebra of $f(X)$ with respect to S.

The proof is completed. □

3.7 Derivations of Weak BCC-Algebras

In the theory of rings, the properties of derivations play an important role. Y.B. Jun and X.L. Xin [95] applied the notion of derivations in ring and near-ring theory to BCI-algebras, and they also introduced a new concept called a regular derivation in BCI-algebras. Ch. Prabpayak and U. Leerawat [124] applied the notion of regular derivation in BCI-algebras to BCC-algebras. In [129], [130] the concept of derivations was generalized to the weak BCC-algebras.

In this section we give the generalized results for derivations in solid weak BCC-algebras.

Definition 3.7.1. Let $\mathfrak{X}$ be a weak BCC-algebra. A map $d : X \longrightarrow X$ is called a *left-right derivation* (briefly, *(l, r)-derivation*) of $\mathfrak{X}$, if it satisfies the identity

$$d(xy) = d(x)y \wedge xd(y), \tag{3.53}$$

where $x \wedge y = y \cdot yx$.

If d satisfies the identity

$$d(xy) = xd(y) \wedge d(x)y, \tag{3.54}$$

then it is called a *right-left derivation* (briefly, *(r, l)-derivation*) of $\mathfrak{X}$. A map d, which is both (l, r)- and (r, l)-derivation is called a *derivation.* Any derivation d with the property $d(0) = 0$ is called *regular.*

We will now give a more general version of the above definition.

Definition 3.7.2. Let f be an endomorphism of a weak BCC-algebra $\mathfrak{X}$. A map $d_f : X \longrightarrow X$ satisfying the identity

$$d_f(xy) = d_f(x)f(y) \wedge f(x)d_f(y) \tag{3.55}$$

is called a *left-right f-derivation* (briefly, *(l, r)-f-derivation*) of $\mathfrak{X}$.

A *right-left f-derivation* (briefly, *(r, l)-f-derivation*) of $\mathfrak{X}$ is defined as follows:

$$d_f(xy) = f(x)d_f(y) \wedge d_f(x)f(y). \tag{3.56}$$

A self-map d_f is called a *f-derivation* of $\mathfrak{X}$ if it is both (l, r)- and (r, l)-f-derivation.

An f-derivation d_f with the property $d_f(0) = 0$ is called *regular.*

Substituting $f(x) = x$ in an f-derivation, we obtain a derivation defined in Definition 3.7.1.

Example 3.7.3. Let $X = \{0, 1, 2, 3\}$ be a weak BCC-algebra with the operation $\cdot$ defined by Table 1.2.13.

Consider two maps d_1, $d_2 : X \longrightarrow X$ defined in the following way:

$$d_1(x) = \begin{cases} 0 & \text{if} \quad x \in \{0, 1, 3\}, \\ 2 & \text{if} \quad x = 2, \end{cases}$$

$$d_2(x) = \begin{cases} 0 & \text{if} \quad x \in \{0, 1\}, \\ 2 & \text{if} \quad x \in \{2, 3\}. \end{cases}$$

Then d_1 is both (l, r)- and (r, l)-derivation of $\mathfrak{X}$. Of course, d_1 is a regular derivation.

We show that d_1 is (r, l)-derivation. The (l, r)-derivation can be proven in a similar way.

We have three cases to prove:

Case 1^0. Let $x = 2$ and $y \in \{0, 1, 3\}$. Then

$$d_1(2 \cdot y) = \begin{cases} d_1(2) & \text{for} \quad y \in \{0, 1\} \\ d_1(0) & \text{for} \quad y \in \{2, 3\} \end{cases} = \begin{cases} 2 & \text{for} \quad y \in \{0, 1\}, \\ 0 & \text{for} \quad y \in \{2, 3\}. \end{cases}$$

On the other hand,

$$2 \cdot d_1(y) \wedge d_1(2) \cdot y = 2 \cdot 0 \wedge 2 \cdot y = 2 \wedge \begin{cases} 2 & \text{for} \quad y \in \{0, 1\} \\ 0 & \text{for} \quad y \in \{2, 3\} \end{cases}$$

$$= \begin{cases} 2 & \text{for} \quad y \in \{0, 1\}, \\ 0 & \text{for} \quad y \in \{2, 3\}. \end{cases}$$

This means that in this case equality (3.54) holds.

Case 2^0. Let $x \in \{0, 1, 3\}$ and $y = 2$. Then

$$d_1(x \cdot 2) = \left\{ \begin{array}{lll} d_1(1) & \text{for} & y \in \{1, 3\} \\ d_1(0) & \text{for} & y \in \{0, 2\} \end{array} \right\} = 0.$$

On the other hand,

$$x \cdot d_1(2) \wedge d_1(x) \cdot 2 = x \cdot 2 \wedge 0 \cdot 2 = x \cdot 2 \wedge 0 = 0 \cdot (0 \cdot (x \cdot 2)) = 0.$$

In this case (3.54) holds, too.

Case 3^0. Let $x, y \in \{0, 1, 3\}$. Then the algebra $\langle\{0, 1, 3\}; \cdot, 0\rangle$ is isomorphic to the improper BCC-algebra with the operation $\cdot$ given in Table 1.2.5. It is also closed under the operation $\cdot$. So, we have

$$d_1(x \cdot y) = 0.$$

On the other hand,

$$x \cdot d_1(y) \wedge d_1(x) \cdot y = x \cdot 0 \wedge 0 \cdot y = x \wedge 0 = 0.$$

Condition (3.54) is satisfied.

We have shown that d_1 is (r, l)-derivation. The map d_2 is a (r, l)-derivation of $\mathfrak{X}$, but it is not a (l, r)-derivation, because

$$d_2(3 \cdot 0) = d_2(3) = 2$$

and

$$d_2(3) \cdot 0 \wedge 3 \cdot d_2(0) = 2 \cdot 0 \wedge 3 \cdot 0 = 2 \wedge 3 = 3 \cdot (3 \cdot 2) = 3 \cdot 1 = 3.$$

Equation (3.55) is not satisfied. □

Example 3.7.4. Let for $X = \{0, 1, 2, 3, 4, 5\}$ the algebra $\mathfrak{X} = \langle X; \cdot, 0\rangle$ be a solid weak BCC-algebra with the operation $\cdot$ defined by the following table:

$\cdot$	0	1	2	3	4	5
0	0	0	2	2	2	2
1	1	0	2	2	2	2
2	2	2	0	0	0	0
3	3	2	1	0	0	0
4	4	2	1	1	0	1
5	5	2	1	1	1	0

Table 3.7.1

We define a map $d_\varphi : X \longrightarrow X$ by

$$d_\varphi(x) = \begin{cases} 2 & if \quad x \in \{0, 1\}, \\ 0 & if \quad x \in \{2, 3, 4, 5\}. \end{cases}$$

Then it is easy to check that d_φ is a derivation of $\mathfrak{X}$, which is not regular.

Moreover, since $\mathfrak{X}$ is solid, by Theorem 2.6.12, the map φ is its endomorphism. The map d_φ is a φ-derivation of $\mathfrak{X}$. □

Example 3.7.5. Let $\mathfrak{X}$ and $d_\varphi = d_f$ be as in the previous example. Then d_f is not a f-derivation for an endomorphism $f(x) = 0$ for $x \in X$ since

$$d_f(2 \cdot 3) = d_f(0) = 2 \neq d_f(2)f(3) \wedge f(2)d_f(3) = 0.$$

□

Example 3.7.6. Let for $X = \{0, 1, 2, 3, 4, 5\}$ the algebra $\mathfrak{X} = \langle X; \cdot, 0\rangle$ be a solid weak BCC-algebra with the operation $\cdot$ defined by Table 3.7.2.

Let us consider the map:

$$f = \begin{pmatrix} 0 & 1 & 2 & 3 & 4 & 5 \\ 0 & 1 & 3 & 2 & 5 & 4 \end{pmatrix}.$$

*	0	1	2	3	4	5
0	0	0	3	2	3	2
1	1	0	5	4	3	2
2	2	2	0	3	0	3
3	3	3	2	0	2	0
4	4	2	1	5	0	3
5	5	3	4	1	2	0

Table 3.7.2

It is not difficult to check that the map f is an endomorphism of $\mathfrak{X}$. f is not a derivation. Indeed,

$$f(1 \cdot 5) = f(2) = 3.$$

On the other hand,

$$f(1) \cdot 5 \wedge 1 \cdot f(5) = 1 \cdot 5 \wedge 1 \cdot 4 = 2 \wedge 3 = 3 \cdot (3 \cdot 2) = 2,$$

i.e., f is not (l, r)-derivation.

Now, we have

$$f(2 \cdot 3) = f(3) = 2.$$

On the other hand,

$$2 \cdot f(3) \wedge f(2) \cdot 3 = 2 \cdot 2 \wedge 3 \cdot 3 = 0 \wedge 0 = 0.$$

So, f is not $(r.l)$-derivation, too.

But, since f is an endomorphism, it is easy to see that the map $d_f = f$ is a f-derivation of $\mathfrak{X}$. □

The next result characterizes the mapping φ in the derivation environment.

Theorem 3.7.7. *The endomorphism φ^2 is a regular derivation of each solid weak BCC-algebra.*

Proof. From Corollary 2.2.3 it follows that the map $d(x) = \varphi^2(x)$ is a regular endomorphism and $d(x) = a$ for all $x \in B(a)$. Thus for $x \in B(a)$, $y \in B(b)$, according to Theorem 2.4.18, we have $d(xy) = ab$.

On the other hand,

$$d(x)y \wedge xd(y) = ay \wedge xb = xb \cdot (xb \cdot ay).$$

Since, by the assumption, a weak BCC-algebra $\mathfrak{X}$ is solid and elements xb, ay are in the same branch

$$(xb \cdot (xb \cdot ay)) \cdot ay = (xb \cdot ay)(xb \cdot ay) = 0,$$

which means that

$$d(x)y \wedge xd(y) = xb \cdot (xb \cdot ay) \leq ay.$$

But, by Lemma 2.6.15, $ay = ab$. Thus $d(x)y \wedge xd(y) \leq ab$. This, in view of Theorem 2.6.13, proves $d(x)y \wedge xd(y) = ab = d(xy)$.

Now,

$$xd(y) \wedge d(x)y = xb \wedge ay = ay \cdot (ay \cdot xb).$$

Since $ay, ab \in B(ab)$, we have

$$(ay \cdot (ay \cdot xb)) \cdot ab = (ay \cdot ab)(ay \cdot xb) \leq by \cdot (ay \cdot xb),$$

from (BCK1) and (1.4). But $by = 0$ and $ay \cdot xb \in B(0)$, which gives

$$by \cdot (ay \cdot xb) = 0 \cdot (ay \cdot xb) = 0.$$

Thus $xd(y) \wedge d(x)y \leq ab$ and this implies $xd(y) \wedge d(x)y = ab = d(xy)$. Therefore, d is a regular derivation of $\mathfrak{X}$. □

Corollary 3.7.8. *A map $d : X \longrightarrow X$ such that $d(x) = a$ for all $x \in B(a)$, is a regular derivation of each solid weak BCC-algebra.*

Observe that the endomorphism φ^2 is a φ^2-derivation, too. Indeed, we have

$$\varphi^2(x)\varphi^2(y) \wedge \varphi^2(x)\varphi^2(y) = \varphi^2(x)\varphi^2(y) = \varphi^2(xy).$$

So, the condition (3.55) is satisfied and φ^2 is (l,r)-φ^2-derivation.

The proof of (r,l)-φ^2-derivation is similar.

But φ^2 does not need to be a φ-derivation. We show it in the next example.

Example 3.7.9. The endomorphism φ^2 is not a φ-derivation. Indeed, φ is an endomorphism of each solid weak BCC-algebra, but for a solid weak BCC-algebra defined in Example 3.7.6 we have

$$\varphi^2(1 \cdot 5) = 2.$$

On the other hand,

$$\begin{aligned} \varphi^2(1)\varphi(5) \wedge \varphi(1)\varphi^2(5) &= \varphi(1)\varphi^2(5) \cdot (\varphi(1)\varphi^2(5) \cdot \varphi^2(1)\varphi(5)) \\ &= (0 \cdot 3) \cdot ((0 \cdot 3) \cdot (0 \cdot 2)) \\ &= 2 \cdot (2 \cdot 3) \\ &= 3. \end{aligned}$$

Hence, φ^2 is not a (l,r)-φ-derivation.

It is also not a (r,l)-φ-derivation because

$$\varphi^2(2 \cdot 5) = 3$$

and

$$\begin{aligned} \varphi(2)\varphi^2(5) \wedge \varphi^2(2)\varphi(5) &= \varphi^2(2)\varphi(5) \cdot (\varphi^2(2)\varphi(5) \cdot \varphi(2)\varphi^2(5)) \\ &= (2 \cdot 2) \cdot ((2 \cdot 2) \cdot (3 \cdot 3)) \\ &= 0. \end{aligned}$$

□

Corollary 3.7.10. *Each endomorphism f of a weak BCC-algebra $\mathfrak{X}$ is its f-derivation.*

Proof. Straightforward. □

Theorem 3.7.11. *Let d_f be a self-map of weak BCC-algebra $\mathfrak{X}$. Then:*

(1) *if d_f is a regular (l,r)-f-derivation of $\mathfrak{X}$, then*
$d_f(x) = d_f(x) \wedge f(x)$,

(2) *if d_f is a (r,l)-f-derivation of $\mathfrak{X}$, then*
$d_f(x) = f(x) \wedge d_f(x)$ *if and only if d_f is regular.*

Proof. (1) Let d_f be a regular (l,r)-f-derivation of $\mathfrak{X}$. Then

$$d_f(x) = d_f(x0) = d_f(x)f(0) \wedge f(x)d_f(0) = d_f(x) \wedge f(x).$$

(2) If d_f is a (r,l)-f-derivation of $\mathfrak{X}$ and $d_f(x) = f(x) \wedge d_f(x)$, then

$$d_f(0) = f(0) \wedge d_f(0) = 0 \wedge d_f(0) = 0.$$

Conversely, if $d_f(0) = 0$, then

$$d_f(x) = d_f(x0) = f(x)d_f(0) \wedge d_f(x)f(0) = f(x) \wedge d_f(x).$$

□

Corollary 3.7.12. *For a derivation d of solid weak BCC-algebra $\mathfrak{X}$, the following conditions are equivalent:*

(1) *d is regular,*

(2) *$d(x) \leq x$ for every $x \in X$,*

(3) *$d(a) = a$ for every $a \in I(X)$.*

Proof. (1) $\Longrightarrow$ (2) For any regular derivation d from Theorem 3.7.11(2) we have

$$d(x)x = (x \wedge d(x))x = (d(x) \cdot d(x)x)x = d(x)x \cdot d(x)x = 0,$$

because elements x and $d(x)$ belong to the same branch (Theorem 2.4.20). This proves (2).

(2) $\Longrightarrow$ (3) It follows directly from Definition 2.2.4.

(3) $\Longrightarrow$ (1) Since $0 \in I(X)$, we have $d(0) = 0$, i.e., d is regular. □

Corollary 3.7.13. *Let d be a regular derivation of solid weak BCC-algebra $\mathfrak{X}$. Then*

(1) $d(x)y \leq xd(y)$,

(2) $d(xy) = d(x)y$

for all $x, y \in X$.

Proof. (1) Applying 1.4 and 1.5 to Corollary 3.7.12 (2) we obtain

$$d(x)y \leq xy \leq xd(y),$$

which proves (1).

(2) From the above

$$d(xy) = xd(y) \wedge d(x)y = d(x)y \cdot (d(x)y \cdot xd(y)) = d(x)y \cdot 0 = d(x)y.$$

This completes the proof. □

Theorem 3.7.14. *If* d_f *is a regular* (r,l)-f-*derivation of a solid weak BCC-algebra* $\mathfrak{X}$, *then* $d_f(B(a)) \subset B(f(a))$.

Proof. Let $x \in B(a)$. Then $a \leq x$, and consequently

$$\begin{aligned} 0 = d_f(0) = d_f(ax) \\ &= f(a)d_f(x) \wedge d_f(a)f(x) \\ &= d_f(a)f(x) \cdot (d_f(a)f(x) \cdot f(a)d_f(x)), \end{aligned}$$

i.e.,

$$d_f(a)f(x) \leq d_f(a)f(x) \cdot f(a)d_f(x).$$

From this, applying (1.4), we obtain

$$\begin{aligned} 0 &= d_f(a)f(x) \cdot d_f(a)f(x) \\ &\leq (d_f(a)f(x) \cdot f(a)d_f(x)) \cdot d_f(a)f(x) \\ &= 0 \cdot f(a)d_f(x), \end{aligned}$$

because $\mathfrak{X}$ is solid. Hence, by Theorem 2.2.2, we have $0 \leq f(a)d_f(x)$. Thus $f(a)d_f(x) \in B(0)$, which means (Theorem 2.4.20) that $f(a)$ and $d_f(x)$ are in the same branch.

But, by Theorem 2.2.5, for every $a \in I(X)$ we have

$$f(a) = f(0 \cdot 0a) = 0 \cdot 0f(a).$$

So, $f(a) \in I(X)$. Consequently, $d_f(x) \in B(f(a))$.
Therefore, $d_f(B(a)) \subset B(f(a))$. □

Theorem 3.7.15. *If a (r,l)-f-derivation of a solid weak BCC-algebra $\mathfrak{X}$ is regular, then $d_f(a) = f(a)$ for every $a \in I(X)$.*

Proof. Since $d_f(B(a)) \subset B(f(a))$ for every $a \in I(X)$, we have

$$f(a) \le d_f(a). \tag{3.57}$$

This means also that the elements $f(a)$ and $d_f(a)$ are in the same branch.

But $d_f(a) = f(a) \wedge d_f(a) = d_f(a) \cdot d_f(a)f(a)$. Hence

$$d_f(a)f(a) = (d_f(a) \cdot d_f(a)f(a))f(a) = d_f(a)f(a) \cdot d_f(a)f(a) = 0.$$

Thus

$$d_f(a) \le f(a). \tag{3.58}$$

Combining (3.57) and (3.58) we get $d_f(a) = f(a)$ for every $a \in I(X)$. □

Theorem 3.7.16. *For any f-derivation d_f of a solid weak BCC-algebra $\mathfrak{X}$ elements $f(x)$ and $d_f(d_f(x))$ are in the same branch.*

Proof. Let us assume $f(y) = d_f(x)$. Then we obtain

$$\begin{aligned} d_f(xy) &= d_f(x)f(y) \wedge f(x)d_f(y) \\ &= f(y)f(y) \wedge f(x)d_f(y) \\ &= 0 \wedge f(x)d_f(y) \\ &= f(x)d_f(y) \cdot (f(x)d_f(y) \cdot 0) \\ &= 0. \end{aligned}$$

Thus $d_f(xy) = 0$ for $f(y) = d_f(x)$. This, together with Theorem 2.2.2, gives

$$\begin{aligned} 0 &= d_f(xy) \\ &= f(x)d_f(y) \wedge d_f(x)f(y) \\ &= f(x)d_f(y) \wedge 0 \\ &= 0(0 \cdot f(x)d_f(y)) \\ &\le f(x)d_f(y). \end{aligned}$$

Therefore, $f(x)d_f(y) \in B(0)$, which shows that the elements $f(x)$ and $d_f(y) = d_f(d_f(x))$ are in the same branch. □

Example 3.7.4 shows that x and $d(x)$ may not be in the same branch if d is not regular. Indeed, by putting $x = 1$ we get $d(1) = 2$, i.e., $x \in B(0)$ and $d(x) \in B(2)$.

Based on this observation, we can formulate the following theorem:

Theorem 3.7.17. *A (r,l)-f-derivation (resp. (l,r)-f-derivation) d_f of a solid weak BCC-algebra $\mathfrak{X}$ is regular if and only if for every $x \in X$ elements $f(x)$ and $d_f(x)$ belong to the same branch.*

Proof. Let d_f be a regular (r,l)-f-derivation of a solid weak BCC-algebra $\mathfrak{X}$. Then for any $x \in X$ we have

$$\begin{aligned} 0 &= d_f(xx) = f(x)d_f(x) \wedge d_f(x)f(x) \\ &= d_f(x)f(x) \cdot (d_f(x)f(x) \cdot f(x)d_f(x)), \end{aligned}$$

which implies

$$d_f(x)f(x) \leq d_f(x)f(x) \cdot f(x)d_f(x).$$

From this we obtain

$$\begin{aligned} 0 &= d_f(x)f(x) \cdot d_f(x)f(x) \\ &\leq (d_f(x)f(x) \cdot f(x)d_f(x)) \cdot d_f(x)f(x) \\ &= (d_f(x)f(x) \cdot d_f(x)f(x)) \cdot f(x)d_f(x) \\ &= 0 \cdot f(x)d_f(x). \end{aligned}$$

So, $f(x)d_f(x) \in B(0)$. This, according to Theorem 2.4.20, means that $f(x)$ and $d_f(x)$ are in the same branch.

Conversely, let for every $x \in X$ elements $f(x)$ and $d_f(x)$ be in the same branch. Then $d_f(x)$, $d_f(d_f(x))$ are in the same branch, too. So, $f(x)d_f(x), d_f(x)f(x)$ and $f(x)d_f(d_f(x))$ are in $B(0)$. Thus

$$0 = 0 \cdot f(x)d_f(x) = 0 \cdot d_f(x)f(x) = 0 \cdot f(x)d_f(d_f(x)).$$

Hence

$$\begin{aligned} d_f(0) &= \\ &= d_f(0 \cdot f(x)d_f(x)) \\ &= (0 \cdot d_f(f(x)d_f(x))) \wedge (d_f(0) \cdot f(x)d_f(x)) \\ &= (0 \cdot ((f(x)d_f(d_f(x))) \wedge (d_f(f(x))d_f(x)))) \wedge (d_f(0) \cdot f(x)d_f(x)) \end{aligned}$$

$$= (0 \cdot (d_f(f(x))d_f(x) \cdot (d_f(f(x))d_f(x) \cdot f(x)d_f(d_f(x))))) \wedge$$
$$(d_f(0) \cdot f(x)d_f(x))$$
$$= ((0 \cdot d_f(f(x))d_f(x))((0 \cdot d_f(f(x))d_f(x))(0 \cdot f(x)d_f(d_f(x))))) \wedge$$
$$(d_f(0) \cdot f(x)d_f(x))$$
$$= ((0 \cdot d_f(f(x))d_f(x))((0 \cdot d_f(f(x))d_f(x))0)) \wedge (d_f(0) \cdot f(x)d_f(x))$$
$$= ((0 \cdot d_f(f(x))d_f(x))(0 \cdot d_f(f(x))d_f(x))) \wedge (d_f(0) \cdot f(x)d_f(x))$$
$$= 0 \wedge (d_f(0) \cdot f(x)d_f(x))$$
$$= (d_f(0) \cdot f(x)d_f(x))((d_f(0) \cdot f(x)d_f(x))0)$$
$$= (d_f(0) \cdot f(x)d_f(x))(d_f(0) \cdot f(x)d_f(x))$$
$$= 0,$$

i.e., d_f is regular.

In a similar way, we can prove the theorem for a (l, r)-f-derivation. □

Corollary 3.7.18. *A f-derivation d_f of a solid weak BCC-algebra $\mathfrak{X}$ is regular if and only if for every $x \in X$ elements $f(x)$ and $d_f(x)$ are in the same branch.*

Theorem 3.7.19. *Let d_f be a self-map of a solid weak BCC-algebra $\mathfrak{X}$ defined by $d_f(x) = \varphi^2(f(x))$ for all $x \in B(a), a \in I(X)$. Then d_f is a branchwise regular f-derivation of $\mathfrak{X}$.*

Proof. First observe that if $x \in B(a)$, then $f(x) \in f(B(a)) = B(f(a))$. And for $x \in B(a)$ we have

$$d_f(x) = \varphi^2(f(x)) = \varphi^2(f(a)) = f(a),$$

i.e.,

$$d_f(x) = f(a). \tag{3.59}$$

Then for $x, y \in B(a)$

$$\begin{aligned} d_f(x)f(y) \wedge f(x)d_f(y) &= f(a)f(y) \wedge f(x)f(a) \\ &= f(x)f(a) \cdot (f(x)f(a) \cdot f(a)f(y)) \\ &= f(x)f(a) \cdot (f(x)f(a) \cdot f(0)) \\ &= f(0) = 0. \end{aligned}$$

On the other hand, $d_f(xy) = \varphi^2(f(xy)) = 0$, because $f(xy) = f(x)f(y) \in B(0)$. Hence, we have

$$d_f(xy) = d_f(x)f(y) \wedge f(x)d_f(y),$$

which proves that d_f is (l,r)-f-derivation of $\mathfrak{X}$.

Now, by (3.59), we have

$$f(x)d_f(y) \wedge d_f(x)f(y) = f(x)f(a) \wedge f(a)f(y) = f(x)f(a) \wedge 0 = 0.$$

Hence $d_f(x)$ is (r,l)-f-derivation of $\mathfrak{X}$. Since $d_f(0) = f(0) = 0$, $d_f(x)$ is regular. □

Theorem 3.7.20. *A solid weak BCC-algebra $\mathfrak{X}$ is group-like if and only if $ker\, d_f = \{0\}$ for every regular f-derivation d_f of $\mathfrak{X}$.*

Proof. If $\mathfrak{X}$ is group-like, then $X = I(X)$ and $B(0) = \{0\}$. From Theorem 3.7.14, $d_f(0) = 0$ for every regular f-derivation d_f of $\mathfrak{X}$.

Let $ker\, d_f = \{0\}$ for every regular f-derivation d_f of $\mathfrak{X}$. Let us define a self-map $d_{f,a}$ of $\mathfrak{X}$, by $d_{f,a}(x) = f(a)$ for $x \in B(a), a \in I(X)$. From Theorem 3.7.19, $d_{f,a}$ is a regular f-derivation of $\mathfrak{X}$ and $ker\, d_{f,a} = \{0\}$. From the definition of the map $d_{f,0}$, for every $x \in B(0)$ $d_{f,0}(x) = 0$, thus $B(0) \subseteq ker\, d_{f,0} = \{0\}$. From Theorem 2.5.2, $\mathfrak{X}$ is group-like. □

Theorem 3.7.21. *If f is an endomorphism of a weak BCC-algebra $\mathfrak{X}$ such that $f(X) = X$, then a f-derivation d_f of $\mathfrak{X}$ is regular if and only if $d_f(A) \subset f(A)$ for all BCC-ideals A of $\mathfrak{X}$.*

Proof. For a regular f-derivation of a solid weak BCC-algebra, by Theorem 3.7.11(2) and Corollary 3.7.18, we have $d_f(x) = f(x) \wedge d_f(x) \leq f(x)$ for all $x \in X$. Let $x \in A$ for some BCC-ideal A of $\mathfrak{X}$. Then, $f(A)$ is a BCC-ideal in $f(X) = X$ and $f(x) \in f(A)$. Hence $d_f(x)f(x) = 0 = f(0) \in f(A)$ and consequently $d_f(x) \in f(A)$. Thus $d_f(A) \subset f(A)$ for any BCC-ideal A of $\mathfrak{X}$.

Conversely, if $d_f(A) \subset f(A)$ for each BCC-ideal A of $\mathfrak{X}$, then also $d_f(\{0\}) \subset f(\{0\}) = \{0\}$. Hence $d_f(0) = 0$, i.e., d_f is regular. □

Corollary 3.7.22. *A derivation d of a solid weak BCC-algebra $\mathfrak{X}$ is regular if and only if $d(A) \subset A$ for all BCK-ideals A of $\mathfrak{X}$.*

Proof. The proof follows from Theorem 2.3.4 and Theorem 3.7.21, by putting $f(x) = x$. □

3.8 Para-Associative Weak BCC-Algebras

As we will show at the beginning of this section, the associative weak BCC-algebra forms a Boolean group. Therefore, it is worth examining weak BCC-algebras with weaker versions of associativity called para-associativity (cf. [23]). We also study quasi-associative weak BCC-algebras that have strong connections to T-type algebras described in Section 4.2.

We start with a theorem on associative weak BCC-algebras and the conclusion that follows from it.

Theorem 3.8.1. *Let $\mathfrak{X}$ be a weak BCC-algebra. Then the following conditions are equivalent:*

(1) *$\mathfrak{X}$ is associative,*

(2) *$0x = x$ for any $x \in X$,*

(3) *$xy = yx$ for any $x, y \in X$,*

(4) *$\mathfrak{X}$ is an associative BCI-algebra.*

Proof. (1) $\Longrightarrow$ (2) Suppose $\mathfrak{X}$ is an associative weak BCC-algebra. From this and by (BCC2) we have

$$0x \cdot x = 0 \cdot xx = 00 = 0.$$

By (BCC2), (BCC3) and associativity

$$0 = xx = x0 \cdot x = x \cdot 0x.$$

Using (BCC5), we get $0x = x$. Hence condition (2) is satisfied.

(2) $\Longrightarrow$ (3) For any $x, y \in X$, by Lemma 2.2.2(1), we have $0 \cdot xy \leqslant yx$. By applying (2), we get $xy \leqslant yx$.
In the same way, we get $yx \leqslant xy$.
Thus $xy = yx$ and (3) holds.

(3) $\Longrightarrow$ (4) By (BCC3), (3) and Lemma 2.2.2(3), we have

$$y \cdot yx = y \cdot xy = y0 \cdot xy = 0y \cdot xy = 0x = x0 = x.$$

Thus the condition

$$y \cdot yx = x$$

is satisfied for any $x, y \in X$.

From Corollary 2.5.13 it follows that $\mathfrak{X}$ is group-like BCI-algebra.
It remains to prove the associativity:

$$\begin{aligned} xy \cdot z &= yx \cdot z && \text{by}(3) \\ &= yz \cdot x && \text{by } (1.6) \\ &= x \cdot yz && \text{by}(3) \end{aligned}$$

This means that $\mathfrak{X}$ is an associative BCI-algebra.

(4) $\Longrightarrow$ (1) It is trivial.

The proof is completed. □

Corollary 3.8.2. *An associative weak BCC-algebra is a Boolean group or in other words a group of exponent* 2.

In this part of the section, we will investigate BCI-algebras, in which one of the conditions, presented in the following definition, is satisfied

Definition 3.8.3. A weak BCC-algebra $\mathfrak{X}$ is called *right alternative, left alternative, flexible*, resp., if the following conditions are satisfied for any $x, y, z \in X$:

(RA) $yx \cdot x = y \cdot xx$, i.e., $yx \cdot x = y$,

(LA) $xx \cdot y = x \cdot xy$, i.e., $0 \cdot y = x \cdot xy$,

(FL) $xy \cdot x = x \cdot yx$.

Lemma 3.8.4. *A right alternative weak BCC-algebra is a Boolean group.*

Proof. Putting $x = y$ in (RA) we obtain $0x = x$.

From Theorem 3.8.1, $\mathfrak{X}$ is an associative weak BCC-algebra. From Corollary 3.8.2, it is a Boolean group. □

Lemma 3.8.5. *A left alternative weak BCC-algebra is a Boolean group.*

Proof. Since for $y = x$ (LA) gives $0x = x$, from Theorem 3.8.1, $\mathfrak{X}$ is an associative weak BCC-algebra. From Corollary 3.8.2, it is a Boolean group. □

Lemma 3.8.6. *A weak BCC-algebra is a Boolean group if and only if it satisfies (at least) one of the following conditions:*

(1) $x \cdot yx = y$,

(2) $xy \cdot x = y$.

Proof. A Boolean group satisfies these identities.

Conversely, if a weak BCC-algebra $\mathfrak{X}$ satisfies (1), then for $x = 0$ we have $0y = y$ for all $y \in X$.

Moreover, the equations $bx = a$ and $xa = b$ have a uniquely determined solution $x \in X$.

Indeed, for every $a, b \in X$ and $x = ab$ we have $bx = b \cdot ab = a$ and $xa = x \cdot bx = b$. If $xa = ya$, then $x = a \cdot xa = a \cdot ya = y$. Thus $\mathfrak{X}$ is a quasi-group.

Hence (BCC1) can be written in the form

$$xy \cdot zy = xz.$$

Therefore,

$$yx \cdot x = yx \cdot 0x = y0 = y \cdot xx$$

for all $x, y \in X$, which proves that $\mathfrak{X}$ is a right alternative weak BCC-algebra. By Lemma 3.8.4, it is a Boolean group.

In the case of (2), the proof is analogous. □

Lemma 3.8.7. *A flexible weak BCC-algebra $\mathfrak{X}$ is a Boolean group.*

Proof. Let $\mathfrak{X}$ be a weak BCC-algebra. Since $\mathfrak{X}$ satisfies (FL), then for $x = y$ we have $0x = x0 = x$. Since $0x = x$, by Theorem 3.8.1, $\mathfrak{X}$ is an associative BCI-algebra. From Corollary 3.8.2, it is a Boolean group. □

Since every group of exponent 2 is an associative BCI-algebra, in which (RA), (LA) and (FL) are satisfied, from the above lemmas it follows:

Theorem 3.8.8. *A BCI-algebra $\mathfrak{X}$ is right alternative, left alternative or flexible if and only if it is a group of exponent* 2.

Definition 3.8.9. A weak BCC-algebra $\mathfrak{X}$ is called *para-associative*, when it is satisfying the (i, j, k)-associative law

$$x_1x_2 \cdot x_3 = x_i \cdot x_jx_k, \tag{3.60}$$

where $\{i, j, k\}$ is a fixed permutation of $\{1, 2, 3\}$.

Theorem 3.8.10. *A para-associative weak BCC-algebra is a Boolean group.*

Proof. Let $\mathfrak{X}$ be a weak BCC-algebra.

It is not difficult to see that we have to consider six cases of the para-associativity.

1^0. The case of the $(1,2,3)$-associativity is described by Lemma 3.8.2.

2^0. Since every $(1,3,2)$-associative groupoid is also right alternative, this case follows from Lemma 3.8.4.

3^0. Every $(2,1,3)$-associative groupoid is left alternative. Thus such weak BCC-algebra is a Boolean group, by Lemma 3.8.5.

4^0. In the case of the $(2,3,1)$-associativity we have

$$x \cdot 0x = xx \cdot 0 = 0$$

and

$$0x \cdot x = x \cdot x0 = xx = 0,$$

which, by (BCC5), gives $0x = x$. From Theorem 3.8.1(2) and Corollary 3.8.2, it follows that $\mathfrak{X}$ is a Boolean group.

5^0. The $(3,2,1)$-associativity implies $0x = x$, too.

Indeed, we have

$$0x \cdot x = x \cdot x0 = 0$$

and

$$x \cdot 0x = xx \cdot 0 = 0.$$

By (BCC5), we get $0x = x$ and again, Theorem 3.8.1(2) and Corollary 3.8.2, complete this part of the proof.

6^0. The $(3,1,2)$-associativity implies

$$xz = x0 \cdot z = z \cdot x0 = zx.$$

Hence $\mathfrak{X}$ is commutative and, by Theorem 3.8.1 and Corollary 3.8.2, it is a Boolean group.

The proof is completed. □

Definition 3.8.11. A weak BCC-algebra $\mathfrak{X}$ is called *quasi-associative* of type (i,j,k), when it is satisfying the following condition:

$$x_1x_2 \cdot x_3 \leq x_i \cdot x_jx_k, \tag{3.61}$$

where $\{i,j,k\}$ is a fixed permutation of $\{1,2,3\}$.

The quasi-associativity for BCI-algebras was studied in [140].

Definition 3.8.12. An element x of a weak BCC-algebra $\mathfrak{X}$ such that

$$\varphi(x) \leq x \tag{3.62}$$

is called a *quasi-unit.*

We now show the connection between a quasi-associative weak BCC-algebra of type $(1,3,2)$ and a weak BCC-algebra containing only quasi-unit elements.

Theorem 3.8.13. *Let $\mathfrak{X}$ be a weak BCC-algebra. Then for every $x \in X$ the condition* (3.62) *is satisfied if and only if $\mathfrak{X}$ is quasi-associative of type* $(1,3,2)$, *i.e.,*

$$xy \cdot z \leq x \cdot zy \tag{3.63}$$

holds for any $x, y, z \in X$.

Proof. Let $\mathfrak{X}$ be a weak BCC-algebra. If it satisfies (3.63), then for any $y \in X$, $x = 0$ and $z = y$, the condition (3.63) has the form

$$\varphi(y) \cdot y \leq 0 \cdot yy = 0,$$

i.e., $\varphi(y) \cdot y = 0$. Hence $\mathfrak{X}$ satisfies (3.62).

On the other hand, if (3.62) is satisfied for every $x \in X$, then for any $y, z \in X$, we obtain

$$\begin{aligned} zy \cdot y &= (zy \cdot y)0 && \text{by (BCC3)} \\ &= (zy \cdot y)(0y \cdot y) && \text{by (3.62)} \\ &\leq zy \cdot 0y && \text{by (BCC1}'\text{)} \\ &\leq z && \text{by (BCC1}'\text{).} \end{aligned}$$

So, we have

$$zy \cdot y \leq z. \tag{3.64}$$

Using (3.64), (1.5) and (BCC1$'$), we get

$$xy \cdot z \leq xy \cdot (zy \cdot y) \leq x \cdot zy.$$

This means that $\mathfrak{X}$ satisfies (3.63).

The proof of the theorem is completed. □

From the above theorem the corollary below follows directly:

Corollary 3.8.14. *Let $\mathfrak{X}$ be a weak BCC-algebra. Then for every $x \in X$ the condition* (3.62) *is satisfied if and only if it satisfies*

$$xy \cdot y \leq x \tag{3.65}$$

for any $x, y \in X$.

Proof. By putting $z = y$ in (3.63) we get condition (3.65). □

3.9 Hyper (Weak) BCC-Algebras

Hyperstructures (also known as multialgebras) were introduced by F. Marty (cf. [114]) at the 8th Scandinavian Congress of Mathematicians.

An algebraic hyperstructure is a generalization of classical algebraic structures. For example, in a classical algebraic structure, the composition of two elements is an element, while in an algebraic hyperstructure, the composition of two elements is a nonempty set. Hyperstructures of different types of algebras have many important applications.

In this chapter, we will present the results concerning hyper BCC-algebras and hyper weak BCC-algebras. The latter are a common generalization of BCC-algebras, hyper BCK-algebras and also hyper BCI-algebras. They have been studied by many authors (cf. for example [12] or [22]).

We start with the appropriate definitions.

Definition 3.9.1. Let X be a nonempty set and $\mathscr{P}(X)$ the set of all nonempty subsets of X.

A map $\circ : X \times X \longrightarrow \mathscr{P}(X)$ is called *(binary) hyperoperation* or *hypercomposition* and $\langle X; \circ \rangle$ is called a *hypergroupoid.*

If $A, B \in \mathscr{P}(X)$ and $x \in X$, then

$$A \circ B = \bigcup_{a \in A, b \in B} a \circ b, \; A \circ x = A \circ \{x\}, \; x \circ B = \{x\} \circ B.$$

In this section we replace the singleton set $\{x\}$ with x.

Definition 3.9.2. An algebra $\mathfrak{X} = \langle X; \circ, 0\rangle$ of type (2,0) with the hyperoperation $\circ$ is called a *hyper BCC-algebra* if it satisfies the following conditions:

(HBCC1) $(x \circ z) \circ (y \circ z) \ll x \circ y$,

(HBCC2) $x \circ 0 = x$,

(HBCC3) $0 \circ x = 0$,

(HBCC4) $x \ll y$ and $y \ll x$ imply $x = y$,

for all $x, y, z \in X$, where $x \ll y$ is defined by

(1) $0 \in x \circ y$,

(2) for every $A, B \subseteq X$ we define

$$A \ll B = \{a \in A \,|\, \exists\, b \in B \text{ such that } a \ll b\}.$$

The relation $\ll$ is called the *hyperorder.*

Definition 3.9.3. An algebra $\mathfrak{X} = \langle X; \circ, 0\rangle$ of type (2,0) with the hyperoperation $\circ$ is called a *hyper BCK-algebra* if it satisfies the following conditions: (HBCC1), (HBCC4) and

(HBCK1) $(x \circ y) \circ z = (x \circ z) \circ y$,

(HBCK2) $0 \circ X \ll x$

for all $x, y, z \in X$.

Definition 3.9.4. An algebra $\mathfrak{X} = \langle X; \circ, 0\rangle$ of type (2,0) with the hyperoperation $\circ$ is called a *hyper BCI-algebra* if it satisfies the following conditions: (HBCC1), (HBCC4), (HBCK1) and

(HBCI1) $x \ll x$,

(HBCI2) $0 \circ (0 \circ x) \ll x$

for all $x \in X$.

Example 3.9.5. Let $\langle X; \star, 0\rangle$ be a BCC-algebra. We define the hyperoperation $\circ$ on X by $x \circ y = \{x \star y\}$ for all $x, y \in X$.
Then $\langle X; \circ, 0\rangle$ is a hyper BCC-algebra. $\square$

Example 3.9.6. The algebra $\langle\{0,1,2,...\};\circ,0\rangle$, with the hyperoperation $\circ$ on X defined in the following way:

$$x \circ y = \begin{cases} \{0, x\} & \text{if } \; x \leq y, \\ x & \text{if } \; x > y \end{cases}$$

for $x, y \in X$ is a hyper BCC-algebra. □

Example 3.9.7. Let $\mathfrak{X} = \langle\{0,1,2,3\};\circ_1,0\rangle$ and $\mathfrak{Y} = \langle\{0,1,2\};\circ_2,0\rangle$ be algebras with the operations $\circ_1$ and $\circ_2$, defined by the following tables:

$\circ_1$	0	1	2	3
0	0	0	0	0
1	1	0	0	0
2	2	2	$\{0,1\}$	2
3	3	$\{1,3\}$	$\{0,1,3\}$	$\{0,1,3\}$

Table 3.9.1

$\circ_2$	0	1	2
0	0	0	0
1	1	0	0
2	2	2	$\{0,1\}$

Table 3.9.2

Then $\mathfrak{X}$ and $\mathfrak{Y}$ are hyper BCC-algebras.

Moreover, since

$$(2 \circ 1) \circ 2 \neq (2 \circ 2) \circ 1$$

for $x \circ y = x \circ_1 y = x \circ_2 y$, i.e., the condition (HBCK1) is not satisfied, they are proper hyper BCC-algebras. □

Lemma 3.9.8. *In a hyper BCC-algebra* $\mathfrak{X}$ *the following conditions are satisfied:*

(1) $x \circ y \ll x$,

(2) $0 \circ 0 = 0$,

(3) $x \ll x$.

Proof. (1) Putting $z = 0$ in (HBCC1) we get $x \circ y \ll x$.

(2) Putting $x = 0$ in (HBCC2) we get $0 \circ 0 = 0$.

(3) Let $y = z = 0$ in (HBCC1). Then, by (2) and (HBCC2), we have

$$(x \circ 0) \circ (0 \circ 0) \ll x \circ 0 \Longleftrightarrow x \circ \{0\} \ll x \Longleftrightarrow x \ll x.$$

□

Now we calculate the number of all nonisomorphic hyper BCC-algebras of order 3.

Definition 3.9.9. We say a hyper BCC-algebra of order 3 is *linear* if all its elements form a chain. That is, if $0 \ll a \ll b$ or $0 \ll b \ll a$. Any other hyper BCC-algebra $\mathfrak{X}$ will be called *nonlinear.*

Theorem 3.9.10. *There are three no proper nonisomorphic nonlinear hyper BCC-algebras of order 3.*

Proof. In a nonlinear hyper BCC-algebra $\mathfrak{X}$, we have the following possibilities:

(1) $x \notin y \circ x$ for all $x, y \in X$ such that $x \neq y$,

(2) $a \circ b = a$ and $b \circ a = b$,

(3) $a \circ a \in \{0, \{0, a\}\}$ and $b \circ b \in \{0, \{0, b\}\}$.

In the sequel, we will prove the following three cases:

(1) If $x \neq y$ and $x \in y \circ x$, then $x \neq 0$ because $x = 0$ implies $y \neq 0$ and $0 \in y \circ 0 = y$, whence $y = 0$, which is impossible. By Lemma 3.9.8(1), $y \circ x \ll y$. So, $x \ll y$. But $x \neq 0$, hence $a \ll b$ or $b \ll a$. From assumption that is impossible. Therefore, condition (1) is satisfied.

(2) From $a \not\ll b$ it follows that $0 \notin a \circ b$. Hence
$a \circ b \notin \{0, \{0, a\}, \{0, b\}, \{0, a, b\}\}$.
Since, by (1), we have $b \notin a \circ b$, $a \circ b \notin \{b, \{a, b\}\}$. Thus $a \circ b = a$. Similarly, we prove that $b \circ a = b$.

(3) It is easy to see that, in this case, $a \circ a \in \{0, \{0, a\}, \{0, b\}, \{0, a, b\}\}$. In the cases, when $a \circ a \in \{\{0, b\}, \{0, a, b\}\}$, by Lemma 3.9.8(1), we get $a \circ a \ll a$. But, from the assumption, $b \in a \circ a$, whence $b \ll a$, which is impossible. Hence $a \circ a \notin \{\{0, a, b\}, \{0, b\}\}$, i.e., $a \circ a \in \{0, \{0, a\}\}$. Similarly, we prove that $b \circ b \in \{0, \{0, b\}\}$.

Therefore, from conditions (1), (2) and (3), it follows that there are four nonlinear hyper BCC-algebras containing three elements.

The following tables show their hyperoperations:

$\circ_1$	0	1	2
0	0	0	0
1	1	0	1
2	2	2	0

Table 3.9.3

$\circ_2$	0	1	2
0	0	0	0
1	1	0	1
2	2	2	$\{0, 2\}$

Table 3.9.4

$\circ_3$	0	1	2
0	0	0	0
1	1	$\{0, 1\}$	1
2	2	2	$\{0, 2\}$

Table 3.9.5

$\diamond$	0	1	2
0	0	0	0
1	1	$\{0, 1\}$	1
2	2	2	0

Table 3.9.6

Let f be a function $f : \langle X; \diamond\rangle \longrightarrow \langle X; \circ_2\rangle$ defined by

$$f(x) = \begin{cases} 0 & \text{if} \quad x = 0, \\ b & \text{if} \quad x = a, \\ a & \text{if} \quad x = b. \end{cases}$$

Then, f is an isomorphism between $\langle X; \diamond\rangle$ and $\langle X; \circ_2\rangle$. The other nonlinear hyper BCC-algebras are not isomorphic.

It can be easily observed that they are not proper hyper BCC-algebras.

The proof is completed. □

Theorem 3.9.11. *Let $\langle X; \circ, 0\rangle$ be a proper hyper BCC-algebra, an element $a \notin X$ and let $X' = X \cup a$. We define the operation $\circ'$ as follows:*

$$x \circ' y = \begin{cases} a & \text{if } \ x = a, \ y = 0, \\ x & \text{if } \ x \in X, \ y = a, \\ \{0, a\} & \text{if } \ x = y = a, \\ 0 & \text{if } \ x = a, \ y \in X \setminus \{0\}, \\ x \circ y & \text{if } \ x, y \in X \end{cases}$$

for all $x, y \in X'$.
Then $\langle X'; \circ', 0\rangle$ is a proper hyper BCC-algebra.

Proof. As it can be easily seen, conditions (HBCC2), (HBCC3) and (HBCC4) are satisfied. The verification of condition (HBCC1) remains. Let $x, y, z \in X'$. We consider two cases for x.

Case 1^0. $x \in X$. Then $y \in X$ or $y = a$.

1^0(a). Let $y \in X$. Then $x \circ' y = v \Longleftrightarrow x \circ y = v$. Since $\circ$ satisfies (HBCC1), it does $\circ'$, too.

1^0(b). Let $y = a$. If $z = 0$, then

$$\begin{aligned} (x \circ' z) \circ' (y \circ' z) &= (x \circ' 0) \circ' (a \circ' 0) \\ &= (x \circ 0) \circ' a && \text{by (HBCC2)} \\ &= x \circ' a \\ &= x \\ &\ll x && \text{by Lemma 3.9.8 (3)} \\ &= x \circ' a \\ &= x \circ' y. \end{aligned}$$

If $z \in X \setminus \{0\}$, then

$$\begin{aligned} (x \circ' z) \circ' (y \circ' z) &= (x \circ' z) \circ' (a \circ' z) \\ &= (x \circ z) \circ' 0 \\ &= x \circ z && \text{by (HBCC2)} \\ &\ll x && \text{by Lemma 3.9.8(1)} \\ &= x \circ' a \\ &= x \circ' y. \end{aligned}$$

If $z = a$, then

$$\begin{aligned}(x \circ' z) \circ' (y \circ' z) &= (x \circ' a) \circ' (a \circ' a)\\ &= a \circ' \{0, a\}\\ &= \{0, a\}\\ &\ll a\\ &= x \circ' a\\ &= x \circ' y.\end{aligned}$$

Case 2^0. $x = a$. Then, again, $y \in X$ or $y = a$.
2^0(a). Let $y \in X$. If $z = 0$, then

$$\begin{aligned}(x \circ' z) \circ' (y \circ' z) &= (a \circ' 0) \circ' (y \circ' 0)\\ &= a \circ' y && \text{by (HBCC2)}\\ &\ll a \circ' y && \text{by Lemma 3.9.8 (3)}\\ &= x \circ' y.\end{aligned}$$

If $z \in X \setminus \{0\}$, then

$$\begin{aligned}(x \circ' z) \circ' (y \circ' z) &= (a \circ' z) \circ' (y \circ' z)\\ &= 0 \circ' (y \circ' z)\\ &= 0 && \text{by (HBCC3)}\\ &\ll a \circ' y\\ &= x \circ' y.\end{aligned}$$

If $z = a$, then

$$\begin{aligned}(x \circ' z) \circ' (y \circ' z) &= (a \circ' a) \circ' (y \circ' a)\\ &= \{0, a\} \circ' y\\ &= \{0\} \cup a \circ' y\\ &\ll a \circ' y\\ &= x \circ' y.\end{aligned}$$

2^0(b). Let $y = a$. If $z = 0$, then

$$\begin{aligned}(x \circ' z) \circ' (y \circ' z) &= (a \circ' 0) \circ' (a \circ' 0)\\ &= a \circ' a\\ &= \{0, a\} \qquad \text{by (HBCC2)}\\ &\ll \{0, a\}\\ &= a \circ' a\\ &= x \circ' y.\end{aligned}$$

If $z \in X \setminus \{0\}$, then

$$\begin{aligned}(x \circ' z) \circ' (y \circ' z) &= (a \circ' z) \circ' (a \circ' z)\\ &= 0 \circ' 0\\ &= 0 \qquad \text{by Lemma 3.9.8(2)}\\ &\ll \{0, a\}\\ &= a \circ' a\\ &= x \circ' y.\end{aligned}$$

If $z = a$, then

$$\begin{aligned}(x \circ' z) \circ' (y \circ' z) &= (a \circ' a) \circ' (a \circ' a)\\ &= \{0, a\}\\ &\ll \{0, a\} \qquad \text{by Lemma 3.9.8(3)}\\ &= a \circ' a\\ &= x \circ' y.\end{aligned}$$

Hence $\langle X'; \circ', 0\rangle$ is a hyper BCC-algebra. It is proper because $\langle X; \circ, 0\rangle$ is proper.

The proof is completed. □

As it can be easily seen, there is no proper hyper BCC-algebra with two elements. But there exists a proper hyper BCC-algebra with three

elements. In Example 3.9.7 we have shown such algebra. This is an interesting fact because it follows from Theorem 1.2.8 that a proper BCC-algebra has at least four elements.
Moreover, from Theorem 3.9.11 it follows that any proper hyper BCC-algebra with n elements (for any $n \geq 3$) can be expanded to a proper hyper BCC-algebra with $n+1$ elements.

The above remarks allow the formulation of the following conclusion:

Corollary 3.9.12. *For any $n \geq 3$ there exists at least one proper hyper BCC-algebra of order n.*

The next theorem is an interesting generalization of the above results to the hyper BCC-algebras with the transfinite cardinal.

Theorem 3.9.13. *Let α be a transfinite cardinal number. Then there exists a hyper BCC-algebra $\mathfrak{X}$ such that* Card(X)$= \alpha$.

Proof. Since α is a transfinite cardinal number, there exists an infinite set A such that Card(A)$= \alpha$.

Now let $C = \{x_0, x_1, ...\}$ be an infinite countable subset of A.
Let hyperoperation $\diamond$ on C be defined as follows:

$$x_i \diamond x_j = \begin{cases} \{x_0, x_i\} & \text{if} \quad i \leq j, \\ x_i & \text{if} \quad i > j \end{cases}$$

for all $x_i \in C$.

Similarly to Example 3.9.6, we can see that $\langle C;\, \diamond, x_0\rangle$ is a hyper BCC-algebra.

If Card(A) $=$ Card(C), the proof is completed.

Let Card(C) $<$ Card(A).
We define the set Ω as follows

$$\Omega = \{\langle X_i;\, \circ_i, x_0\rangle \mid \langle X_i;\, \circ_i, x_0\rangle \text{ is a hyper BCC-algebra}, \\ C \subseteq X_i \subseteq A,\, \circ_i|_C = \diamond\}.$$

Since $\langle C;\, \diamond, x_0\rangle \in \Omega$, Ω is nonempty.
We define the relation $\subseteq$ on Ω^2 in the following way:

$$(\langle X_i;\, \circ_i, x_0\rangle, \langle X_j;\, \circ_j, x_0\rangle) \in \subseteq \Longleftrightarrow X_i \subseteq X_j \text{ and } \circ_j|_{X_i} = \circ_i.$$

$(\Omega; \subseteq)$ is a partially ordered set.
Indeed, the reflexivity is obvious. Now,

$$(\langle X_i;\, \circ_i, x_0\rangle, \langle X_j;\, \circ_j, x_0\rangle) \in \subseteq \text{ and } (\langle X_j;\, \circ_j, x_0\rangle, \langle X_i;\, \circ_i, x_0\rangle) \in \subseteq,$$

then

$$X_i = X_j$$

and

$$\circ_i = \circ_j \mid_{X_i} = (\circ_i \mid_{X_j}) \mid_{X_i} = (\circ_i \mid_{X_j}) \mid_{X_j} = \circ_i \mid_{X_j} = \circ_j.$$

So, $\langle X_i;\, \circ_i, x_0\rangle = \langle X_j;\, \circ_j, x_0\rangle$ and we have proven the antisymmetry of $\subseteq$.

We will show the transitivity of $\subseteq$. Let

$$(\langle X_i;\, \circ_i, x_0\rangle, \langle X_j;\, \circ_j, x_0\rangle) \in \subseteq \text{ and } (\langle X_j;\, \circ_j, x_0\rangle, \langle X_k;\, \circ_k, x_0\rangle) \in \subseteq,$$

then of course $X_i \subseteq X_k$. And hence

$$\circ_i = \circ_j \mid_{X_i} = (\circ_k \mid_{X_j}) \mid_{X_i} = \circ_k \mid_{X_i}.$$

So, the relation $\subseteq$ is transitive and $(\Omega; \subseteq)$ is partially ordered set.

Let $\{\langle X_k;\, \circ_k, x_0\rangle\}$ be a totally ordered set in Ω and $X^{'} = \bigcup_{k \in I} X_k$. For all $x, y \in X^{'}$, there exists $X_{t_{xy}} \in \{\langle X_k;\, \circ_k, x_0\rangle\}_{k \in I}$ such that $x, y \in X_{t_{xy}}$ and we can define hyperoperation $\circ$ on $X^{'}$ by

$$x \circ y = x \circ_{t_{xy}} y.$$

It is easy to prove that $\langle X^{'};\, \circ, x_0\rangle$ is a hyper BCC-algebra and so $X^{'} \in \Omega$.

Hence, by Zorn's lemma, there exists a maximal element in Ω. Let $\langle X;\, \circ^{'}, x_0\rangle$ be a maximal element in Ω.

If $X = A$, then the proof is completed.

If $X \neq A$, then there exists an element $a \in A$ such that $a \notin X$.

The definition of the hyperoperation $\circ$ on $X \cup \{a\}$ now looks like this:

$$x \circ y = \begin{cases} a & \text{if } \; x = y,\, y = x_0, \\ x & \text{if } \; x \in X,\, y = a, \\ \{x_0, a\} & \text{if } \; x = y = a, \\ x_0 & \text{if } \; x = a,\, y \in X \setminus \{x_0\}, \\ x \circ^{'} y & \text{if } \; x, y \in X \end{cases}$$

for all $x, y \in X$.

Then by Theorem 3.9.11, $\langle X \cup \{a\};\, \circ, x_0\rangle$ is a hyper BCC-algebra and so it belongs to Ω, which is a contradiction because $\langle X;\, \circ^{'}, x_0\rangle$ is a maximal element in Ω. Hence $X = A$ and so Card(X) $= \alpha$. □

We summarize the theorem with the following conclusion:

Corollary 3.9.14. *For every transfinite cardinal number α, there exists an infinite hyper BCC-algebra of order α.*

Proof. By Theorem 3.9.13, there exists a hyper BCC-algebra $\mathfrak{X}$ such that Card(X) $= \alpha$. Let $A = \{a_1, a_2, a_3, ...\}$ be an infinite set such that $X \cap A = \emptyset$. From Theorem 3.9.11 we have a hyper BCC-algebra $X_1 = X \cup \{a_1\}$ with order $\alpha+1$. Since α is a transfinite cardinal number, $\alpha + 1 = \alpha$, i.e., Card(X_1) $= \alpha$.

Similarly, $X_2 = X_1 \cup \{a_2\}$ is a hyper BCC-algebra of order $\alpha+1 = \alpha$. By continuing this process, we get infinite hyper BCC-algebras of order α. □

We now give a definition of a hyper weak BCC-algebra that is a generalization of hyper BCC-algebra.

Definition 3.9.15. An algebra $\langle X; \circ, 0\rangle$ of type (2,0) with the hyperoperation $\circ$ satisfying the following axioms: (HBCC1), (HBCC4) and

(HBCC5) $x \ll x$,

(HBCC6) $0 \circ (0 \circ x) \ll x$,

(HBCC7) $x \ll x \circ 0$

for all $x, y, z \in X$ is called a *hyper* weak BCC-algebra.

Remark 3.9.16. Let us observe that every weak BCC-algebra is a hyper weak BCC-algebra and that the classes of all hyper BCK-, hyper BCI- and hyper BCC-algebras are included in the class of all hyper weak BCC-algebras.

Example 3.9.17. Let $\langle \{0, 1, 2, 3\}; \circ, 0\rangle$ be algebra with the operation $\circ$ defined by the following table:

$\circ$	0	1	2	3
0	0	0	2	2
1	1	$\{0,1\}$	3	3
2	2	2	0	0
3	3	3	1	$\{0,1\}$

Table 3.9.7

Then $\langle\{0,1,2,3\};\circ,0\rangle$ is a hyper weak BCC-algebra. It is not a hyper BCC-algebra, because $0\cdot 3=2\neq 0$, i.e., the axiom (HBCC3) is not satisfied. □

Example 3.9.18. The proper weak BCC-algebra presented in Table 2.6.3 of Example 2.6.5 is, by Remark 3.9.16, a hyper weak BCC-algebra. It is not a hyper BCI-algebra. □

From the above two examples and from Example 3.9.7, it follows that there are proper hyper (weak) BCC-algebras.

Lemma 3.9.19. *Let $\mathfrak{X}$ be a hyper weak BCC-algebra. Then for all $x,y,z\in X$ and for nonempty subsets A and B of X, the following conditions are satisfied:*

(1) $x\ll 0$ implies $x=0$,

(2) $0\circ(x\circ y)\ll y\circ x$,

(3) $A\ll A$,

(4) $A\subseteq B$ implies $A\ll B$,

(5) $A\ll 0$ implies $A=0$,

(6) $0\circ 0=0$,

(7) $(0\circ x)\circ(0\circ x)=0$,

(8) $0\circ x$ is a singleton set,

(9) $x\circ y=0$ implies $(x\circ z)\circ(y\circ z)=0$ and $x\circ z\ll y\circ z$,

(10) $A\circ 0=0$ implies $A=0$,

(11) $x\ll y$ implies $0\ll y\circ x$,

(12) $0\circ(0\circ(0\circ x))\ll 0\circ x$,

(13) $x\circ x=0$ implies $Card(y\circ z)=1$.

Proof. (1) Let us assume $x\ll 0$. Then $0\in x\circ 0$ and, by (HBCC1) and (HBCC5), we have

$$0\in 0\circ 0\subseteq 0\circ(x\circ 0)\subseteq(0\circ 0)\circ(x\circ 0)\ll 0\circ x,$$

i.e., $0\ll 0x$. Now, by (HBCC6), we have $0\in 0\circ(0\circ x)\ll x$. Then $0\ll x$, which, by assumption, gives $x=0$.

(2) (HBCC1) and (HBCC5) imply

$$0\circ(x\circ y)\subseteq(y\circ y)\circ(x\circ y)\ll y\circ x,$$

i.e., $0 \circ (x \circ y) \ll y \circ x$.

(3) Of course, by (HBCC5), we have $x \ll x$ and $0 \in x \circ x$, for any $x \in A$. Then $A \ll A$.

(4) From assumption, $a \in A$ implies $a \in B$. Now, by (HBCC5), $x \ll x$ and $0 \in x \circ x$. Hence $A \ll B$.

(5) Let $a \in A$. Then $a \ll 0$ and so $a = 0$. This gives $A = \{0\}$.

(6) By (BCC6), $0 \circ (0 \circ 0) \ll 0$, then $0 \circ (0 \circ 0) = 0$. By (BCC5),

$$0 \in 0 \circ 0 \subseteq 0 \circ (0 \circ 0) = 0.$$

Hence $0 \circ 0 = 0$.

(7) From (HBCC1), it follows

$$(0 \circ x) \circ (0 \circ x) \ll 0 \circ 0 = 0, \text{ i.e., } (0 \circ x) \circ (0 \circ x) = 0.$$

(8) Let a, b be arbitrary and $a, b \in 0 \circ x$ such that $a \neq b$. Then

$$a \circ b \subseteq (0 \circ x) \circ (0 \circ x) = 0$$

and

$$b \circ a \subseteq (0 \circ x) \circ (0 \circ x) = 0.$$

Hence $a \ll b$ and $b \ll a$. From (HBCC4), it follows $a = b$. This means that $0 \circ x$ is a singleton set.

(9) By (HBCC1) and by assumption, we get

$$(x \circ z) \circ (y \circ z) \ll x \circ y = 0.$$

Condition (5) implies $(x \circ z) \circ (y \circ z) = 0$. So, $x \circ z \ll y \circ z$.

(10) Let us assume that $A \circ 0 = 0$, then $A \ll 0$. Hence $A = 0$.

(11) Let us assume that $x \ll y$. Then $0 \in x \circ y$ and

$$0 \in 0 \circ 0 \subseteq (y \circ y) \circ (x \circ y) \ll y \circ x.$$

Hence $0 \ll y \circ x$.

(12) It follows directly from (HBCC6). Indeed, by putting $a = 0 \circ x$ we get

$$0 \circ (0 \circ (0 \circ x) = 0 \circ (0 \circ a) \ll a = 0 \circ x, \text{ i.e., } 0 \circ (0 \circ (0 \circ x) \ll 0 \circ x.$$

(13) Let $x \circ x = 0$, for any $x \in X$. Let us assume $Card(y \circ z) > 1$. Let now $a, b \in y \circ z$ and let $a \neq b$. Then

$$a \circ b \subseteq (y \circ z) \circ (y \circ z) \ll y \circ y = 0$$

and

$$b \circ a \subseteq (y \circ z) \circ (y \circ z) \ll y \circ y = 0.$$

Thus $a \circ b \ll 0$ and $b \circ a \ll 0$. So, $a \ll a$ and $b \ll a$. Hence, by (HBCC4), $a = b$ and $Card(y \circ z) = 1$. □

Let us consider the condition $x \circ 0 = x$ for $x \in X$, where $\mathfrak{X}$ is a hyper weak BCC-algebra. This condition is satisfied in some hyper weak BCC-algebras, but there are hyper weak BCC-algebras where it is not satisfied. In order to make use of this condition, the concept of a standard hyper weak BCC-algebra was introduced.
We give its definition.

Definition 3.9.20. A hyper weak BCC-algebra $\mathfrak{X}$ is called *standard* hyper weak BCC-algebra if for any $x \in X$ the condition $x \circ 0 = x$ is satisfied.

Lemma 3.9.21. *In any standard hyper weak BCC-algebra $\mathfrak{X}$, the following conditions are satisfied:*

(SHBCC1) $\quad A \circ 0 = A$,

(SHBCC2) $\quad x \ll y$ implies $z \circ y \ll z \circ x$

for all $x, y \in X$ and for nonempty subset A of X.

Proof. (SHBCC1) For any $a \in A$, clearly, $a \circ 0 = a$. Hence $A \circ 0 = A$.

(SHBCC2) Let us assume that $x \ll y$. Then $0 \in x \circ y$. We write the axiom (HBCC1) in the form $(z \circ y) \circ (x \circ y) \ll z \circ x$. For any $a \in z \circ y$, from the assumption and Definition 3.9.20 we have

$$a = a \circ 0 \subseteq (z \circ y) \circ (x \circ y) \ll z \circ x.$$

It is easy to see that $z \circ y \ll z \circ x$ is satisfied. □

Corollary 3.9.22. *Every hyper weak BCC-algebra $\mathfrak{X}$ is a standard hyper weak BCC-algebra.*

Proof. The corollary follows from Definition 3.9.2. □

The converse statement is not true. Indeed, the following example shows that a hyper BCI-algebra is not necessarily a standard hyper weak BCC-algebra.

Example 3.9.23. The algebra $\langle\{0, 1, 2, 3\}; \circ\rangle$ with the operation $\circ$, defined by Table 3.9.8 is a hyper BCI-algebra. Since $1 \cdot 0 = \{1, 2\} \neq 1$, it is not standard hyper weak BCC-algebra.

$\circ$	0	1	2	3
0	0	0	0	3
1	$\{1,2\}$	$\{0,2\}$	0	3
2	2	2	0	3
3	3	3	3	0

Table 3.9.8

□

Definition 3.9.24. A hyper weak BCC-algebra $\mathfrak{X}$ is said to be a *transitive* hyper weak BCC-algebra if the relation $\ll$ is transitive.

Lemma 3.9.25. *A transitive hyper weak BCC-algebra $\mathfrak{X}$ satisfies the following conditions:*

$$\text{(THBCC1)} \quad A \ll B \text{ and } B \ll C \text{ imply } A \ll C,$$

$$\text{(THBCC2)} \quad x \circ y \ll z \text{ implies } (x \circ u) \circ (y \circ u) \ll z$$

for all $x, y, z, u \in X$ and for nonempty subsets A, B, C of X.

Proof. (THBCC1) Let $a \in A$, Then, from assumption, there is an element $b \in B$ such that $a \ll b$. Similarly, for an element $b \in B$ there is $c \in C$ such that $b \ll c$. Hence, by transitivity of $\ll$, $a \ll c$. Then, for any $a \in A$ there exists $c \in C$ such that $a \ll c$, which means $A \ll C$.

(THBCC2) From (HBCC1) and assumption, we have

$$(x \circ u) \circ (y \circ u) \ll x \circ y \ll z, \text{ i.e., } (x \circ u) \circ (y \circ u) \ll z.$$

□

Example 3.9.26. The algebra $\langle\{0,1,2,3\}; \circ, 0\rangle$ with the operation $\circ$, defined by the following table:

$\circ$	0	1	2	3
0	0	0	0	0
1	1	$\{0,1\}$	0	0
2	2	2	$\{0,2\}$	$\{0,2\}$
3	3	2	$\{1,2\}$	$\{0,1,2\}$

Table 3.9.9

is a transitive hyper weak BCC-algebra. □

Example 3.9.27. The algebra $\langle\{0,1,2,3\};\circ,0\rangle$ with the operation $\circ$, defined by the following table:

$\circ$	0	1	2	3
0	0	0	0	0
1	$\{1,2\}$	$\{0,2\}$	0	0
2	2	2	0	0
3	3	3	2	0

Table 3.9.10

is a transitive, but not standard hyper weak BCC-algebra. □

In the following, we show properties of hyper weak BCC-subalgebra.

We define a hyper weak BCC-subalgebra $\mathfrak{S}$ of $\langle X;\circ,0\rangle$ as a nonempty subset $S\subseteq X$ with $0\in S$, which is a hyper weak BCC-algebra with respect to the operation $\circ$.

Lemma 3.9.28. *Let $\langle X;\circ,0\rangle$ be a hyper weak BCC-algebra and let $S\subseteq X$ and $S\neq\emptyset$. If for any $x,y\in S$, $x\circ y\subseteq S$ then $0\in S$.*

Proof. Let us assume that $x\circ y\subseteq S$ for $x,y\in S$. Let $a\in S$. Since $a\ll a$, we have $0=a\circ a\subseteq S$. □

Theorem 3.9.29. *Let $\mathfrak{X}=\langle X;\circ,0\rangle$ be a hyper weak BCC-algebra and let $S\subseteq X$ and $S\neq\emptyset$. Then $\mathfrak{S}$ is a hyper weak BCC-subalgebra of $\mathfrak{X}$ if and only if $x\circ y\subseteq S$, for all $x,y\in S$.*

Proof. Let $x\circ y\subseteq S$ for $x,y\in S$. Then, by Lemma 3.9.28, $0\in S$. From assumption $x\circ z\subseteq S$ and $y\circ z\subseteq S$. This implies

$$(x\circ z)\circ(y\circ z)=\bigcup_{a\in x\circ z,b\in y\circ z} a\circ b\subseteq S.$$

The assumption $S\subseteq X$ implies $(x\circ z)\circ(y\circ z)\ll x\circ y\subseteq S$. Hence the axiom (HBCC1) holds in $\mathfrak{S}$. In a similar way we can prove that the axioms (HBCC4), (HBCC5), (HBCC6), (HBCC7) are also satisfied in $\mathfrak{S}$. Thus $\mathfrak{S}$ is a hyper subalgebra of $\mathfrak{X}$.

The converse is obvious. □

Theorem 3.9.30. *Let $\mathfrak{X}=\langle X;\circ,0\rangle$ be a hyper weak BCC-algebra. Then the set*

$$S_I=\{x\in X\,|\,x\circ x=0\}$$

is a hyper subalgebra of $\mathfrak{X}$ and for any $x,y\in S_I$, the element $x\circ y$ is a singleton set.

Proof. Let $x, y \in S_I$ and $a \in x \circ y$. Then, by (HBCC1),

$$(x \circ y) \circ (x \circ y) \ll x \circ x = 0.$$

Hence

$$(x \circ y) \circ (x \circ y) = 0$$

and

$$a \circ a \subseteq (x \circ y) \circ (x \circ y) = 0.$$

Therefore, $x \circ y \subseteq S_I$. From Theorem 3.9.29, $\mathfrak{S}_I$ is a hyper subalgebra of $\mathfrak{X}$.

According to Lemma 3.9.19(13), for $x, y \in S_I$, the element $x \circ y$ is a singleton set. □

However, $\mathfrak{S}_I$ is not necessarily a weak BCC-algebra. We have namely the following example:

Example 3.9.31. The algebra $\mathfrak{X} = \langle\{0,1,2,3\};\circ\rangle$ with the operation $\circ$ defined by the following table is a hyper weak BCC-algebra:

$\circ$	0	1	2	3
0	0	0	0	0
1	$\{1,2\}$	$\{0,2\}$	0	0
2	2	2	0	2
3	2	2	0	0

Table 3.9.11

The algebra $\mathfrak{S}_I = \langle\{0,2,3\};\circ\rangle$ is a hyper BCC-subalgebra of $\mathfrak{X}$. But, since $3 \circ 0 = 2 \neq 3$, it is not a weak BCC-algebra. □

Theorem 3.9.32. *Let $\mathfrak{X} = \langle X;\circ,0\rangle$ be a hyper weak BCC-algebra. Then the set*

$$B(0) = \{x \in X \,|\, 0 \circ x = 0\},$$

is a hyper BCC-subalgebra of $\mathfrak{X}$.

Proof. Let $x, y \in B(0)$ and let $a = x \circ y$. Then

$$0 \circ (x \circ y) = (0 \circ y) \circ (x \circ y) \ll 0 \circ x = 0, \text{ i.e., } 0 \circ (x \circ y) = 0.$$

Thus $x \circ y \subseteq B(0)$. From Theorem 3.9.29 it follows that $B(0)$ is a hyper subalgebra of $\mathfrak{X}$. Since for any $x \in B(0)$, (HBCC2) holds, $B(0)$ is a hyper BCC-algebra. □

3.10 Group-Like Hyper Weak BCC-Algebras

We begin this section by presenting the relationships between hyper weak BCC-algebras and semigroups. Next, we will examine the relationships between hyper weak BCC-algebras and hypersemigroups. Moreover, we introduce the concepts of hyper group-like weak BCC-algebra and generalized hyper group-like weak BCC-algebra.

Let $\mathfrak{X} = \langle X; \circ, 0\rangle$ be a hyper weak BCC-algebra. For any $a, x \in X$ we define a map:

$$\rho_a : X \longrightarrow \mathscr{P}(X)$$

as

$$\rho_a(x) = x \circ a.$$

Let $a, b \in X$. Then for any $x \in X$ we define the *composition* operation of mappings $\rho_a \diamond \rho_b$ as follows

$$\rho_a \diamond \rho_b(x) = \bigcup_{\forall y \in \rho_b(x)} \rho_a(y).$$

Let $M(X)$ denote a set of all results of compositions of finite mappings ρ_a for all $a \in X$.

We provide proofs of the following two theorems:

Theorem 3.10.1. *M(X) is a semigroup.*

Proof. Let $a, b, c \in X$ and $x \in X$. Then for any $s \in (\rho_a \diamond \rho_b) \diamond \rho_c(x)$, there exists $y \in \rho_c(x)$ such that $s \in \rho_a \diamond \rho_b(y)$ and there exists $u \in \rho_b(y)$ such that $u \in \rho_b(\rho_c(x)) = \rho_b \diamond \rho_c(x)$ and $s \in \rho_a(u)$.

Then

$$s \in \rho_a \diamond (\rho_b \diamond \rho_c)(x)$$

and

$$(\rho_a \diamond \rho_b) \diamond \rho_c(x) \subseteq \rho_a \diamond (\rho_b \diamond \rho_c)(x). \tag{3.66}$$

On the other hand, for any $t \in \rho_a \diamond (\rho_b \diamond \rho_c)(x)$, there exists $m \in \rho_b \diamond \rho_c(x)$ such that $t \in \rho_a(m)$ and there exists $n \in \rho_c(x)$ such that $m \in \rho_b(n)$ and $t \in \rho_a(\rho_b(n)) = \rho_a \diamond \rho_b(n)$.

Then

$$t \in (\rho_a \diamond \rho_b) \diamond \rho_c(x)$$

and

$$\rho_a \diamond (\rho_b \diamond \rho_c)(x) \subseteq (\rho_a \diamond \rho_b) \diamond \rho_c(x) \tag{3.67}$$

By combining of (3.66) and (3.67), we get

$$(\rho_a \diamond \rho_b) \diamond \rho_c(x) = \rho_a \diamond (\rho_b \diamond \rho_c)(x),$$

which means that the set $M(X)$ satisfies the associative law. □

Theorem 3.10.2. *Let $\langle X; \circ, 0\rangle$ be a standard hyper weak BCC-algebra. Then $\langle M(X); \diamond\rangle$ is a semigroup with the identity ρ_0.*

Proof. From Definition 3.9.20 it follows $x \circ 0 = x$ for all $x \in X$. Then for any $a \in X$, we have

$$\rho_0 \diamond \rho_a(x) = \rho_0(x \circ a) = (x \circ a) \circ 0 = x \circ a = \rho_a(x),$$

and

$$\rho_a \diamond \rho_0(x) = \rho_a(x \circ 0) = (x \circ 0) \circ a = x \circ a = \rho_a(x).$$

It follows that ρ_0 is the identity element. □

Example 3.10.3. Let $\mathfrak{X} = \langle\{0, 1, 2, 3\}; \circ, 0\rangle$ be an algebra with the operation $\circ$ defined as follows:

$\circ$	0	1	2	3
0	0	0	2	2
1	1	$\{0,1\}$	2	2
2	2	2	0	0
3	3	3	1	$\{0,1\}$

Table 3.10.1

Then $\mathfrak{X}$ is a hyper weak BCC-algebra and for $\rho_2^2 = \rho_2 \diamond \rho_2$ and $\rho_3^3 = \rho_3 \diamond \rho_3 \diamond \rho_3$ the set $M(X) = \{\rho_0, \rho_1, \rho_2, \rho_3, \rho_2^2, \rho_3^3\}$.

We have 36 following cases to verify. The verification is long but easy.

$\rho_0 \diamond \rho_0 = \rho_0, \rho_0 \diamond \rho_1 = \rho_1, \rho_0 \diamond \rho_2 = \rho_2, \rho_0 \diamond \rho_3 = \rho_3, \rho_0 \diamond \rho_2^2 = \rho_2^2,$
$\rho_0 \diamond \rho_3^3 = \rho_3^3;$
$\rho_1 \diamond \rho_0 = \rho_1, \rho_1 \diamond \rho_1 = \rho_1, \rho_1 \diamond \rho_2 = \rho_3, \rho_1 \diamond \rho_3 = \rho_3, \rho_1 \diamond \rho_2^2 = \rho_2^2,$
$\rho_1 \diamond \rho_3^3 = \rho_3^3;$
$\rho_2 \diamond \rho_0 = \rho_2, \rho_2 \diamond \rho_1 = \rho_2, \rho_2 \diamond \rho_2 = \rho_2^2, \rho_2 \diamond \rho_3 = \rho_2^2, \rho_2 \diamond \rho_2^2 = \rho_3^3,$

$\rho_2 \diamond \rho_3^3 = \rho_2^2$;
$\rho_3 \diamond \rho_0 = \rho_3, \rho_3 \diamond \rho_1 = \rho_3, \rho_3 \diamond \rho_2 = \rho_2^2, \rho_3 \diamond \rho_3 = \rho_2^2, \rho_3 \diamond \rho_2^2 = \rho_3^3$,
$\rho_3 \diamond \rho_3^3 = \rho_2^2$;
$\rho_2^2 \diamond \rho_0 = \rho_2^2, \rho_2^2 \diamond \rho_1 = \rho_2^2, \rho_2^2 \diamond \rho_2 = \rho_3^3, \rho_2^2 \diamond \rho_3 = \rho_3^3, \rho_2^2 \diamond \rho_2^2 = \rho_2^2$,
$\rho_2^2 \diamond \rho_3^3 = \rho_3^3$;
$\rho_3^3 \diamond \rho_0 = \rho_3^3, \rho_3^3 \diamond \rho_1 = \rho_3^3, \rho_3^3 \diamond \rho_2 = \rho_2^2, \rho_3^3 \diamond \rho_3 = \rho_2^2, \rho_3^3 \diamond \rho_2^2 = \rho_3^3$,
$\rho_3^3 \diamond \rho_3^3 = \rho_2^2$.

As an example, we verify the case $\rho_2 \diamond \rho_2^2 = \rho_3^3$.
We have

$$\begin{aligned}\rho_2 \diamond \rho_2^2(x) &= \rho_2 \diamond \rho_2 \diamond \rho_2(x)\\ &= \rho_2(\rho_2(\rho_2))(x)\\ &= \rho_2(\rho_2(x \circ 2))\\ &= \rho_2((x \circ 2) \circ 2) =\end{aligned}$$

$$= ((x \circ 2) \circ 2) \circ 2 = \begin{cases} 2 & \text{for} \quad x \in \{0, 1\}, \\ 0 & \text{for} \quad x \in \{2, 3\}. \end{cases}$$

On the other hand,

$\rho_3^3(x) = \rho_3 \diamond \rho_3 \diamond \rho_3(x) = \rho_3(\rho_3(\rho_3))(x) =$

$$= ((x \circ 3) \circ 3) \circ 3 = \begin{cases} 2 & \text{for} \quad x \in \{0, 1\}, \\ 0 & \text{for} \quad x \in \{2, 3\}. \end{cases}$$

The above case is satisfied.

The appropriate Cayley table looks like this:

$\diamond$	ρ_0	ρ_1	ρ_2	ρ_3	ρ_2^2	ρ_3^3
ρ_0	ρ_0	ρ_1	ρ_2	ρ_3	ρ_2^2	ρ_3^3
ρ_1	ρ_1	ρ_1	ρ_3	ρ_3	ρ_2^2	ρ_3^3
ρ_2	ρ_2	ρ_2	ρ_2^2	ρ_2^2	ρ_3^3	ρ_2^2
ρ_3	ρ_3	ρ_3	ρ_2^2	ρ_2^2	ρ_3^3	ρ_2^2
ρ_2^2	ρ_2^2	ρ_2^2	ρ_3^3	ρ_3^3	ρ_2^2	ρ_3^3
ρ_3^3	ρ_3^3	ρ_3^3	ρ_2^2	ρ_2^2	ρ_3^3	ρ_2^2

Table 3.10.2

From Theorem 3.10.1 it follows that $\langle M(X); \diamond\rangle$ is a semigroup. But since

$$\rho_1 \diamond \rho_2 = \rho_3$$

and

$$\rho_2 \diamond \rho_1 = \rho_2,$$

it is not commutative. □

Definition 3.10.4. A hyper weak BCC-algebra $\mathfrak{X} = \langle X; \circ, 0\rangle$ is called a hyper *group-like* weak BCC-algebra if it is standard and it satisfies

$$(x \circ z) \circ (y \circ z) = x \circ y.$$

Theorem 3.10.5. *In any hyper group-like weak BCC-algebra the condition*

$$0 \circ (0 \circ x) = x \tag{3.68}$$

is satisfied for all $x \in X$.

Proof. For any $x \in X$ we have $(x \circ x) \circ (0 \circ x) = x \circ 0 = x$ and $0 \in x \circ x$. This implies $0 \circ (0 \circ x) \subseteq (x \circ x) \circ (0 \circ x) = x \circ 0 = x$. Hence the condition (3.68) is satisfied. □

Theorem 3.10.6. *Let a hyper weak BCC-algebra $\mathfrak{X}$ satisfy the condition $0 \circ (0 \circ x) = x$ for all $x \in X$. Then the following conditions hold:*

(1) $x \circ x = 0,$

(2) $x \circ 0 = x,$

(3) $Card(x \circ y) = 1$

for all $x, y \in X$.

Proof. (1) Since $0 \circ x$ is a singleton set, we get

$$x \circ x = (0 \circ (0 \circ x)) \circ (0 \circ (0 \circ x)) \ll 0 \circ 0 = 0.$$

So, $x \circ x = 0$.

(2) By (HBCC7), $x \ll x \circ 0$. On the other hand,

$$x \circ 0 = (0 \circ (0 \circ x)) \circ ((0 \circ x) \circ (0 \circ x)) \ll 0 \circ (0 \circ x) = x.$$

So, $x \ll x \circ 0 \ll x$, i.e., $x \circ 0 = x$.

(3) From Lemma 3.9.19(13), for any $x, y \in X$, $Card(x \circ y) = 1$. □

Theorem 3.10.7. *A hyper weak BCC-algebra $\mathfrak{X}$ satisfies $0\circ(0\circ x) = x$ for all $x \in X$ if and only if it satisfies the condition $x \ll y \Longrightarrow x = y$ for all $x, y \in X$.*

Proof. ($\Rightarrow$) Let us assume that $x \ll y$. Then $0 = 0\circ 0 \subseteq 0\circ(x\circ y) \ll y\circ x$. For $0 \in x \circ y$ there exists $m \in y \circ x$ such that $0 \ll m$.

If $m \neq 0$, $Card(0\circ m) = 1$ and $0 = 0\circ m$, then $0\circ(0\circ m) = 0\circ 0 \neq m$. Hence $m = 0$ and so, $y \ll x$. This, by (HBCC4), implies $x = y$.

($\Leftarrow$) Let $m = 0 \circ (0 \circ x)$. Since $m \ll x$, we get $m = x$. That is, $0 \circ (0 \circ x) = x$. □

Theorem 3.10.8. *In any hyper weak BCC-algebra $\mathfrak{X} = \langle X; \circ, 0\rangle$ the following conditions are equivalent:*

(1) *$\mathfrak{X}$ satisfies $0 \circ (0 \circ x) = x$,*

(2) *$\mathfrak{X}$ satisfies $x \circ x = 0$ and $(x \circ z) \circ (y \circ z) = x \circ y$,*

(3) *$\mathfrak{X}$ is a group-like weak BCC-algebra*

for any $x, y, z \in X$.

Proof. (1) $\Longrightarrow$ (2) From Theorem 3.10.6 it follows that $\mathfrak{X}$ satisfies $x \circ x = 0$. Now, by (HBCC1), $(x \circ z) \circ (y \circ z) \ll x \circ y$. From Theorem 3.10.6(3) it follows $Card(x \circ y) = 1$ and Theorem 3.10.7 implies

$$(x \circ z) \circ (y \circ z) = x \circ y.$$

(2) $\Longrightarrow$ (3) Let us assume $Card(x \circ y) > 1$ for any $x, y \in X$. Then there exist $a, b \in x \circ y$ such that $a \neq b$, and we have

$$a \circ b \subseteq (x \circ y) \circ (x \circ y) \ll x \circ x = 0$$

and

$$b \circ a \subseteq (x \circ y) \circ (x \circ y) \ll x \circ x = 0.$$

Thus, by (HBCC4),

$$(a \circ b \ll 0 \text{ and } b \circ a \ll 0) \Longrightarrow (a \ll b \text{ and } b \ll a) \Longrightarrow a = b.$$

This implies $Card(x \circ y) = 1$.

Let now $x \circ 0 = y$ and $y \neq x$. Then, by (HBCC7), we get

$$x \ll x \circ 0 = y, \text{ i.e., } x \ll y$$

and

$$y \circ x = (y \circ y) \circ (x \circ y) = 0 \circ 0 = 0, \text{ i.e., } y \ll x.$$

Hence $x = y$ and

$$0 \circ (0 \circ x) = (x \circ x) \circ (0 \circ x) = x \circ 0 = x.$$

Therefore, $\mathfrak{X}$ is a group-like weak BCC-algebra.

(3) $\Longrightarrow$ (1) It follows from Theorem 3.10.5. □

By combining Theorem 3.10.5 and Theorem 3.10.8, we get the following corollary:

Corollary 3.10.9. *Every hyper group-like weak BCC-algebra is a group-like weak BCC-algebra.*

We will now look at generalized hyper group-like weak BCC-algebras. First, we will define them.

Definition 3.10.10. A hyper weak BCC-algebra $\langle X; \circ, 0\rangle$ satisfying for all $x, y, z \in X$ the condition

$$(x \circ (0 \circ y)) \circ (0 \circ z) = x \circ (0 \circ (y \circ (0 \circ z))) \tag{3.69}$$

is called a *generalized* hyper group-like weak BCC-algebra.

Example 3.10.11. Let $\mathfrak{X} = \langle X; \circ, 0\rangle$ be an algebra with the operation defined by Table 3.9.7. Then $\mathfrak{X}$ is a generalized hyper group-like weak BCC-algebra. But, since $0 \circ 3 = 2 \neq 0$, it is not a hyper BCC-algebra. This is also not a hyper BCI-algebra, because

$$(1 \circ 1) \circ 2 = \{2, 3\} \neq 3 = (1 \circ 2) \circ 1.$$

□

Later in this section we will use the expression hypersemigroup. First, however, we will recall its definition.

Definition 3.10.12. An algebra $\langle X; \circ\rangle$ with the binary hyperoperation $\circ$ is called a *hypersemigroup* if it satisfies the associativity condition:

$$(x \circ y) \circ z = x \circ (y \circ z)$$

for all $x, y, z \in X$.

That is,

$$\bigcup_{u \in x \circ y} u \circ z = \bigcup_{v \in y \circ z} x \circ v.$$

Of course, if $\langle X; \circ \rangle$ is a hypersemigroup, then $(A \circ B) \circ C = A \circ (B \circ C)$ for all $A, B, C \in \mathscr{P}(X)$.

Theorem 3.10.13. *Let $\langle X; \circ, 0 \rangle$ be a generalized hyper group-like weak BCC-algebra. We define the operation $\oplus$ on X as follows:*

$$x \oplus y = x \circ (0 \circ y).$$

Then $\langle X; \oplus \rangle$ is a hypersemigroup.

Proof. Let $x, y, z \in X$. Then, by (3.69), we have

$$\begin{aligned}(x \oplus y) \oplus z &= (x \circ (0 \circ y)) \circ (0 \circ z) \\ &= x \circ (0 \circ (y \circ (0 \circ z))) \\ &= x \oplus (y \oplus z).\end{aligned}$$

So, $\oplus$ is associative and $\langle X; \oplus \rangle$ is a hypersemigroup. □

Theorem 3.10.14. *Let $\langle X; \circ, 0 \rangle$ be a hyper weak BCC-algebra with the condition $0 \circ x = 0$ for all $x \in X$. We define the operation $\oplus$ on X as follows:*

$$x \oplus y = x \circ (0 \circ y).$$

Then $\langle X; \oplus \rangle$ is a hypersemigroup.

Proof. Let $x, y, z \in X$. Then we have

$$(x \oplus y) \oplus z = (x \circ (0 \circ y)) \circ (0 \circ z) = (x \circ 0) \circ 0 = x \circ 0.$$

On the other hand,

$$x \oplus (y \oplus z) = x \oplus (y \circ (0 \circ z)) = x \circ (0 \circ (y \circ (0 \circ z))) = x \circ 0.$$

So, $(x \oplus y) \oplus z = x \oplus (y \oplus z)$, i.e., $\langle X; \oplus \rangle$ is a hypersemigroup. □

Theorem 3.10.15. *Let $\langle X; \circ, 0 \rangle$ be a hyper BCC-algebra. We define the operation $\oplus$ on X as follows:*

$$x \oplus y = x \circ (0 \circ y).$$

Then $\langle X; \oplus \rangle$ is a hypersemigroup. Moreover, every element of X is a right identity with respect to the operation $\oplus$.

Proof. It is easy to show that $\langle X; \oplus\rangle$ is a hypersemigroup. The second part of the statement of the theorem follows from (HBCC3) and (HBCC2), respectively. Indeed, we have

$$x \oplus y = x \circ (0 \circ y) = x \circ 0 = x,$$

i.e., $x \oplus y = x$. □

The following interesting corollary follows directly from Theorem 2.5.7.

Corollary 3.10.16. *Every (hyper) group-like weak BCC-algebra is a generalized hyper group-like weak BCC-algebra.*

Example 3.10.17. Let $\mathfrak{X} = \langle\{0, 1, 2, 3\}; \circ_1, 0\rangle$ be a proper hyper BCC-algebra with the hyperoperation $\circ_1$ defined in Table 3.9.1.

According to Theorem 3.10.15 we get a hypersemigroup and any element of $\{0, 1, 2, 3\}$ is a right identity. This hypersemigroup is presented in the next table:

$\oplus$	0	1	2	3
0	0	0	0	0
1	1	1	1	1
2	2	2	2	2
3	3	3	3	3

Table 3.10.3 □

Example 3.10.18. Let $\mathfrak{X} = \langle\{0, 1, 2, 3\}; \circ, 0\rangle$ be a proper hyper BCC-algebra with the hyperoperation $\circ$ defined in the following table:

$\circ$	0	1	2	3
0	0	0	0	0
1	$\{1, 2\}$	$\{0, 2\}$	0	2
2	2	2	0	2
3	3	3	0	0

Table 3.10.4

$\mathfrak{X}$ is a hyper BCC-algebra.

According to Theorem 3.10.15 we get a hypersemigroup and any element of $\{0,1,2,3\}$ is a right identity. This hypersemigroup is presented in the next table:

$\oplus$	0	1	2	3
0	0	0	0	0
1	$\{1,2\}$	$\{1,2\}$	$\{1,2\}$	$\{1,2\}$
2	2	2	2	2
3	3	3	3	3

Table 3.10.5 □

In the following, we present a method of construction of a hyper weak BCC-algebra by a hyper BCC-algebra and a standard generalized hyper group-like weak BCC-algebra.

Theorem 3.10.19. *Let* $\langle X_1; \circ_1, 0\rangle$ *be a hyper BCC-algebra,* $\langle X_2; \circ_2, 0\rangle$ *be a standard generalized hyper group-like weak BCC-algebra and let* $X_1 \cap X_2 = \{0\}$.

We define the operation $\star$ *in the following way:*

$$x \star y = \begin{cases} x \circ_1 y & \text{for} \quad x, y \in X_1, \\ x \circ_2 y & \text{for} \quad x, y \in X_2, \\ 0 \circ_2 y & \text{for} \quad x \in X_1,\ y \in X_2 \setminus \{0\}, \\ x & \text{for} \quad x \in X_2 \setminus \{0\},\ y \in X_1. \end{cases}$$

Then $\langle X_1 \cup X_2; \star, 0\rangle$ *is a hyper weak BCC-algebra.*

Proof. For the proof of (HBCC1) we have to consider seven cases.

1^0. Let $x, y, z \in X_k$ for $k \in \{1,2\}$. Then, by the assumption that X_k is a hyper (weak) BCC-algebra, we have

$$(x \star z) \star (y \star z) = (x \circ_k z) \circ_k (y \circ_k z) \ll x \circ_k y = x \star y.$$

2^0. Let $x, y \in X_1$ and $z \in X_2 \setminus \{0\}$, then

$$\begin{aligned} ((x \star z) \star (y \star z)) \star (x \star y) &= ((0 \circ_2 z) \circ_2 (0 \circ_2 z)) \star (x \circ_1 y) \\ &= 0 \star (x \circ_1 y) \\ &= 0. \end{aligned}$$

Thus $(x \star z) \star (y \star z) \ll x \star y$.

3^0. Let $x, y \in X_2 \setminus \{0\}$ and $z \in X_1$. Then

$$(x \star z) \star (y \star z) = x \star y = x \circ_2 y \ll x \circ_2 y = x \star y.$$

4^0. Let $x, z \in X_1$ and $y \in X_2 \setminus \{0\}$, Then

$$\begin{aligned}((x \star z) \star (y \star z)) \star (x \star y) &= ((x \circ_1 z) \star y) \star (0 \circ_2 y) \\ &= (0 \circ_2 y) \star (0 \circ_2 y) \\ &= (0 \circ_2 y) \circ_2 (0 \circ_2 y) \\ &= 0.\end{aligned}$$

and so, $(x \star z) \star (y \star z) \ll x \star y$.

5^0. Let $x, z \in X_2 \setminus \{0\}$ and $y \in X_1$. Then

$$\begin{aligned}(x \star z) \star (y \star z) &= (x \circ_2 z) \star (0 \circ_2 y) \\ &= (x \circ_2 z) \circ_2 (0 \circ_2 y) \\ &\ll x \circ_2 0 \\ &= x.\end{aligned}$$

But $x \star y = x$. Hence $(x \star z) \star (y \star z) \ll x \star y$.

6^0. Let $x \in X_1$ and $y, z \in X_2 \setminus \{0\}$. Then

$$\begin{aligned}(x \star z) \star (y \star z) &= (0 \circ_2 z) \star (y \circ_2 z) \\ &= (0 \circ_2 z) \circ_2 (y \circ_2 z) \\ &\ll 0 \circ_2 y.\end{aligned}$$

On the other hand,

$$x \star y = 0 \circ_2 y.$$

Thus

$$(x \star z) \star (y \star z) \ll x \star y.$$

7^0. Let $x \in X_2 \setminus \{0\}$ and $y, z \in X_1$. Then

$$(x \star z) \star (y \star z) = x \star (y \circ_1 z) = x.$$

But $x \star y = x$. So, we have

$$(x \star z) \star (y \star z) = x \ll x = x \star y.$$

We have shown that $\langle X_1 \cup X_2; \star, 0\rangle$ satisfies (HBCC1).

Now we show condition (HBCC4). If $x, y \in X_k$ for $k \in \{1, 2\}$, then the case is obvious. Let $x \in X_1$ and $y \in X_2$. Then $0 \in x \star y = 0 \circ_2 y$, because $Card(0 \circ_2 x) = 1$. Hence $0 \circ_2 y = 0$ and $y \star x = y = 0$. Then $x \ll 0$ and $0 \ll x$, i.e., $x = 0$. This implies $x = y$.

As it can be easily seen, condition (HBCC5) is satisfied, too. Indeed, if $x \in X_1 \cup X_2$, then $x \in X_k$ for $k \in \{1, 2\}$. But $\langle X_k; \circ_k, 0\rangle$ satisfies the condition (HBCC5) and so it is also satisfied in $\langle X_1 \cup X_2; \star, 0\rangle$.

Since $\langle X_k; \circ_k, 0\rangle$ satisfies (HBCC6) and (HBCC7) for $k \in \{1, 2\}$, they are also satisfied in $\langle X_1 \cup X_2; \star, 0\rangle$.

We have shown that $\langle X_1 \cup X_2; \star, 0\rangle$ is a hyper weak BCC-algebra. □

We conclude this chapter with an example.

Example 3.10.20. Let $\mathfrak{X} = \langle\{0, 1, 2\}; \circ_2, 0\rangle$ be a hyper BCC-algebra with the operation $\circ_2$ defined by Table 3.9.2.

Let $\mathfrak{Y} = \langle\{0, 3, 4, 5\}; \circ, 0\rangle$ be a standard generalized hyper group-like weak BCC-algebra with the operation $\circ$ defined by Table 3.10.6.

Using the construction method shown in Theorem 3.10.19, we get a hyper weak BCC-algebra $\langle\{0, 1, 2\} \cup \{0, 3, 4, 5\}; \star, 0\rangle$ with the operation $\star$ defined Table 3.10.7.

$\circ$	0	3	4	5
0	0	0	4	4
3	3	$\{0, 3\}$	5	5
4	4	4	0	0
5	5	5	3	$\{0, 3\}$

Table 3.10.6

$\star$	0	1	2	3	4	5
0	0	0	0	0	4	4
1	1	0	0	0	4	4
2	2	2	$\{0, 1\}$	0	4	4
3	3	3	3	$\{0, 3\}$	5	5
4	4	4	4	4	0	0
5	5	5	5	5	3	$\{0, 3\}$

Table 3.10.7 □

3.11 Soft BCC-Algebras

To solve complicated economics, engineering, and environmental problems, we cannot successfully use classical methods because of various uncertainties typical for those problems. Uncertainties cannot be handled using traditional mathematical tools but may be dealt with using a wide range of existing theories such as probability theory, theory of (intuitionistic) fuzzy sets, theory of vague sets and theory of interval mathematics or rough sets. However, all these theories have their own difficulties which Molodtsov points out [121]. Also, in [113] was suggested that one reason for these difficulties might be the inadequacy of the parametrization tool of the theory. To overcome these difficulties, Molodtsov introduced the concept of a soft set as a new mathematical tool for dealing with uncertainties and as a generalization of the fuzzy set theory that is free from the difficulties that have troubled the usual theoretical approaches. He pointed out several directions for the applications of soft sets. After that, the work on the soft set theory progressed rapidly.

Jun et al. (cf. [89]) applied the notion of soft sets to BCC-algebras. In this section, we present their notion and results on soft BCC-algebras.

Let U be an initial universe set and E be a set of parameters. Let $P(U)$ denote the power set of U and $A \subset E$.

We begin with definitions concerned with a soft set.

Definition 3.11.1. A pair (δ, A) is said to be a *soft set* over U, where δ is a mapping defined as

$$\delta : A \longrightarrow P(U).$$

We can say that a soft set over U is a parameterized family of subsets of the universe U. For $\varepsilon \in A$ we can consider $\delta(\varepsilon)$ as the set of ε-approximate elements of the soft set (δ, A).

As it can be easily seen, a soft set is not a set.

Definition 3.11.2. Let (δ, A) and (γ, B) be two soft sets over a common universe U. The intersection of (δ, A) and (γ, B) is the soft set (ρ, C) for which the following conditions hold:

(1) $C = A \cap B$,

(2) $(\forall x \in C)$ $(\rho(x) = \delta(x)$ or $\rho(x) = \gamma(x)$, (as both are same sets)).

In this case, we write

$$(\delta, A)\tilde{\cap}(\gamma, B) = (\rho, C).$$

Definition 3.11.3. Let (δ, A) and (γ, B) be two softs sets over a common universe U. The union of (δ, A) and (γ, B) is the soft set (ρ, C) satisfying the following conditions:

(1) $C = A \cup B$,

(2) for all $x \in C$,

$$\rho(x) = \begin{cases} \delta(x) & \text{if } x \in A \setminus B, \\ \gamma(x) & \text{if } x \in B \setminus A, \\ \delta(x) \cup \gamma(x) & \text{if } x \in A \cap B. \end{cases}$$

In this case, the notation looks like this:

$$(\delta, A)\tilde{\cup}(\gamma, B) = (\rho, C).$$

Definition 3.11.4. Let (δ, A) and (γ, B) be two soft sets over a common universe U. Then we define a logical operator $(\delta, A)\,\text{AND}\,(\gamma, B)$ by

$$(\delta, A)\tilde{\wedge}(\gamma, B) = (\rho, A \times B),$$

where $\rho(x, y) = \delta(x) \cap \gamma(y)$ for all $(x, y) \in A \times B$.

It will be denoted by

$$(\delta, A)\tilde{\wedge}(\gamma, B).$$

Definition 3.11.5. Let (δ, A) and (γ, B) be two soft sets over a common universe U. Then we define a logical operator $(\delta, A)\,\text{OR}\,(\gamma, B)$ by

$$(\delta, A)\tilde{\vee}(\gamma, B) = (\rho, A \times B),$$

where $\rho(x, y) = \delta(x) \cup \gamma(y)$ for all $(x, y) \in A \times B$.

It will be denoted by

$$(\delta, A)\tilde{\vee}(\gamma, B).$$

Definition 3.11.6. Let (δ, A) and (γ, B) be two softs sets over a common universe U. We define a *soft subset* (δ, A) of (γ, B) as $(\delta, A)\tilde{\subset}(\gamma, B)$, if it satisfies the conditions:

(1) $A \subset B$,

(2) for every $\varepsilon \in A$, $\delta(\varepsilon)$ and $\gamma(\varepsilon)$ are identical approximations.

Now we define the soft set over a BCC-algebra $\mathfrak{X}$ and a soft BCC-algebra over $\mathfrak{X}$.

Let θ be an arbitrary binary relation between an element of a nonempty set A and an element of X, where $\mathfrak{X} = \langle X; \cdot, 0\rangle$ is a BCC-algebra. Then $\theta \subseteq A \times X$. We define a set-valued function $\delta : A \longrightarrow \mathscr{P}(X)$ as

$$\delta(x) = \{y \in X \mid (x, y) \in \theta\}$$

for all $x \in A$.

Then the pair (δ, A) is a *soft set* over a BCC-algebra $\mathfrak{X}$.

Definition 3.11.7. Let (δ, A) be a soft set over a BCC-algebra $\mathfrak{X}$. Then (δ, A) is said to be a *soft BCC-algebra* over $\mathfrak{X}$ if $\langle\delta(x); \cdot, 0\rangle$ is a subalgebra of $\mathfrak{X}$ for all $x \in A$.

Let us illustrate the definition, using the following examples:

Example 3.11.8. The algebra $\langle\{0, 1, 2, 3, 4\}; \cdot, 0\rangle$ with the operation $\cdot$ defined by Table 3.11.1 is an improper BCC-algebra (cf. [146], algebra B_{5-1-28} on page 330).

·	0	1	2	3	4
0	0	0	0	0	0
1	1	0	1	1	1
2	2	2	0	2	2
3	3	3	3	0	3
4	4	4	4	4	0

Table 3.11.1

We will show two cases of soft BCC-algebras over $\mathfrak{X}$.

Case 1^0. Let (δ, A) be a soft set over $\mathfrak{X}$, where $A = X$ and $\delta : A \longrightarrow \mathscr{P}(X)$ is a set-valued function defined by

$$\delta(x) = \Big\{y \in X \mid y \cdot yx \in \{0, 1\}\Big\}$$

for all $x \in A$.

It is easy to see that

$$\delta(x) = \begin{cases} X & \text{for} \quad x = 0, \\ X & \text{for} \quad x = 1, \\ \{0,1,3,4\} & \text{for} \quad x = 2, \\ \{0,1,2,4\} & \text{for} \quad x = 3, \\ \{0,1,2,3\} & \text{for} \quad x = 4. \end{cases}$$

The values of $\delta(x)$ are subalgebras.

Indeed, $\langle \delta(x); \cdot, 0 \rangle$ for $x \in \{1,2,3\}$ are isomorphic to the BCK-algebra B_{4-1-8} in [146] on page 316, i.e., they are all subalgebras of $\mathfrak{X}$.

Therefore, (δ, A) is a soft BCC-algebra over $\mathfrak{X}$.

Case 2^0. Let (ρ, B) be a soft set over $\mathfrak{X}$, where $B = \{1,3,4\}$ and $\rho : B \longrightarrow \mathscr{P}(X)$ is a set-valued function defined for all $x \in B$ as follows:

$$\rho(x) = \Big\{ y \in X \,|\, yx \cdot x \in \{0,2\} \Big\}.$$

Then, as it is easy to see, we have

$$\delta(x) = \begin{cases} \{0,1,2\} & \text{for} \quad x = 1, \\ \{0,2,3\} & \text{for} \quad x = 3, \\ \{0,2,4\} & \text{for} \quad x = 4. \end{cases}$$

The algebras $\langle \delta(x); \cdot, 0 \rangle$ for $x \in \{1,3,4\}$ are isomorphic to Table 1.2.6, i.e., they are all subalgebras of $\mathfrak{X}$.

Therefore, (ρ, B) is a soft BCC-algebra over $\mathfrak{X}$. □

Example 3.11.9. The algebra $\langle \{0,1,2,3\}; \cdot, 0 \rangle$ with the operation $\cdot$ given by Table 1.2.14 is a BCC-algebra.

Let (γ, A) be a soft set over $\mathfrak{X}$, where $A = \{1,2\}$ and $\gamma : A \longrightarrow \mathscr{P}(X)$ is a set-valued function defined by

$$\gamma(x) = \Big\{ y \in X \,|\, yx \cdot x \in \{0,1\} \Big\}$$

for all $x \in A$.

Then $\langle \gamma(1); \cdot, 0 \rangle$ and $\langle \gamma(2); \cdot, 0 \rangle$, where $\gamma(1) = \{0,1\}$ and $\gamma(2) = X$ are subalgebras of $\mathfrak{X}$. Hence (γ, A) is a soft BCC-algebra over $\mathfrak{X}$. □

Theorem 3.11.10. *Let (δ, A) be a soft BCC-algebra over $\mathfrak{X}$. If $B \subset A$, then $(\delta_{|B}, B)$ is a soft BCC-algebra over $\mathfrak{X}$.*

Proof. Straightforward. □

In the next example we show that there exists a soft set (ρ, A) that is not a soft BCC-algebra over $\mathfrak{X}$ and we show also that there exists a subset $B \subset A$ such as in the last theorem.

Example 3.11.11. Let $\mathfrak{X} = \langle\{0,1,2,3,4\}; \cdot, 0\rangle$ with the operation defined by Table 3.11.2 (cf. [146], algebra B_{5-3-5} on page 320) be an improper BCC-algebra.

$\cdot$	0	1	2	3	4
0	0	0	0	0	0
1	1	0	0	0	0
2	2	2	0	0	0
3	3	2	1	0	1
4	4	4	2	2	0

Table 3.11.2

Let (ρ, A) be a soft set over $\mathfrak{X}$, where $A = X$ and $\rho : A \longrightarrow \mathscr{P}(X)$ is a set-valued function defined by

$$\rho(x) = \Big\{y \in X \mid yx \in \{0,3,4\}\Big\}$$

for all $x \in A$.

But (ρ, A) is not a soft BCC-algebra over $\mathfrak{X}$ because $\rho(0) = \{0,3,4\}$ equipped with the operation $\cdot$ is not a subalgebra of $\mathfrak{X}$.

Indeed, the elements $3, 4 \in \rho(0)$. But $3 \cdot 4 = 1 \neq \rho(0)$.

When we take $B \subseteq A \setminus \{0\}$, then $(\rho_{|B}, B)$ is also a soft BCC-algebra over $\mathfrak{X}$. □

Theorem 3.11.12. *Let (δ, A) and (γ, B) be two soft BCC-algebras over $\mathfrak{X}$. If $A \cap B \neq \emptyset$, then the intersection $(\delta, A)\tilde{\cap}(\gamma, B)$ is a soft BCC-algebra over $\mathfrak{X}$.*

Proof. From Definition 3.11.2 it follows that $(\delta, A)\tilde{\cap}(\gamma, B) = (\rho, C)$, where $C = A \cap B$ and $\rho(x) = \delta(x)$ or $\rho(x) = \gamma(x)$ for $x \in C$. In this case $\rho : C \longrightarrow \mathscr{P}(X)$ is a mapping and therefore (ρ, C) is a soft set over $\mathfrak{X}$. But because (δ, A) and (γ, B) are soft BCC-algebras over $\mathfrak{X}$, it follows that $\langle\rho(x); \cdot, 0\rangle = \langle\delta(x); \cdot, 0\rangle$ is a subalgebra of $\mathfrak{X}$ or $\langle\rho(x); \cdot, 0\rangle = \langle\gamma(x); \cdot, 0\rangle$ is a subalgebra of $\mathfrak{X}$ for all $x \in C$. This implies, $(\rho, C) = (\delta, A)\tilde{\cap}(\gamma, B)$ is a soft BCC-algebra over $\mathfrak{X}$. □

From Theorem 3.11.12 immediately the following corollary follows:

Corollary 3.11.13. *Let (δ, A) and (γ, A) be two soft BCC-algebras over $\mathfrak{X}$. Then the intersection $(\delta, A)\tilde{\cap}(\gamma, A)$ is a soft BCC-algebra over $\mathfrak{X}$.*

Theorem 3.11.14. *Let (δ, A) and (γ, B) be two soft BCC-algebras over $\mathfrak{X}$. If A and B are disjoint, then the union $(\delta, A)\tilde{\cup}(\gamma, B)$ is a soft BCC-algebra over $\mathfrak{X}$.*

Proof. From Definition 3.11.3 it follows $(\delta, A)\tilde{\cup}(\gamma, B) = (\rho, C)$, where $C = A \cup B$. Moreover, for any $x \in C$ we have

$$\rho(x) = \begin{cases} \delta(x) & \text{if } \ x \in A \setminus B, \\ \gamma(x) & \text{if } \ x \in B \setminus A, \\ \delta(x) \cup \gamma(x) & \text{if } \ x \in A \cap B. \end{cases}$$

From the assumption $A \cap B = \emptyset$. Then either $x \in A \setminus B$ or $x \in B \setminus A$ for all $x \in C$. In the case, when $x \in A \setminus B$, then $\langle \rho(x); \cdot, 0\rangle = \langle \delta(x); \cdot, 0\rangle$ is a subalgebra of $\mathfrak{X}$ since (δ, A) is a soft BCC-algebra over $\mathfrak{X}$. If $x \in B \setminus A$, then $\langle \rho(x); \cdot, 0\rangle = \langle \gamma(x); \cdot, 0\rangle$ is also a subalgebra of $\mathfrak{X}$ because (γ, B) is a soft BCC-algebra over $\mathfrak{X}$.

Hence $(\rho, C) = (\delta, A)\tilde{\cup}(\gamma, B)$ is a soft BCC-algebra over $\mathfrak{X}$. □

It turns out that the assumption in Theorem 3.11.14 that the sets A and B are disjoint, is necessary. We show it in the next examples.

Example 3.11.15. Let us consider the BCC-algebra $\mathfrak{X}$ in Example 3.11.11. Let (δ, A) and (γ, B) be two soft sets over $\mathfrak{X}$, where $A = \{0, 1, 2, 3\}, B = \{0, 3, 4\}$ and, $\delta : A \longrightarrow \mathscr{P}(X)$ and $\gamma : B \longrightarrow \mathscr{P}(X)$ are set-valued functions defined in the following way:

$$\delta(a) = \Big\{ y \in X \,|\, ya \in \{0, 3\} \Big\}$$

and

$$\gamma(b) = \Big\{ y \in X \,|\, yb \in \{0, 4\} \Big\}$$

for all $a \in A$ and $b \in B$, respectively. Then (δ, A) and (γ, B) are soft BCC-algebras over $\mathfrak{X}$.

Let us observe that A and B are not disjoint and their union $(\delta, A)\tilde{\cup}(\gamma, B)$ is not a soft BCC-algebra.

Indeed, $\langle \delta(0) \cup \gamma(0); \cdot, 0\rangle = \langle \{0, 3, 4\}; \cdot, 0\rangle$ is not a subalgebra. □

Example 3.11.16. Let us consider the BCC-algebra $\mathfrak{X}$ in Example 3.11.9. Let us put $A = \{1\}$ and let $\delta : A \longrightarrow \mathscr{P}(X)$ be a set-valued function defined by

$$\delta(x) = \Big\{y \in X \,|\, xy \in \{1\}\Big\}$$

for all $x \in A$. Then (δ, A) is a soft BCC-algebra over $\mathfrak{X}$ since $\langle\delta(1);\, \cdot\,, 0\rangle$, where $\delta(1) = \{0, 2\}$ is a subalgebra of $\mathfrak{X}$.

Now let $B = \{0, 1\}$ and let $\gamma : B \longrightarrow \mathscr{P}(X)$ be a set-valued function defined by

$$\gamma(x) = \Big\{y \in X \,|\, y \cdot xy \in \{0, 3\}\Big\}$$

for all $x \in B$.

Then $\langle\gamma(0);\, \cdot\,, 0\rangle = \langle\gamma(1);\, \cdot\,, 0\rangle$, where $\gamma(0) = \gamma(1) = \{0, 3\}$ are subalgebras of $\mathfrak{X}$. Hence (γ, B) is a soft BCC-algebra over $\mathfrak{X}$.

Clearly, $A \cap B \neq \emptyset$. So, when we put $(\rho, C) = (\delta, A)\tilde{\cup}(\gamma, B)$, then $\rho(1) = \delta(1) \cup \gamma(1) = \{0, 2, 3\}$ equipped with the operation $\cdot$ is not a subalgebra of $\mathfrak{X}$. Indeed, $2 \cdot 3 = 1 \notin \rho(1)$.

This means that $(\rho, C) = (\delta, A)\tilde{\cup}(\gamma, B)$ is not a soft BCC-algebra over $\mathfrak{X}$. □

Theorem 3.11.17. *If (δ, A) and (γ, B) are soft BCC-algebras over $\mathfrak{X}$, then $(\delta, A)\tilde{\wedge}(\gamma, B)$ is a soft BCC-algebra over $\mathfrak{X}$.*

Proof. By Definition 3.11.4, we know that $(\delta, A)\tilde{\wedge}(\gamma, B) = (\rho, A \times B)$, where $\rho(x, y) = \delta(x) \cap \gamma(y)$ for all $(x, y) \in A \times B$. From the fact that $\delta(x)$ and $\gamma(y)$ equipped with the operation $\cdot$ are subalgebras of $\mathfrak{X}$ it follows that their intersection $\delta(x) \cap \gamma(y)$ is a subalgebra of $\mathfrak{X}$, too. Hence $\langle\rho(x, y);\, \cdot\,, 0\rangle$ is a subalgebra of $\mathfrak{X}$ for all $(x, y) \in A \times B$. Therefore, $(\delta, A)\tilde{\wedge}(\gamma, B) = (\rho, A \times B)$ is a soft BCC-algebra over $\mathfrak{X}$. □

Theorem 3.11.17 in the version for $\tilde{\vee}$ is not true. In the following examples we show two soft BCC-algebras (δ, A) and (γ, B) over $\mathfrak{X}$ such that $(\delta, A)\tilde{\vee}(\gamma, B)$ is not a soft BCC-algebra over $\mathfrak{X}$.

Example 3.11.18. Let us consider the BCC-algebra $\mathfrak{X}$ in Example 3.11.11. Let (δ, A) and (γ, B) be two soft sets over $\mathfrak{X}$, where $A = \{0, 3\}, B = \{2\}$ and $\delta : A \longrightarrow \mathscr{P}(X)$ and $\gamma : B \longrightarrow \mathscr{P}(X)$ are set-valued functions defined as follows:

$$\delta(a) = \Big\{x \in X \,|\, xa \in \{0, 3\}\Big\}$$

and

$$\gamma(b) = \Big\{x \in X \mid xb = 2\Big\}$$

for all $a \in A$ and $b \in B$, respectively.

Then $\delta(0) = \{0, 3\}, \delta(3) = \{0, 1, 2, 3\}$ and $\gamma(2) = \{0, 4\}$ all equipped with the operation $\cdot$ are subalgebras of $\mathfrak{X}$, i.e., (δ, A) and (γ, B) are soft BCC-algebras over $\mathfrak{X}$. But $(\delta, A)\tilde{\vee}(\gamma, B)$ is not a soft BCC-algebra. Indeed, the algebra $\langle \delta(0) \cup \gamma(2); \cdot, 0\rangle$ for $\delta(0) \cup \gamma(2) = \{0, 3, 4\}$ is not a subalgebra of $\mathfrak{X}$. □

Example 3.11.19. Let (δ, A) and (γ, B) be two soft BCC-algebras over $\mathfrak{X}$, which are described in Example 3.11.16. In this case, $(\rho, A \times B) = (\delta, A)\tilde{\vee}(\gamma, B)$ is not a soft BCC-algebra over $\mathfrak{X}$ because

$$\rho(1, 1) = \delta(1) \cup \gamma(1) = \{0, 2, 3\}$$

is not a subalgebra of $\mathfrak{X}$ due to the operation $\cdot$. □

Now we provide definitions and an example of trivial and whole soft BCC-algebra over $\mathfrak{X}$.

Definition 3.11.20. A soft BCC-algebra (δ, A) over $\mathfrak{X}$ is called *trivial* if $\delta(x) = \{0\}$ for all $x \in A$. Analogously, we call (δ, A) *whole* if $\delta(x) = X$ for all $x \in A$.

Example 3.11.21. Let $\mathfrak{X}$ be the BCC-algebra described in Example 3.11.8. Let $A = X$ and $\delta : A \longrightarrow \mathscr{P}(X)$ be a set-value function defined as follows:

$$\delta(x) = \Big\{y \in X \mid xy \in \{0, x\}\Big\}$$

for all $x \in A$. Then we have $\delta(0) = \delta(1) = \delta(2) = \delta(3) = \delta(4) = X$. As it can be easily seen, (δ, A) is a whole soft BCC-algebra over $\mathfrak{X}$.

Let $\gamma : A \longrightarrow \mathscr{P}(X)$ be a set-valued function defined by

$$\gamma(x) = \Big\{y \in X \mid y \cdot xy \in \{0\}\Big\}$$

for all $x \in A$. Then we have $\gamma(0) = \gamma(1) = \gamma(2) = \gamma(3) = \gamma(4) = \{0\}$, and so (γ, A) is an example of a trivial soft BCC-algebra over $\mathfrak{X}$. □

Let us take a look at homomorphisms in the environment of soft BCC-algebras.

Let $f : X \longrightarrow Y$ be a mapping of BCC-algebras. If (δ, A) is a soft set over $\mathfrak{X}$, then $(f(\delta), A)$ is a soft set over $\mathfrak{Y}$, where $f(\delta) : A \longrightarrow P(Y)$ is defined by $f(\delta)(x) = f(\delta(x))$ for all $x \in A$.

Lemma 3.11.22. *Let $\mathfrak{X}$ and $\mathfrak{Y}$ be BCC-algebras and let $f : X \longrightarrow Y$ be a homomorphism. If (δ, A) is a soft BCC-algebra over $\mathfrak{X}$, then $(f(\delta), A)$ is a soft BCC-algebra over $\mathfrak{Y}$.*

Proof. For every $x \in A$, $f(\delta)(x) = f(\delta(x))$ is a subalgebra of $\mathfrak{Y}$ due to the operation $\cdot$. Indeed, we know that $\langle \delta(x); \cdot, 0\rangle$ is a subalgebra of $\mathfrak{X}$ and, of course, its homomorphic image is also a subalgebra of $\mathfrak{Y}$. Hence $(f(\delta), A)$ is a soft BCC-algebra over $\mathfrak{Y}$. □

Example 3.11.23. Let $\mathfrak{X}$ and $\mathfrak{Y}$ be the BCC-algebras that are described in Example 3.11.8 and in Example 3.11.9, respectively. We define a map $f : X \longrightarrow Y$ as follows:

$$f(x) = \begin{cases} 0 & \text{for} \quad x = 0, \\ 2 & \text{for} \quad x = 1, \\ 1 & \text{for} \quad x = 2, \\ 0 & \text{for} \quad x = 3, \\ 0 & \text{for} \quad x = 4. \end{cases}$$

It is not difficult to verify that f is a homomorphism of BCC-algebras. Let (δ, A) be a soft BCC-algebra over $\mathfrak{X}$, which is given in Example 3.11.8. Then

$$f(\delta)(0) = f(\delta)(1) = f(\delta)(3) = f(\delta)(4) = \{0, 1, 2\} \text{ and } f(\delta)(2) = \{0, 2\}$$

are subalgebras of $\mathfrak{Y}$ due to the operation $\cdot$. Indeed, the improper algebra $\langle\{0, 1, 2\}; \cdot, 0\rangle$ with the operation $\cdot$ defined by Table 1.2.6 is a subalgebra of a BCC-algebra $\mathfrak{Y} = \langle\{0, 1, 2, 3\}; \cdot, 0\rangle$ with the operation defined in Table 1.2.14. And, obviously, $\langle\{0, 2\}; \cdot, 0\rangle$ is an improper subalgebra of $\mathfrak{Y}$.

This means that $(f(\delta), A)$ is a soft BCC-algebra over $\mathfrak{Y}$.

When we take a soft BCC-algebra (ρ, B) which is defined in Example 3.11.8, then $(f(\rho); B)$ is a soft BCC-algebra over $\mathfrak{Y}$. Indeed, we have $f(\rho)(1) = \{0, 1, 2\}$ and $f(\rho)(3) = f(\rho)(4) = \{0, 1, \}$. Clearly, the algebras $\langle\{0, 1, 2\}; \cdot, 0\rangle$ and $\langle\{0, 1\}; \cdot, 0\rangle$ are subalgebras of $\mathfrak{Y}$. □

Theorem 3.11.24. *Let $\mathfrak{X}$ and $\mathfrak{Y}$ be BCC-algebras. Let $f : X \longrightarrow Y$ be a homomorphism and (δ, A) be a soft BCC-algebra over $\mathfrak{X}$. Then the following statements are true:*

(1) *If $\delta(x) \subseteq kerf$ for all $x \in A$, then $(f(\delta), A)$ is a trivial soft BCC-algebra over $\mathfrak{Y}$,*

(2) *If f is onto and (δ, A) is whole, then $(f(\delta), A)$ is a whole soft BCC-algebra over $\mathfrak{Y}$.*

Proof. (1) Let us assume that $\delta(x) \subseteq kerf$ for all $x \in A$. Then $f(\delta)(x) = f(\delta(x)) = \{0_Y\}$ for all $x \in A$. Hence, by Lemma 3.11.22 and Definition 3.11.20, $(f(\delta), A)$ is a trivial soft BCC-algebra over $\mathfrak{Y}$.

(2) Let us suppose that f is onto and (δ, A) is whole. Then $\delta(x) = X$ for all $x \in A$. We get $f(\delta)(x) = f(\delta(x)) = f(X) = Y$ for all $x \in A$. It follows from Lemma 3.11.22 and Definition 3.11.20 that $(f(\delta), A)$ is a whole soft BCC-algebra over $\mathfrak{Y}$. □

Definition 3.11.25. Let (δ, A) and (γ, B) be two soft BCC-algebras over $\mathfrak{X}$. Then (δ, A) satisfying the following conditions:

(1) $A \subset B$,

(2) $\langle \delta(x); \cdot, 0\rangle$ is a subalgebra of $\langle \gamma(x); \cdot, 0\rangle$ for all $x \in A$

is called a *soft BCC-subalgebra* of (γ, B), denoted by $(\delta, A)\tilde{<}(\gamma, B)$.

Example 3.11.26. Let (δ, A) be a soft BCC-algebra over $\mathfrak{X}$, which is given in Case 1^0 of Example 3.11.8. Let $B = \{1, 3, 4\}$ be a subset of A and let $\gamma : B \longrightarrow \mathscr{P}(X)$ be a set-valued function defined in the following way:

$$\gamma(x) = \Big\{y \in X \,|\, y \cdot yx \in \{0, 1\}\Big\}$$

for all $x \in B$. Then we have

$$\gamma(x) = \begin{cases} X & \text{for} \quad x = 1, \\ \{0, 1, 2, 4\} & \text{for} \quad x = 3, \\ \{0, 1, 2, 3\} & \text{for} \quad x = 4. \end{cases}$$

That are subalgebras of $\delta(1), \delta(3)$ and $\delta(4)$ due to $\cdot$, respectively. As a conclusion, we have that (γ, B) is a soft BCC-subalgebra of (δ, A). □

Theorem 3.11.27. *Let $\mathfrak{X}$ be a BCC-algebra. Let (δ, A) and (γ, A) be two soft BCC-algebras over $\mathfrak{X}$. Then the following conditions are satisfied:*

(1) *If $\delta(x) \subset \gamma(x)$ for all $x \in A$, then $(\delta, A)\tilde{<}(\gamma, A)$,*

(2) *If $B = \{0\}$ and $(\rho, B), (\rho, X)$ are soft BCC-algebras over $\mathfrak{X}$, then $(\rho, B)\tilde{<}(\rho, X)$.*

Proof. Straightforward. □

Theorem 3.11.28. *Let $\mathfrak{X}$ be a BCC-algebra. Let (δ, A) be a soft BCC-algebra over $\mathfrak{X}$ and let (γ_1, B_1) and (γ_2, B_2) be soft BCC-subalgebras of (δ, A). Then the following conditions are true:*

(1) $(\gamma_1, B_1)\tilde{\cap}(\gamma_2, , B_2)\tilde{<}(\delta, A)$,

(2) $B_1 \cap B_2 = \emptyset \Longrightarrow (\gamma_1, B_1)\tilde{\cup}(\gamma_2, B_2)\tilde{<}(\delta, A)$.

Proof. (1) From Definition 3.11.2 it follows that $(\gamma_1, B_1)\tilde{\cap}(\gamma_2, B_2) = (\gamma, B)$, where $B = B_1 \cap B_2$ and $\gamma(x) = \gamma_1(x)$ or $\gamma(x) = \gamma_2(x)$ for all $x \in B$. Obviously, $B \subset A$. And for $x \in B$ we have $x \in B_1$ and $x \in B_2$. Hence, because $x \in B_1$, then $\gamma(x) = \gamma_1(x)$ is a subalgebra due to the operation $\cdot$ of $\langle \delta(x); \cdot, 0\rangle$. Indeed, we have $(\gamma_1, B_1)\tilde{<}(\delta, A)$.

Since $x \in B_2$, $\gamma(x) = \gamma_2(x)$. It, equipped with the operation $\cdot$, is a subalgebra of $\langle \delta(x); \cdot, 0\rangle$ because $(\gamma_2, B_2)\tilde{<}(\delta, A)$.

So, we have

$$(\gamma_1, B_1)\tilde{\cap}(\gamma_2, B_2) = (\gamma, B)\tilde{<}(\delta, A).$$

To prove (2) let us assume that $B_1 \cap B_2 = \emptyset$. From Definition 3.11.3 it follows that we can write $(\gamma_1, B_1)\tilde{\cup}(\gamma_2, B_2) = (\gamma, B)$. So, we have

$$\gamma(x) = \begin{cases} \gamma_1(x) & \text{for } \ x \in B_1 \setminus B_2, \\ \gamma_2(x) & \text{for } \ x \in B_2 \setminus B_1, \\ \gamma_1(x) \cup \gamma_2(x) & \text{for } \ x \in B_1 \cap B_2 \end{cases}$$

for all $x \in B$.

Since for $i \in \{1, 2\}$ we have $(\gamma_i, B_i)\tilde{<}(\delta, A)$, $B = B_1 \cup B_2 \subset A$ and $\langle \gamma_i(x); \cdot, 0\rangle$ is a subalgebra of $\langle \delta(x); \cdot, 0\rangle$ for all $x \in B_i$. From assumption $B_1 \cap B_2 = \emptyset$, it follows $\langle \gamma(x); \cdot, 0\rangle$ is a subalgebra of $\langle \delta(x); \cdot, 0\rangle$ for all $x \in B$. So, we have proven that

$$(\gamma_1, B_1)\tilde{\cup}(\gamma_2, B_2) = (\gamma, B)\tilde{<}(\delta, A).$$

The proof is completed. □

In Theorem 3.11.28(2) the assumption that B_1 and B_2 are disjoint is necessary. In the following example, we show that without that assumption that result is not valid.

Example 3.11.29. Let $\mathfrak{X}$ be a BCC-algebra $\langle \{0, 1, 2, 3, 4\}; \cdot, 0\rangle$ with the operation defined by Table 3.11.2. For $A = X$ we define a set-valued function $\delta : A \longrightarrow \mathscr{P}(X)$ in the following way:

$$\delta(x) = \Big\{ y \in X \,|\, xy \cdot x \in \{0\} \Big\}$$

for all $x \in A$. Then $\delta(x) = X$ for all $x \in A$, and so (δ, A) is a whole soft BCC-algebra over $\mathfrak{X}$.

Let (γ_1, B_1) be a soft set over $\mathfrak{X}$, where $B_1 = \{3\} \subset A$ and let $\gamma_1 : B_1 \longrightarrow \mathscr{P}(X)$ be a set-valued function defined by

$$\gamma_1(x) = \Big\{ y \in X \mid y \cdot yx \in \{0, 2\} \Big\}$$

for all $x \in B_1$. In this case $\gamma_1(3) = \{0, 2, 4\}$ equipped with the operation $\cdot$ is a subalgebra of $\langle \delta(3);\ \cdot, 0\rangle$, where, as it can be easily seen, $\delta(3) = X$. This means $(\gamma_1, B_1)\tilde{<}(\delta, A)$.

Now, we put $B_2 = \{0, 3\} \subset A$ that is not disjoint with B_1 and we define a set-valued function $\gamma_2 : B_2 \longrightarrow \mathscr{P}(X)$ by

$$\gamma_2(x) = \Big\{ y \in X \mid xy \in \{0, 3\} \Big\}$$

for all $x \in B_2$. Then we get $(\gamma_2, B_2)\tilde{<}(\delta, A)$ since $\gamma_2(0) = X$ and $\gamma_2(3) = \{0, 3\}$ are subalgebras of $\mathfrak{X}$ due to the operation $\cdot$. Obviously, we have $\delta(0) = \delta(3) = X$.

But $(\gamma_1, B_1)\tilde{\cup}(\gamma_2, B_2)$ is not a soft BCC-subalgebra of (δ, A) because $\gamma_1(3) \cup \gamma_2(3) = \{0, 2, 3, 4\}$ and $\langle \{0, 2, 3, 4\};\ \cdot, 0\rangle$ is not a subalgebra of $\mathfrak{X}$, where, as mentioned above, $X = \delta(3)$.

So, the assumption about B_1 and B_2 in Theorem 3.11.28 is necessary. □

Theorem 3.11.30. *Let $\mathfrak{X}$ and $\mathfrak{Y}$ be BCC-algebras. Let $f : X \longrightarrow Y$ be a homomorphism of BCC-algebras and let (δ, A) and (γ, B) be soft BCC-algebras over $\mathfrak{X}$. Then*

$$(\delta, A)\tilde{<}(\gamma, B) \Longrightarrow (f(\delta), A)\tilde{<}(f(\gamma), B).$$

Proof. Let us assume that $(\delta, A)\tilde{<}(\gamma, B)$. Let $x \in A$. Then $A \subset B$ and $\langle \delta(x);\ \cdot, 0\rangle$ is a subalgebra of $\langle \gamma(x);\ \cdot, 0\rangle$. Since f is a homomorphism, $f(\delta)(x) = f(\delta(x))$ is a subalgebra of $f(\gamma(x)) = f(\gamma)(x)$, both due to $\cdot$. Therefore, $(f(\delta), A)\tilde{<}(f(\gamma), B)$. □

Theorem 3.11.31. *For every fuzzy subalgebra μ of a BCC-algebra $\mathfrak{X}$, there exists a soft BCC-algebra (δ, A) over $\mathfrak{X}$.*

Proof. Let μ be a fuzzy subalgebra of $\mathfrak{X}$. Then we define

$$U(\mu;t) = \{x \in X \mid \mu(x) \geq t\}.$$

Then $\langle U(\mu;t); \cdot, 0\rangle$ is a subalgebra of $\mathfrak{X}$ for all $t \in \mathrm{Im}(\mu)$. Let us put $A = \mathrm{Im}(\mu)$ and let $\delta : A \longrightarrow \mathscr{P}(X)$ be a set-valued function defined as $\delta(t) = U(\mu;t)$ for all $t \in A$. Then (δ, A) is a soft BCC-algebra over $\mathfrak{X}$. □

The converse of the above result is obvious. We formulate it in the next theorem.

Theorem 3.11.32. *For any fuzzy set μ in a BCC-algebra $\mathfrak{X}$ if a soft BCC-algebra (δ, A) over $\mathfrak{X}$ is given by $A = \mathrm{Im}(\mu)$ and $\delta(t) = U(\mu;t)$ for all $t \in A$, then μ is a fuzzy subalgebra of $\mathfrak{X}$.*

Let μ be fuzzy set in a BCC-algebra $\mathfrak{X}$ and let (δ, A) be a soft set over $\mathfrak{X}$ in which $A \subseteq [0,1]$ and $\delta : A \longrightarrow \mathscr{P}(X)$ is a set-valued function defined by

$$(\forall t \in A)(\delta(t) = \{x \in X \mid \mu(x) + t > 1\}) \tag{3.70}$$

Then there exists $t \in A$ such that $\langle \delta(t); \cdot, 0\rangle$ is not a subalgebra of $\mathfrak{X}$.

We show it in the next example.

Example 3.11.33. Let $\mathfrak{X} = \langle\{0,1,2,3,4\}; \cdot, 0\rangle$ be an improper BCC-algebra with the operation $\cdot$ defined by the following table (cf. [146], algebra B_{5-2-8} on page 323):

$\cdot$	0	1	2	3	4
0	0	0	0	0	0
1	1	0	1	0	0
2	2	2	0	0	2
3	3	2	1	0	2
4	4	1	4	1	0

Table 3.11.3

We define a fuzzy set $\mu : X \longrightarrow [0,1]$ as follows:

$$\mu(x) = \begin{cases} 0.6 & \text{if} \quad x = 0, \\ 0.3 & \text{if} \quad x = 1, \\ 0.1 & \text{if} \quad x = 2, \\ 0.7 & \text{if} \quad x = 3, \\ 0.8 & \text{if} \quad x = 4 \end{cases}$$

If we take the set $A = \{0.1, 0.3, 0.6, 0.7, 0.8\}$, then

$$\delta(0.6) = \{x \in X \mid \mu(x) + 0.6 > 1\} = \{0, 3, 4\}.$$

It is not a subalgebra of $\mathfrak{X}$ due the operation $\cdot$ because $3 \cdot 4 = 2 \notin \delta(0.6)$.

□

Theorem 3.11.34. *Let $\mathfrak{X}$ be a BCC-algebra and μ be a fuzzy set in $\mathfrak{X}$. Let (δ, A) be a soft set over $\mathfrak{X}$ in which $A = [0, 1]$ and $\delta : A \longrightarrow \mathscr{P}(X)$ is given by (3.70). Then the following statements are equivalent:*

(1) *μ is a fuzzy subalgebra of $\mathfrak{X}$,*

(2) *(δ, A) is a soft BCC-algebra over $\mathfrak{X}$.*

Proof. Let μ be a fuzzy subalgebra of $\mathfrak{X}$. Let $t \in A$ and let $x, y \in \delta(t)$. Then $\mu(x) + t > 1$ and $\mu(y) + t > 1$. From the fact that μ is a fuzzy subalgebra of $\mathfrak{X}$ it follows that $\mu(xy) \geq \min\{\mu(x), \mu(y)\}$ thus

$$\begin{aligned} \mu(xy) + t &\geq \min\{\mu(x), \mu(y)\} + t \\ &= \min\{\mu(x) + t, \mu(y) + t\} > 1. \end{aligned}$$

It is easy to see that $xy \in \delta(t)$ and hence $\langle \delta(t); \cdot, 0 \rangle$ is a subalgebra of $\mathfrak{X}$ for all $t \in A$. So, (δ, A) is a soft BCC-algebra over $\mathfrak{X}$.

Conversely, let us suppose that (δ, A) is a soft BCC-algebra over $\mathfrak{X}$. Let $x_0, y_0 \in X$ be such that $\mu(x_0y_0) < \min\{\mu(x_0), \mu(y_0)\}$.
Let us take $t_0 \in A$ such that

$$\mu(x_0y_0) + t_0 < 1 < \min\{\mu(x_0), \mu(y_0)\} + t_0.$$

Then

$$\mu(x_0) + t_0 > 1 \text{ and } \mu(y_0) + t_0 > 1.$$

This implies that $x_0, y_0 \in \delta(t_0)$. But $x_0y_0 \notin \delta(t_0)$, which is contrary to our assumption. Hence

$$\mu(xy) \geq \min\{\mu(x), \mu(y)\}$$

for all $x, y \in X$.
Therefore, μ is a fuzzy subalgebra of $\mathfrak{X}$. □

4

Ideal Theory of Weak BCC-Algebras

Ideals are one of the most important concepts in the theory of BCC-algebras. Various mathematicians studied them from different points of view (see Bibliography). Different ideals are defined and studied in different subclasses of the class of weak BCC-algebras. The results of these studies together form a single whole, the elements of which match and complement each other. Therefore, we can extract the theory of ideals from the general theory of weak BCC-algebras.

4.1 Closed Ideals

The next results and some results in the following chapters, concern the so-called closed BCC-ideals. These interesting subsets of base sets of weak BCC-algebras coincide in some subclasses of weak BCC-algebras with their subalgebras.

We start with the following definition

Definition 4.1.1. A BCC-ideal (BCK-ideal) A of a weak BCC-algebra $\mathfrak{X}$ is called *closed* if $\varphi(A) \subseteq A$, i.e., $\varphi(x) \in A$ for every $x \in A$.

In BCC-algebras all BCC-ideals are closed, but in weak BCC-algebras there are BCC-ideals which are not closed.

Example 4.1.2. Let $\mathfrak{X} \times \mathfrak{Z}$ be a weak BCC-algebra defined in Example 1.1.9. Then $S = A \times \mathbb{N}$, where A is a BCC-ideal of $\mathfrak{X}$ and $\mathbb{N}$ is the set of nonnegative integers, is a BCC-ideal of this weak BCC-algebra. This BCC-ideal is not closed since $\varphi(x, m) = (0x, 0 - m) \notin S$.
But, as it can be easily seen, the BCC-ideal $\{0\} \times \mathbb{Z}$ is closed. □

Corollary 4.1.3. *$B(0)$ is a closed BCC-ideal of each weak BCC-algebra.*

Proof. $B(0)$ is a BCC-ideal, by Theorem 2.3.7. It is closed since for every $x \in B(0)$ we have $0x = 0 \in B(0)$. □

Theorem 4.1.4. *A BCK-ideal A of a weak BCC-algebra $\mathfrak{X}$ is closed if and only if $\mathfrak{A}$ is a subalgebra of $\mathfrak{X}$.*

Proof. Assume that A is a BCK-ideal and $\mathfrak{A}$ is a subalgebra of $\mathfrak{X}$. Then $0 \in A$ and $0x \in A$ for every $x \in A$. So, $\varphi(x) \in A$, i.e., an BCK-ideal A is closed.

Conversely, if a BCK-ideal A is closed, then, by (BCC3), (BCC1), for any $x, y \in A$, we have

$$(xy \cdot 0y) \cdot x = (xy \cdot 0y) \cdot x0 = 0 \in A,$$

which, by the condition $(\mathrm{I2})_{BCK}$ from Definition 2.3.1, implies $xy \cdot 0y \in A$ and $xy \in A$, because $0y \in A$. This proves that $\mathfrak{A}$ is a subalgebra of $\mathfrak{X}$. □

Corollary 4.1.5. *If a BCC-ideal A of a weak BCC-algebra $\mathfrak{X}$ is a subalgebra of $\mathfrak{X}$, then it is closed.*

Proof. The proof follows from the first part of the proof of the above theorem. □

Corollary 4.1.6. *Let A be a nonempty subset of X. Then A is a closed BCC-ideal of group-like weak BCC-algebra $\mathfrak{X}$ if and only if $\mathfrak{A}$ is a normal subgroup of the corresponding group.*

Proof. Let A be a closed BCC-ideal of a group-like weak BCC-algebra $\mathfrak{X} = \langle X; \cdot, 0\rangle$. Then, according to Theorem 2.5.2(2) and (BCC1), for all $a \in A$ and $x \in X$ we have

$$((x \cdot 0a)a)x = ((x \cdot 0a)(0 \cdot 0a)) \cdot x0 = 0 \in A,$$

and consequently in the corresponding group $\langle X; *, 0\rangle$

$$x * a * x^{-1} = (x \cdot 0a)x \in A.$$

So, $\mathfrak{A}$ is a normal subgroup.

On the other hand, if $\langle A; *, 0\rangle$ is a normal subgroup, then for every $x \in X$ and $a \in A$ there is $c \in A$ such that $x * a = c * x$.

Let $a \in A$ and $(xa)y \in A$. Hence $(xa)y = x * a^{-1} * y^{-1} = b \in A$, whence $b * y = x * a^{-1} = d * x$ for some $d \in A$. Therefore, $xy = x * y^{-1} = d^{-1} * b \in A$, which proves that A is a BCC-ideal.

If $x \in A$, then $0x = 0 * x^{-1} = x^{-1} \in A$. So, A is closed. □

Theorem 4.1.7. *A BCC-ideal A of a weak group-like BCC-algebra $\mathfrak{X}$ is closed if and only if*

$$xy, yx \in A \quad \text{for all} \quad x, y \notin A. \tag{4.1}$$

Proof. Let A be a closed ideal of a weak group-like BCC-algebra $\mathfrak{X}$. If the condition (4.1) is not satisfied, then there are $a \in A$ and $x \notin A$ such that $ax \in A$. Then $0 \cdot ax \in A$. But, by Theorem 2.5.2(5), we obtain

$$0 \cdot ax = \varphi(ax) = xa.$$

Thus $xa \in A$, and consequently $x \in A$. Obtained contradiction proves that the condition (4.1) is satisfied.

Conversely, suppose that there exists an ideal A satisfying (4.1), which is not closed. Then $0x \notin A$ for some $x \in A$. But in this case, for $y = 0x$, by Theorem 2.5.2(2), we have

$$0y = 0 \cdot 0x = \varphi^2(x) = x \in A.$$

Thus $0y \in A$ for $y \notin A$.

This contradicts the assumption that A satisfies (4.1) and the ideal A is closed. □

Theorem 4.1.8. *A BCK-ideal of a weak group-like BCC-algebra $\mathfrak{X}$ is closed if and only if it is a subalgebra of $\mathfrak{X}$.*

Proof. If A is a closed BCK-ideal, then $0y \in A$ for every $y \in A$. Let $b = xy$, where x and y are arbitrary elements of A.

Multiplying the last equality by $0y \in A$ and using Theorem 2.5.7 we see that, by (BCC1),

$$b \cdot 0y = xy \cdot 0y = x \in A,$$

which implies $b \in A$. So, $xy \in A$ for all $x, y \in A$. Hence A is a subalgebra.

Conversely, let A be a subalgebra. Then $0 \in A$ and $0y \in A$ for every $y \in A$. If y and xy are in A, then also $xy \cdot 0y$ is in A. But, as in the previous part of this proof, $xy \cdot 0y = x$. So, $y \in A$ together with $xy \in A$ imply $x \in A$. This means that A is a BCK-ideal. □

As a simple consequence of the above theorem, we obtain:

Corollary 4.1.9. *A subset containing* 0 *is a closed* BCK*-ideal of a weak group-like BCC-algebra* $\mathfrak{X}$ *if and only if it is a subgroup of the corresponding group.*

Corollary 4.1.10. *If a weak BCC-algebra* $\mathfrak{X}$ *is group-like, then each commutative subgroup of the corresponding group* $\mathfrak{X}$ *is a closed* BCC*-ideal of* $\mathfrak{X}$.

Theorem 4.1.11. *A* BCC*-ideal* A *of a weak BCC-algebra* $\mathfrak{X}$ *is closed if and only if* $\varphi^2(A)$ *is a closed* BCC*-ideal of a weak* BCC*-algebra* $\varphi^2(\mathfrak{X})$.

Proof. Let A be a closed BCC-ideal of $\mathfrak{X}$. Then, by Lemma 2.2.2(2) and (2.7), $\varphi^2(A) \subseteq A$.

Let $xy \cdot z \in \varphi^2(A)$, where $y \in \varphi^2(A)$ and $x, z \in \varphi^2(X)$.

Then $xy \cdot z \in A$ and $y \in A$, consequently $xz \in A$.

Since $x = \varphi^2(a)$, $z = \varphi^2(b)$ for some $a, b \in X$, by Lemma 2.2.2(6) and Corollary 2.2.9, we obtain

$$xz = \varphi^2(a)\varphi^2(b) = \varphi^2(ab) \in \varphi^2(X) = I(X).$$

Hence, by Theorem 2.5.2(2), $xz \in \varphi^2(A)$. So, $\varphi^2(A)$ is a BCC-ideal of $\varphi^2(X)$. It is closed because

$$\varphi(\varphi^2(A)) = \varphi^2(\varphi(A)) \subseteq \varphi^2(A).$$

Conversely, let A be a BCC-ideal of $\mathfrak{X}$. If a BCC-ideal $\varphi^2(A)$ is closed then, by Lemma 2.2.2(5), Theorem 2.3.4 and Lemma 2.3.14, we have

$$\varphi(A) = \varphi^3(A) = \varphi(\varphi^2(A)) \subseteq \varphi^2(A) = A \cap I(X) \subseteq A,$$

which means that the BCC-ideal A is closed. □

Corollary 4.1.12. *A* BCK*-ideal* A *of a weak BCC-algebra* $\mathfrak{X}$ *is closed if and only if* $\varphi^2(A)$ *is a closed* BCK*-ideal of* $\varphi^2(X)$.

Proof. The proof follows from Theorem 2.3.4. □

Theorem 4.1.13. *Let* $\mathfrak{X}$ *be a weak BCC-algebra. A nonempty subset* A *of* X *such that* $\varphi(X) \subseteq A$ *is a closed* BCC*-ideal of* $\mathfrak{X}$ *if and only if*

(1) $ay \in A$ *for all* $a \in A$ *and* $y \in X$,

(2) $x(xa \cdot b) \in A$ *for* $a, b \in A$ *and* $x \in X$.

Proof. If A is a closed BCC-ideal of $\mathfrak{X}$ and $a, b \in A$, $x, y \in X$, then from $aa \cdot y = \varphi(y) \in A$ we conclude $ay \in A$. This proves that A satisfies (1).

Similarly, from $xa \cdot xa = 0 \in A$ we obtain $x \cdot xa \in A$. But, by (BCC1$'$),

$$xb \cdot (xa \cdot b) \leq x \cdot xa,$$

whence, by Lemma 2.3.13, we get

$$xb \cdot (xa \cdot b) \in A.$$

Consequently, from Definition 2.3.3, $x(xa \cdot b) \in A$. This proves that A satisfies (2).

Conversely, for any subset A of X satisfying (1) and (2) an element

$$b = x \cdot xc, \tag{4.2}$$

where $x \in X$, $c \in A$, belongs to A because

$$x \cdot xc = x(xc \cdot 0) \in A,$$

by (2). If

$$xc \cdot y = a \in A, \tag{4.3}$$

then, applying (BCC1$'$), we get

$$xy \cdot (xc \cdot y) \leq x \cdot xc.$$

Hence

$$(xy \cdot (xc \cdot y))(x \cdot xc) = 0. \tag{4.4}$$

By, respectively, (4.4), (4.3), (4.2) and (2) we have

$$xy = xy \cdot 0 = xy \cdot ((xy \cdot (xc \cdot y))(x \cdot xc)) = xy \cdot ((xy \cdot a)b) \in A,$$

which completes the proof. □

As a simple consequence, we obtain the following characterization of BCC-ideals:

Corollary 4.1.14. *A subset A of X containing 0 is a BCC-ideal of a BCC-algebra $\mathfrak{X}$ if and only if it satisfies conditions* (1) *and* (2) *of the above theorem.*

Now we will show some properties of maximal BCC-ideals of BCC-algebras. We begin with a definition.

Definition 4.1.15. A proper BCC-ideal (BCK-ideal) of a BCC-algebra $\mathfrak{X}$ is called *maximal* if it is not contained in any proper BCC-ideal (BCK-ideal).

Theorem 4.1.16. *Any BCC-algebra may be viewed as a maximal BCC-ideal of some BCC-algebra.*

Proof. If $\mathfrak{Y}$ is a BCC-algebra and $e \notin Y$, then $\mathfrak{X} = \langle Y \cup \{e\}; \star, 0\rangle$ with the operation $\star$ defined by

$$x \star y = \begin{cases} xy & for \quad x, y \in Y, \\ 0 & for \quad x \in X, y = e, \\ e & for \quad x = e, y \in Y \end{cases}$$

is a BCC-algebra and e is the greatest element of X.
If $\mathfrak{Y}$ is proper, then $\mathfrak{X}$ is proper, too. $\mathfrak{Y}$ is a subalgebra of $\mathfrak{X}$.

It is not difficult to verify that Y is a maximal BCC-ideal of $\mathfrak{X}$. Indeed, let $x \in Y$ and $e \star z = e \notin Y$. Then $(e \star x) \star z = e \notin Y$. From Theorem 2.3.10 it follows that Y is a BCC-ideal of $\mathfrak{X}$. It is not properly contained in any proper BCC-ideal.

So, Y is a maximal BCC-ideal of $\mathfrak{X}$. □

Corollary 4.1.17. *If the base set X of a BCC-algebra $\mathfrak{X}$ includes an element e such that $xy = e$ if and only if $x = e, y \neq e$, then $X \setminus \{e\}$ is the maximal BCC-ideal of $\mathfrak{X}$.*

Proof. Assume $xy \cdot z \neq e$ for some $y \neq e$. Then $xz \neq e$. If not, then $xz = e$, by the assumption, implies $x = e$, $z \neq e$. Hence $xy \cdot z = ey \cdot z = ez = e$, which is impossible. □

Corollary 4.1.18. *If in a BCC-algebra $\mathfrak{X}$, the base set X includes an element e such that $X \setminus \{e\}$ is a BCC-ideal (BCK-ideal), then $ey = e$ for all $y \neq e$ and e is the maximal element of $(X, \leq)$.*

Definition 4.1.19. A BCC-algebra without proper BCC-ideals (BCK-ideals) is called *BCC-simple* (*BCK-simple*).

Obviously any BCK-simple BCC-algebra is BCC-simple. The converse is not true.

A BCC-algebra $\mathfrak{X}$ given in Example 2.3.2 is BCC-simple, but it is not BCK-simple, because it has two maximal BCK-ideals $A = \{0, 1\}$ and $B = \{0, 2\}$.

As a conclusion, we have:

Corollary 4.1.20. *A BCC-simple BCC-algebra has only two regular congruences.*

Theorem 4.1.21. *Let A be a proper BCC-ideal of a BCC-algebra $\mathfrak{X}$. Then A is a maximal BCC-ideal of $\mathfrak{X}$ if and only if $\mathfrak{X}/\mathfrak{A}$ is a BCC-simple BCC-algebra.*

Proof. Let $\mathfrak{X}/\mathfrak{A}$ be a BCC-simple BCC-algebra. If A is not a maximal BCC-ideal, then there exists a proper BCC-ideal B such that $A \subset B \subset X$. Obviously B/A is properly contained in X/A. It is easy to see that B/A has at least two elements. Indeed, for all $x \in B \setminus A$ we have $x \in C_x$ and $C_x \in B/A$ and obviously $A = C_0 \in B/A$.

Moreover, if $C_y \in B/A$ and $C_{xy \cdot z} = C_x C_y \cdot C_z \in B/A$, then $y, xy \cdot z \in B$, which, by (I2) (Definition 2.3.3), implies $xz \in B$. Therefore, $C_x C_z \in B/A$. Thus B/A is a proper BCC-ideal of $\mathfrak{X}/\mathfrak{A}$, i.e., $\mathfrak{X}/\mathfrak{A}$ is not simple. This contradiction proves that A is a maximal BCC-ideal.

Conversely, if A is a maximal BCC-ideal of $\mathfrak{X}$ and $\mathfrak{X}/\mathfrak{A}$ is not BCC-simple, then there exists a proper BCC-ideal D of $\mathfrak{X}/\mathfrak{A}$. Then $f^{-1}(D)$, where $f(x) = C_x$ is the canonical homomorphism from $\mathfrak{X}$ onto $\mathfrak{X}/\mathfrak{A}$, is a proper BCC-ideal of $\mathfrak{X}$. Moreover $A = C_0 \subsetneq f^{-1}(D)$, which contradicts our hypothesis. Hence $\mathfrak{X}/\mathfrak{A}$ is simple.

The proof is completed. □

Finally, we present a very interesting result about connection between initial segments and maximal ideals of a BCC-algebra.

Theorem 4.1.22. *Any initial segment $[0, c]$ is isomorphic to a maximal ideal of some BCC-algebra.*

Proof. The proof is a consequence of Corollary 3.5.23. □

Let $\{X_i\}_{i \in I}$ be a nonempty family of BCC-chains (or BCC-algebras) such that $X_i \cap X_j = \{0\}$ for any distinct $i, j \in I$.

In $\{X_i\}_{i \in I}$ we define a new operation identifying it with an operation in any X_i, and putting $xy = x$ if $x \in X_i$ and $y \in X_j$ for $i, j \in I$

and $i \neq j$.

Direct computations show that the union $\bigcup_{i\in I} X_i$ is a BCC-algebra. It is called *the disjoint union of* $\{X_i\}_{i\in I}$ and it is denoted by $\sum_{i\in I} X_i$. The BCC-algebra X_i is called *a component of* $\sum_{i\in I} X_i$. It is easy to show that *any component* X_i *is a BCC-ideal of* $\sum_{i\in I} X_i$.

In a general case where $\{X_i\}_{i\in I}$ is an arbitrary nonempty family of BCC-chains (BCC-algebras), we consider $\{X_i \times \{i\}\}_{i\in I}$ and identify all $(0_i, i)$, where 0_i is a constant of X_i. By identifying each $x_i \in X_i$ with (x_i, i), the assumption of the above definition is satisfied.

Consequently, we can define the disjoint union of an arbitrary family of BCC-chains.

Let $\prod X_i$ be the direct product of a nonempty family of BCC-algebras X_i. For any fixed $i \in I$, let x_i be an element of $\prod X_i$ such that $x_i(j) = 0$ for any $i \neq j$ and $x_i(i) = x \in X_i$. Then $X_i^* = \{x_i \mid x \in X_i\}$ is a subalgebra of $\prod X_i$, which is naturally isomorphic to X_i.

If $i \neq j$, then $x_i x_j = x_i$ and $X_i^* \cap X_j^* = \{0\}$. Hence $\bigcup_{i\in I} X_i^* = \sum_{i\in I} X_i^*$ and in the consequence, $\bigcup_{i\in I} X_i$ is a subalgebra of $\prod X_i$.

Since $\bigcup X_i^*$ is isomorphic to $\sum X_i$, we obtain

Theorem 4.1.23. $\sum X_i$ *is a subdirect product of* X_i.

By the identification X_i with X_i^* we get

Corollary 4.1.24. $\sum X_i$ *is the minimal subalgebra of* $\prod X_i$ *containing all* X_i.

It is clear that if in the above construction, all X_i are BCK-algebras, then $\sum X_i$ and $\prod X_i$ are BCK-algebras.

And, if at least one BCC-algebra X_i is proper, then $\sum X_i$ and $\prod X_i$ are proper BCC-algebras.

Now we present results concerned with the horizontal BCC-ideals of weak BCC-algebras.

Definition 4.1.25. A *BCC*-ideal A of a weak BCC-algebra $\mathfrak{X}$ is called *horizontal* if it satisfies the following condition:

$$A \cap ker\,\varphi = \{0\}.$$

Theorem 4.1.26. *Any closed horizontal ideal of a weak BCC-algebra $\mathfrak{X}$ is contained in $I(X)$.*

Proof. Let A be a closed horizontal ideal of a weak BCC-algebra $\mathfrak{X}$. From Definition 4.1.25 it follows that there are no comparable elements in A. So all branches contain only the initial element. Hence $A \subseteq I(X)$. □

Theorem 4.1.27. *Let $\mathfrak{X}$ be a weak BCC-algebra. A subalgebra $\mathfrak{S}$ with $S \subseteq I(X)$ is a BCC-ideal, which is a horizontal ideal of $\mathfrak{X}$ if and only if the following conditions are satisfied:*

(1) $y \in S$ *implies* $x \cdot xy \in S$,

(2) $xb = ab$ *implies* $x = a$ *for* $a, b \in S$.

Proof. From Corollary 2.3.23 it follows that S is a *BCC*-ideal if and only if it satisfies the conditions (1) and (2).

It remains to prove that S is a horizontal ideal.

As it can be easily seen, $S \cap ker\,\varphi = B(0)$. But $S \subseteq I(X)$ and in this case $B(0) = \{0\}$.

The proof is completed. □

4.2 T-Ideals of T-Type Weak BCC-Algebras

In this section we provide the notion of T-type weak BCC-algebra as a generalizations of T-type BCI-algebra introduced in [158] and the notion of T-ideal in weak BCC-algebra. Since we base T-ideal on *BCC*-ideal, it is an object specific to weak BCC-algebras.

The generalized T-type BCI-algebras have been first researched for BZ-algebras (cf. [163]). The results given in this section are transferred to the environment of weak BCC-algebras.

Before we begin, a few notes on the notation used in this section. Let $\mathfrak{X}$ be a weak BCC-algebra. For $x \in X$ and $n \in \mathbb{N}$ we denote

$$\varphi(x^n) = ((\varphi(x)x)...)x,$$

in which x occurs n times.

Definition 4.2.1. A weak BCC-algebra $\mathfrak{X}$ is called *T-type* if all its elements are quasi-unite, i.e., the condition (3.62) is satisfied for all $x \in X$.

Let us observe that a weak BCC-algebra $\mathfrak{X}$ is T-type if and only if it satisfies the condition

$$\varphi(x^2) = 0. \tag{4.5}$$

Indeed, we have

$$\varphi(x^2) = 0 \Longleftrightarrow \varphi(x)x = 0 \Longleftrightarrow \varphi(x) \leq x.$$

Corollary 4.2.2. *In a T-type weak BCC-algebra $\mathfrak{X}$ for $a \in I(X)$ we have $0a = a$.*

Proof. Let $a \in I(X)$, then, by (3.62), $0a \leq a$. But $\mathfrak{I}(\mathfrak{X})$ is a group-like weak BCC-algebra with discrete order. Hence $0a = a$. □

Corollary 4.2.3. *In a T-type weak BCC-algebra $\mathfrak{X}$ for any $a \in I(X)$ we have $B(0a) = B(a)$.*

Example 4.2.4. Let $\mathfrak{X}$ be a weak BCC-algebra. We define the set:

$$Z_0 = \{x \in X \mid x0 = 0x\} = \{x \in X \mid x = \varphi(x)\}.$$

The set Z_0 is called the *centralizer* of 0. The $\mathfrak{Z}_0 = \langle Z_0; \cdot, 0\rangle$ is not a proper BCC-subalgebra, it is a group-like BCI-subalgebra. It is either trivial or a Boolean group (cf. [36]).

As it can be easily seen, $\mathfrak{Z}_0$ is a T-type weak BCC-algebra.

For $a \neq 0$ the set

$$Z_a = \{x \in X \mid xa = ax\}$$

is not closed under $\cdot$. Indeed, in the improper BCC-algebra $\mathfrak{X}$ defined by Table 3.3.5 we have $1 \cdot 1 \notin Z_1 = \{1, 2\}$. □

Example 4.2.5. Let $\mathfrak{X} = \langle\{0, 1, 2, 3, 4\}; \cdot, 0\rangle$ be an algebra with the operation $\cdot$ defined by Table 4.2.1.

Then $\mathfrak{X}$ is a weak BCC-algebra. In fact, it can be easily seen that in the operation $\cdot$ the axioms (BCC3) and (BCC5) are satisfied. Simple but long computations show that the algebra $\mathfrak{X}$ satisfies (BCC1), too.

It is easy to verify, by Definition 4.2.1, that the weak BCC-algebra $\mathfrak{X}$ is a T-type weak BCC-algebra.

$\cdot$	0	1	2	3	4
0	0	0	2	2	0
1	1	0	2	2	0
2	2	2	0	0	2
3	3	3	1	0	3
4	4	4	3	3	0

Table 4.2.1

□

Example 4.2.6. Let $X = \{0, 1, 2, 3, 4, 5, 6\}$ and let the operation $\cdot$ be defined in the following Cayley table:

$\cdot$	0	1	2	3	4	5	6
0	0	0	0	0	4	6	5
1	1	0	0	0	4	6	5
2	2	2	0	0	4	6	5
3	3	3	1	0	4	6	5
4	4	4	4	4	0	5	6
5	5	5	5	5	6	0	4
6	6	6	6	6	5	4	0

Table 4.2.2

It can be verified that $\mathfrak{X} = \langle X; \cdot, 0\rangle$ is a weak BCC-algebra. But since $(0 \cdot 5) \cdot 5 = 6 \cdot 5 = 4 \neq 0$, it is not a T-type weak BCC-algebra. □

From Theorem 3.8.13, we get the following:

Theorem 4.2.7. *Let $\mathfrak{X}$ be a weak BCC-algebra. Then $\mathfrak{X}$ is T-type if and only it satisfies*

$$xy \cdot z \leq x \cdot zy$$

for any $x, y, z \in X$.

The following result follows immediately from Corollary 3.8.14:

Corollary 4.2.8. *Let $\mathfrak{X}$ be a weak BCC-algebra. Then $\mathfrak{X}$ is T-type if and only if it satisfies*

$$xy \cdot y \leq x$$

for any $x, y \in X$.

Theorem 4.2.9. *Let $\mathfrak{X}$ be a weak BCC-algebra. Then $\mathfrak{X}$ is T-type if and only if it satisfies*

$$\varphi^2(x) = \varphi(x) \tag{4.6}$$

for any $x \in X$.

Proof. Let $\mathfrak{X}$ be a weak BCC-algebra. If $\mathfrak{X}$ satisfies (4.6), then for any $x \in X$ we have

$$\varphi(x)x = \varphi^2(x)x = 0,$$

by using Lemma 2.2.2 (2), which proves that (3.62) holds in $\mathfrak{X}$, and so $\mathfrak{X}$ is T-type.

Conversely, if $\mathfrak{X}$ is T-type, then for any $x \in X$, by (3.62), we get

$$\varphi^2(x)\varphi(x) = 0, \text{ i.e., } \varphi^2(x) \leq \varphi(x).$$

But we also have $\varphi(x) \leq x$, which by (1.5) implies $\varphi(x) \leq \varphi^2(x)$.

From the above and (BCC5) it follows that (4.6) holds.

The proof is completed. □

Later in the chapter, we consider the ideals of T-algebras.

Definition 4.2.10. Let $\mathfrak{X}$ be a weak BCC-algebra. A BCC-ideal A of $\mathfrak{X}$ is called a T-ideal of X if it satisfies

(T1) $0 \in A$,

(T2) $x \cdot yz \in A$ and $y \in A$ imply $xz \in A$ for any $x, y, z \in X$.

Lemma 4.2.11. *Let $\mathfrak{X}$ be a weak BCC-algebra. If A is a T-ideal of $\mathfrak{X}$, then the following condition holds:*

$$0x \cdot x \in A \tag{4.7}$$

for any $x \in X$.

Proof. Putting $y = 0, x = 0x, z = x$ in (T2) we get (4.7). □

In the next results, we give some characterizations of T-ideals.

Theorem 4.2.12. *Let $\mathfrak{X}$ be a T-type weak BCC-algebra. Then any BCC-ideal A of $\mathfrak{X}$ is also a T-ideal.*

Proof. Let A be a BCC-ideal of $\mathfrak{X}$ and let $x \cdot yz \in A$ and $y \in A$. From Theorem 4.2.7 it follows $xy \cdot z \leq x \cdot zy$. By (2.7) $xy \cdot z \in A$. Hence $xz \in A$, i.e., A is a T-ideal of $\mathfrak{X}$. □

Corollary 4.2.13. *Let $\mathfrak{X}$ be a weak BCC-algebra and let A be a T-ideal of $\mathfrak{X}$. Then we have*

$$x \cdot 0z \in A \Longrightarrow xz \in A \tag{4.8}$$

for any $x, z \in X$.

The above theorem is proven for T-algebras. But what conditions have to be satisfied for a BCC-ideal to be a T-ideal of any weak BCC-algebra ?

In the following results we provide the answers.

Theorem 4.2.14. *Let $\mathfrak{X}$ be a weak BCC-algebra and let A be a BCC-ideal of $\mathfrak{X}$. If A satisfies (4.8), then A is a T-ideal.*

Proof. Since A is a BCC-ideal, the condition (T1) is satisfied. We will prove that (T2) holds for A. Let $x \cdot yz \in A$ and $y \in A$ for any $x, y, z \in X$. Then $xz \in A$.

Indeed, by (BCC1$'$) we get

$$(x \cdot 0z)(yz \cdot 0z) \leq x \cdot yz,$$

which, from assumption and (2.7) implies

$$(x \cdot 0z)(yz \cdot 0z) \in A. \tag{4.9}$$

On the other hand,

$$yz \cdot 0z \leq y0 = y,$$

by (BCC1$'$), (BCC3). The fact that $y \in A$ and (2.7) implies

$$yz \cdot 0z \in A. \tag{4.10}$$

From Theorem 2.3.4 A is a BCK-ideal. Then from (4.9) and (4.10), it follows $x \cdot 0z \in A$, which by (4.8) gives that $xz \in A$.

The proof is completed. □

Corollary 4.2.13 and Theorem 4.2.14 imply the following corollary

Corollary 4.2.15. *Let $\mathfrak{X}$ be a weak BCC-algebra and A a BCC-ideal of $\mathfrak{X}$. Then A is a T-ideal if and only if A satisfies* (4.8).

Theorem 4.2.16. *Let $\mathfrak{X}$ be a weak BCC-algebra and A a BCC-ideal of $\mathfrak{X}$. Then A is a T-ideal if and only if A satisfies*

$$x \cdot 0z \in A \Longrightarrow x(0 \cdot 0z) \in A \tag{4.11}$$

for any $x, z \in X$.

Proof. Let $\mathfrak{X}$ be a weak BCC-algebra and A be a BCC-ideal of $\mathfrak{X}$. To show Theorem 4.2.16, it is sufficient to prove only that (4.11) and (4.8) are equivalent.

First, we show that (4.11) implies (4.8). Indeed, let $x \cdot 0z \in A$. Then for any $x, z \in X$ we have $0 \cdot 0z = \varphi^2(z) \leq z$, by Lemma 2.2.2(2), and thus, by (1.5), $xz \leq x(0 \cdot 0z)$. Whence from assumption and (2.7) $xz \in A$. Thus (4.8) holds.

Now we show that (4.8) implies (4.11). Let $x \cdot 0z \in A$ for any $x, z \in X$. We get $x \cdot 0z = x\varphi(z) = x\varphi^3(z) = x(0(0 \cdot 0z))$ by Lemma 2.2.2(5), and so $x(0(0 \cdot 0z)) \in A$. By (4.8) we have $x(0 \cdot 0z) \in A$, i.e., (4.8) implies (4.11).

That completes the proof of theorem. □

In Lemma 4.2.11 we have shown that in a weak BCC-algebra (4.7) is the necessary condition for a BCC-ideal to be a T-ideal. We show that the condition is also sufficient.

Theorem 4.2.17. *Let $\mathfrak{X}$ be a weak BCC-algebra and let A be a BCC-ideal of $\mathfrak{X}$. If A satisfies* (4.7), *then A is a T-ideal.*

Proof. Let $x \cdot 0z \in A$ for any $x, z \in X$. We show $xz \in A$.

Indeed, by (BCC1$'$) $xz \cdot (0z \cdot z) \leq x \cdot 0z$ and, by (2.7), $xz \cdot (0z \cdot z) \in A$. The assumption and Theorem 2.3.4 imply $xz \in A$. So, the condition (4.8) is satisfied. Finally, from Theorem 4.2.14 we know that A is a T-ideal of $\mathfrak{X}$. □

Corollary 4.2.18. *Let $\mathfrak{X}$ be a weak BCC-algebra and let A be a BCC-ideal of $\mathfrak{X}$. Then A is a T-ideal if and only if A satisfies* (4.7).

Corollary 4.2.19. *Let $\mathfrak{X}$ be a weak BCC-algebra and let A be a T-ideal of $\mathfrak{X}$. If B is a BCC-ideal of $\mathfrak{X}$ and $A \subseteq B$, then B is also a T-ideal.*

In other words,

Corollary 4.2.20. *Any ideal containing T-ideal is a T-ideal.*

The next corollary follows from the corollary above and from the fact that $\{0\}$ is the zero element in the lattice of ideals of a weak BCC-algebra $\mathfrak{X}$, and therefore it is included in every BCC-ideal of $\mathfrak{X}$.

Corollary 4.2.21. *Let $\mathfrak{X}$ be a weak BCC-algebra. If $\{0\}$ is a T-ideal of $\mathfrak{X}$, then every BCC-ideal of $\mathfrak{X}$ is also a T-ideal.*

Corollary 4.2.22. *Let $\mathfrak{X}$ be a weak BCC-algebra. Then $\mathfrak{X}$ is T-type if and only if every BCC-ideal of $\mathfrak{X}$ is a T-ideal.*

Corollary 4.2.23. *A weak BCC-algebra is T-type if and only if $\{0\}$ is its T-ideal.*

Proof. The proof follows immediately from Corollary 4.2.21 and Corollary 4.2.22. □

Theorem 4.2.24. *Let $\mathfrak{X}$ be a weak BCC-algebra. If A is a BCC-ideal of $\mathfrak{X}$, then the quotient algebra $\mathfrak{X}/\mathfrak{A} = \langle X/A; \cdot, C_0\rangle$ is T-type if and only if A is a T-ideal.*

Proof. Suppose A is a T-ideal of $\mathfrak{X}$. Then for any $x \in X$ we have $0x \cdot x \in A$, by Corollary 4.2.18. So, $(0x \cdot x)0 \in A$.

On the other hand,

$$\begin{aligned}
0(0x \cdot x) &= \varphi(0x \cdot x) \\
&= \varphi^2(x \cdot 0x) && \text{by Lemma 2.2.2(8)} \\
&= \varphi^2(x)\varphi^2(0x) && \text{by Lemma 2.2.2(6)} \\
&= \varphi^2(x)\varphi^3(x) \\
&= \varphi^2(x)\varphi(x) && \text{by Lemma 2.2.2(5)} \\
&= (0 \cdot 0x) \cdot 0x \in A && \text{by (4.7).}
\end{aligned}$$

We have shown that both elements $(0x \cdot x)0$ and $0(0x \cdot x)$ are in A, which means that $(0x \cdot x, 0) \in \theta$, where θ is a congruence relation defined by (2.8). That is, for its equivalence classes we have $C_{0x \cdot x} = C_0$, i.e., $C_0 C_x \cdot C_x = C_0$. Thus (3.62) is satisfied and $\mathfrak{X}/\mathfrak{A}$ is a T-type weak BCC-algebra.

Conversely, if $\mathfrak{X}/\mathfrak{A}$ is a T-type weak BCC-algebra, then for any $x \in X$, we have $C_0C_x \cdot C_x = C_0$, that is $C_{0x\cdot x} = C_0$, and therefore, $0x \cdot x = (0x \cdot x)0 \in A$. Hence the condition (4.7) is satisfied. From Theorem 4.2.17 we know that A is a T-ideal of $\mathfrak{X}/\mathfrak{A}$. $\square$

Theorem 4.2.25. *A BCC-ideal A of a weak BCC-algebra $\mathfrak{X}$ is a T-ideal if and only if one of the following conditions is satisfied:*

(1) *$x\varphi(z) \in A$ implies $xz \in A$ for all $x, z \in X$,*

(2) *$x\varphi(z) \in A$ implies $x\varphi^2(z) \in A$ for all $x, z \in X$,*

(3) *$\varphi(x)x \in A$ for all $x \in X$.*

Proof. Putting $y = 0$ in the condition (T2) of the definition of T-ideal, we obtain (1).

To prove the converse, assume that (1) holds and $x \cdot yz \in A$ for some $y \in A$. Since

$$yz \cdot 0z \leq y \text{ and } x\varphi(z) \cdot (yz \cdot \varphi(z)) \leq x \cdot yz,$$

by BCC1$'$, Lemma 2.3.13 implies

$$yz \cdot \varphi(z) \in A \text{ and } (x\varphi(z))(yz \cdot \varphi(z)) \in A,$$

whence $x\varphi(z) \in A$. So, $xz \in A$.

(2) is a consequence of (1) and Lemma 2.2.2(5).

We have

$$x\varphi(z) = x\varphi^3(z) = x\varphi(\varphi^2(z)) \in A.$$

Then (1) implies $x\varphi^2(z)$.

Conversely, let us assume (2). We have $x\varphi^2(z) \in A$. From Lemma 2.2.2(2) we have $\varphi^2(z) \leq z$ and from (1.5) $xz \leq x\varphi^2(z)$, which by Lemma 2.3.13 gives $xz \in A$.

To prove (3), observe that $0x \cdot 0x = 0 \in A$ for all $x \in X$. Whence, according to the definition of a T-ideal, we conclude $0x \cdot x \in A$. This proves (3).

Conversely, if (3) holds and $x\varphi(z) \in A$, then, by (BCC1$'$), we have

$$xz \cdot \varphi(z)z \leq x\varphi(z),$$

which, by Lemma 2.3.13, gives $xz \cdot \varphi(z)z \in A$. So, $xz \in A$, i.e., (3) implies (1). Therefore, an ideal A satisfying (3) is a T-ideal. $\square$

In this part of the section, we will generalize the concept of T-algebras and some related objects. The following results were obtained for BZ-algebras (cf. [97]), but here we provide them for weak BCC-algebras.

So, we can start with the definition of the general form of T-type algebras.

Definition 4.2.26. A weak BCC-algebra $\mathfrak{X}$ is said to be $T(m)$-type, where $m \in \mathbb{N}$, if it satisfies

$$\varphi(x^{m+1}) = 0 \tag{4.12}$$

for all $x \in X$.

It is easy to see that a T-type weak BCC-algebra is a $T(m)$-type for $m = 1$. The converse is not true. We will show it in the following example:

Example 4.2.27. Let us consider the weak BCC-algebra defined in Example 4.2.6. It is easy to observe that if $x \in \{0, 1, 2, 3\}$, then $\varphi(x^m) = 0$, for every $m \in \mathbb{N}$. For $x = 4$ observe that $\varphi(4^2) = 0$. Hence $\varphi(4^4) = 0$. For $x = 5$ we have

$$\varphi(5^4) = ((\varphi(5)5)5)5 = ((6 \cdot 5)5)5 = (4 \cdot 5)5 = 5 \cdot 5 = 0.$$

Similarly, for $x = 6$ we get $\varphi(6^4) = 0$. Hence $\mathfrak{X}$ is a $T(3)$-type.

But from Example 4.2.6 it follows that $\mathfrak{X}$ is not a T-algebra. So, there is a $T(m)$-type weak BCC-algebra, which is not T-type. □

Theorem 4.2.28. *Every T-type weak BCC-algebra $\mathfrak{X}$ is $T(m)$-type for every odd number m.*

Proof. If $m \in \mathbb{N}$ is an odd number, then $m + 1$ is even and we have

$$\begin{aligned}
\varphi(x^{m+1}) &= (\varphi(x)x)x^{m+1-2} \\
&= 0x^{m+1-2} \\
&= (\varphi(x)x)x^{m+1-4} \\
&= \dots \\
&= \varphi(x)x \\
&= 0,
\end{aligned}$$

because from assumption $\mathfrak{X}$ is T-type. □

Theorem 4.2.29. *Let $m \in \mathbb{N}$. Then a weak BCC-algebra $\mathfrak{X}$ is $T(m)$-type if and only if it satisfies the following inequality:*

$$xy^m \cdot z \leq x \cdot zy \tag{4.13}$$

for all $x, y, z \in X$.

Proof. Let $\mathfrak{X}$ be $T(m)$-type. Then

$$\begin{aligned}
zy^{m+1} &= (zy^m \cdot y)0 \\
&= (zy^m \cdot y) \cdot 0y^{m+1} && \text{by Definition 4.2.26} \\
&= (zy^m \cdot y)(0y^m \cdot y) \\
&\leq zy^m \cdot 0y^m && \text{by BCC1}' \\
&= (zy^{m-1} \cdot y)(0y^{m-1} \cdot y) \\
&\leq zy^{m-1} \cdot 0y^{m-1} \\
&\leq \ldots \leq z0 = z,
\end{aligned}$$

i.e., $zy^{m+1} \leq z$.

Now, using (1.5) we get

$$\begin{aligned}
xy^m \cdot z \leq\ & xy^m \cdot zy^{m+1} \\
&= (xy^{m-1} \cdot y)(zy^m \cdot y) \\
&\leq xy^{m-1} \cdot zy^m && \text{by (BCC1}'\text{)} \\
&\leq \ldots \leq x \cdot zy.
\end{aligned}$$

So, the inequality (4.13) is satisfied.

Suppose that $\mathfrak{X}$ satisfies the inequality (4.13). Then for $x = 0$ and $z = y$ in (4.13) we have

$$\varphi(x^{m+1}) = 0y^{m+1} = 0y^m \cdot y \leq 0 \cdot yy = 0,$$

i.e., $\varphi(x^{m+1}) = 0$. Therefore, $\mathfrak{X}$ is $T(m)$-type.

This completes the proof. □

Theorem 4.2.30. *Let $m \in \mathbb{N}$. Then $\mathfrak{X}$ is $T(m)$-type if and only if it satisfies the following inequality:*

$$xy^{m+1} \leq x \tag{4.14}$$

for all $x, y \in X$.

Proof. Let $\mathfrak{X}$ be $T(m)$-type. By putting $z = y$ in (4.13) and, by (BCC2) and (BCC3), we get

$$xy^{m+1} = xy^m \cdot y \leq x \cdot yy = x0 = x.$$

So, the condition (4.14) is satisfied.

Conversely, let us suppose that $\mathfrak{X}$ satisfies (4.14). By putting $x = 0$ in (4.14) we have $0y^{m+1} \leq 0$, which implies $0y^{m+1} = 0$. Hence $\mathfrak{X}$ is $T(m)$-type. □

Before we define and examine $T(m,n)$-ideals, we give the notation of multiplication $m(x) \cdot y$, for a natural m.

Let $\mathfrak{X}$ be a weak BCC-algebra. For any elements $x, y \in X$ we denote a equation

$$m(x) \cdot y = x(...(x \cdot xy)...),$$

in which x occurs m times.

Definition 4.2.31. Let $\mathfrak{X}$ be a weak BCC-algebra. A subset A of X is called a *T(m,n)-ideal* of $\mathfrak{X}$ if there are $m, n \in \mathbb{N}$ such that

(T1) $0 \in A$,

(T2) $m(x) \cdot yz \in A, y \in A \Longrightarrow n(x) \cdot z \in A$.

Example 4.2.32. Let $\mathfrak{X} = \langle\{0,1,2,3,4\}; \cdot, 0\rangle$ be a weak BCC-algebra with the operation $\cdot$ defined by Table 4.2.3.

$\cdot$	0	1	2	3	4
0	0	0	2	2	0
1	1	0	3	2	0
2	2	2	0	0	2
3	3	2	1	0	2
4	4	4	3	3	0

Table 4.2.3

It is not difficult to check that $A = \{0, 1\}$ is a $T(1,3)$-ideal of $\mathfrak{X}$. □

Example 4.2.33. Let $\mathfrak{X} = \langle\{0, a, b, c, d, e, f, 1\}; \cdot, 0\rangle$ be a weak BCC-algebra with the operation $\cdot$ defined by the following table:

$\cdot$	0	a	b	c	d	e	f	1
0	0	0	0	0	0	0	0	0
a	a	0	e	e	c	c	a	0
b	b	f	0	f	c	b	c	0
c	c	0	0	0	c	c	c	0
d	d	f	e	d	0	f	e	0
e	e	0	e	e	0	0	e	0
f	f	f	0	f	0	f	0	0
1	1	f	e	d	c	b	a	0

Table 4.2.4

The subset $A = \{0, c, f\}$ is a $T(3, 2)$-ideal and a $T(5, 4)$-ideal of $\mathfrak{X}$. A is not a $T(4, 2)$-ideal, because

$$\begin{aligned} 4(d) \cdot (0 \cdot a) &= d(d(d(d \cdot (0 \cdot a)))) \\ &= d(d(d(d \cdot 0))) \\ &= d(d(d \cdot d)) \\ &= d(d \cdot 0) \\ &= d \cdot d \\ &= 0 \in A. \end{aligned}$$

But $2(d) \cdot a = d(d \cdot a) = d \cdot f = e \notin A$.

Similarly, A is not a $T(4, 3)$-ideal of $\mathfrak{X}$, too. We have namely

$$4(d) \cdot (0 \cdot b) = 0 \in A, \text{ but } 3(d) \cdot b = e \notin A.$$

□

Theorem 4.2.34. *Let m be an odd number. If A is a $T(m, 1)$-ideal of $\mathfrak{X}$, then $\varphi(x^2) \in A$ for all $x \in X$.*

Proof. Putting $y = 0, x = 0x, z = x$ in $(\mathrm{I2})_T$ we have

$$m(\varphi(x)) \cdot \varphi(x) = 0 \in A$$

because m is odd. It follows from $(\mathrm{I2})_T$ that $\varphi(x^2) = 0x \cdot x \in A$. □

The following example shows that a $BCC(m,n)$-ideal may not be a $T(m,n)$-ideal for $m,n \in \mathbb{N}$.

Example 4.2.35. In Example 2.3.25, $A = \{0,4\}$ is a $BCC(2,4)$-ideal of $\mathfrak{X}$, but it is not a $T(2,4)$-ideal of $\mathfrak{X}$. Indeed, we have

$$3 \cdot (3 \cdot (0 \cdot 1)) = 0 \in A \text{ but } 3 \cdot (3 \cdot (3 \cdot (3 \cdot 1))) = 1 \notin A.$$

□

The converse statement is true. In the next example, we show that a $T(m,n)$-ideal may not be a $BCC(m,n)$-ideal for $m,n \in \mathbb{N} \setminus \{1\}$.

Example 4.2.36. Let $\mathfrak{X}$ be a weak BCC-algebra in Example 4.2.33. The subset $A = \{0,c,f\}$ of X is a $T(3,2)$-ideal and a $T(5,4)$-ideal of $\mathfrak{X}$. But it is not a $BCC(3,2)$-ideal. Indeed,

$$\begin{aligned}(b \cdot c) \cdot e^3 &= (((b \cdot c) \cdot e) \cdot e) \cdot e \\ &= ((f \cdot e) \cdot e) \cdot e \\ &= (f \cdot e) \cdot e \\ &= f \cdot e \\ &= f \in A, c \in A.\end{aligned}$$

But $b \cdot e^2 = (b \cdot e) \cdot e = b \cdot e = b \notin A$.
A is not a $BCC(5,4)$-ideal, too. Indeed,

$$\begin{aligned}(b \cdot c) \cdot e^5 &= (((((b \cdot c) \cdot e) \cdot e) \cdot e) \cdot e) \cdot e \\ &= ((((f \cdot e) \cdot e) \cdot e) \cdot e) \cdot e \\ &= (((f \cdot e) \cdot e) \cdot e) \cdot e \\ &= ((f \cdot e) \cdot e) \cdot e \\ &= (f \cdot e) \cdot e \\ &= f \cdot e \\ &= f \in A, c \in A.\end{aligned}$$

But

$$b \cdot e^4 = (((b \cdot e) \cdot e) \cdot e) \cdot e = b \notin A.$$

□

Theorem 4.2.37. *Let A be a $T(m,n)$-ideal of $\mathfrak{X}$. Then for any $x, y, z \in X$*

$$m(x) \cdot 0z \in A \Longrightarrow n(x) \cdot z \in A.$$

Proof. By putting $y = 0$ in $(\mathrm{I}2)_T$, we get the desired result. □

It follows from Theorem 4.2.12 that in a T-type weak BCC-algebra every BCC-ideal is a T-ideal. In the following example, we show that in a T-type weak BCC-algebra, a $BCC(m,n)$-ideal may not be a $T(m,n)$-ideal for some $m, n \in \mathbb{N} \setminus \{1\}$.

Example 4.2.38. Let $\mathfrak{X}$ be the weak BCC-algebra in Example 2.3.25. It is not difficult to show that $\mathfrak{X}$ is $T(2)$-type and $A = \{0, 4\}$ is a $BCC(2,4)$-ideal, which is not a $T(2,4)$-ideal of $\mathfrak{X}$.

Indeed, putting $x = 1, y = 4, z = 1$ we have

$$1 \cdot (1 \cdot (4 \cdot 1)) = 1 \cdot (1 \cdot 4) = 1 \cdot 1 = 0 \in A, 4 \in A.$$

On the other hand,

$$1 \cdot (1 \cdot (1 \cdot (1 \cdot 1))) = 1 \notin A.$$

□

4.3 Anti-Grouped Ideals

In this section we describe the so-called anti-grouped BCC-ideals in weak BCC-algebras. The class of these BCC-ideals is specific for group-like weak BCC-algebras (cf. [62] and [143]).

The results provided below are proven for BCC-ideals, but, by the Theorem 2.3.4, they are true also for BCK-ideals.

We begin with the definition of an anti-grouped BCC-ideal.

Definition 4.3.1. A BCC-ideal A of a weak BCC-algebra $\mathfrak{X}$ is called *anti-grouped* if it satisfies the following condition:

$$\varphi^2(x) \in A \Longrightarrow x \in A \tag{4.15}$$

for $x \in X$.

Theorem 4.3.2. *Any anti-grouped BCC-ideal A of a weak BCC-algebra $\mathfrak{X}$ containing an element x contains also the whole branch containing this element.*

Proof. Let $z \in A \cap B(a)$, where A is an anti-grouped BCC-ideal of $\mathfrak{X}$ and $B(a)$ is a branch initiated by $a \in I(X)$. Then $a \leq z$. Thus, by Lemma 2.3.13, $a \in A$.

Moreover, for an arbitrary element $x \in B(a)$, from $a \leq x$, by Lemma 2.2.2(7), we obtain $\varphi(x) = \varphi(a)$. Hence, $\varphi^2(x) = \varphi^2(a) = a \in A$. Therefore, $\varphi^2(x) \in A$ and from Definition 4.3.1 $x \in A$. Since x is an arbitrary element of $B(a)$, the whole $B(a) \subseteq A$. □

Theorem 4.3.3. *In weak group-like BCC-algebras all BCC-ideals are anti-grouped.*

Proof. The proof follows directly from Theorem 2.2.8(3). □

Corollary 4.3.4. *Let $\mathfrak{X}$ be a weak BCC-algebra. Then $ker\,\varphi$ is a closed anti-grouped BCC-ideal of $\mathfrak{X}$.*

Proof. From Theorem 2.2.11, it follows that $ker\,\varphi = B(0)$ and, by Theorem 2.3.7, $B(0)$ is a BCC-ideal. So, $ker\,\varphi$ is a BCC-ideal.

Now we show that the BCC-ideal $ker\,\varphi$ is anti-grouped.
Let $\varphi^2(x) \in ker\,\varphi$ for any $x \in X$. Then

$$0 = 0\varphi^2(x) = \varphi^3(x) = \varphi(x) = 0x.$$

Hence, $x \in ker\,\varphi$ and the BCC-ideal $ker\,\varphi$ is anti-grouped.

According to Corollary 4.1.3, $B(0)$ is closed BCC-ideal, so is $ker\,\varphi$, too.

The proof is completed. □

Theorem 4.3.5. *For a BCC-ideal A of a weak BCC-algebra $\mathfrak{X}$, the following conditions are equivalent:*

(1) *A is anti-grouped,*

(2) *$x \leq y$, $x \in A$ imply $y \in A$,*

(3) *$xz \cdot yz \in A$, $y \in A$ imply $x \in A$,*

(4) *$xz \cdot \varphi(z) \in A$ implies $x \in A$.*

Proof. (1) $\Longrightarrow$ (2) Let $x \le y$ and $x \in A$. Then

$$\varphi^2(yx) = \varphi(xy) = \varphi(0) = 0 \in A,$$

by Lemma 2.2.2(8). This, according to the definition of an anti-grouped BCC-ideal, implies $yx \in A$. From this, by Theorem 2.3.4, we get $y \in A$. This means that (2) holds.

(2) $\Longrightarrow$ (3) Since $xz \cdot yz \in A$ and $xz \cdot yz \le xy$, by assumption we have $xy \in A$, whence, by Theorem 2.3.4, we obtain $x \in A$. This means that (3) holds.

(3) $\Longrightarrow$ (4) Putting $y = 0$ in (3), we get (4).

(4) $\Longrightarrow$ (1) Putting $z = x$ in (4), we get (1). □

Theorem 4.3.6. *A BCC-ideal A of a weak BCC-algebra $\mathfrak{X}$ is closed and anti-grouped if and only if for every $x \in X$ both x and $\varphi(x)$ belong or do not belong to A.*

Proof. Let A be a closed anti-grouped BCC-ideal of a weak BCC-algebra $\mathfrak{X}$. Then $x \in A$ implies $\varphi(x) \in A$. Similarly, $\varphi(x) \in A$ implies $\varphi^2(x) \in A$. But A is anti-grouped, whence $x \in A$. Therefore, both x and $\varphi(x)$ belong or do not belong to A.

Conversely, any BCC-ideal A of $\mathfrak{X}$ with the property that both x and $\varphi(x)$ belong or do not belong to A, is obviously closed.

Moreover, if $\varphi^2(x) = \varphi(\varphi(x)) \in A$, then also $\varphi(x) \in A$, whence $x \in A$. So, a BCC-ideal A of $\mathfrak{X}$ is anti-grouped. □

Theorem 4.3.7. *For a BCC-ideal A of a weak BCC-algebra $\mathfrak{X}$, the following statements are equivalent:*

(1) *A is an anti-grouped BCC-ideal,*

(2) *$B(\varphi^2(x)) \subseteq A$ for $x \in A$,*

(3) *$A = \bigcup\{B(a) \mid a \in \varphi^2(A)\}$.*

Proof. (1) $\Longrightarrow$ (2) Let $x \in A$. Then, according to the Lemma 2.2.2(2), $\varphi^2(x) \le x$ and so $x \in B(\varphi^2(x))$. From Theorem 4.3.2, it follows that $B(\varphi^2(x)) \subseteq A$.

(2) $\Longrightarrow$ (3) The proof is obvious.

(3) $\Longrightarrow$ (1) If $\varphi^2(x) \in A$, then $a \le \varphi^2(x) \le x$ for some $a \in \varphi^2(A)$. Hence $a \le x$, i.e., $x \in B(a) \subseteq A$. So, a BCC-ideal A is anti-grouped. □

Definition 4.3.8. A BCC-ideal A of a weak BCC-algebra $\mathfrak{X}$ is called *strong* if it satisfies the following condition:

$$ax \notin A \text{ for all } a \in A \text{ and } x \notin A.$$

Theorem 4.3.9. *$B(0)$ is the smallest strong BCC-ideal of X.*

Proof. From Theorem 2.3.7 it follows that $B(0)$ is a BCC-ideal. It is strong. Indeed, let $a \in B(0)$ and $x \notin B(0)$. Hence $x \in B(b)$, for $b \in I(X), b \neq 0$. As a consequence,

$$ax \in B(0)B(b) = B(0b).$$

Since $I(X)$ is group-like weak BCC-algebra, $0b \neq 0$ implies $B(0b) \neq B(0)$. So, $B(0)$ is strong.

To prove that $B(0)$ is the smallest strong BCC-ideal of $\mathfrak{X}$, let us consider an arbitrary strong BCC-ideal $A \subset X$. Obviously, $0 \in A \cap B(0)$.

If $B(0)$ is not contained in A, then there exists x_0 which belongs to $B(0) \setminus A$. For this x_0 we have $0x_0 = 0 \in A$, which means that A is not strong. So, $B(0) \subseteq A$. □

Theorem 4.3.10. *A BCC-ideal A of a group-like weak BCC-algebra $\mathfrak{X}$ is strong if and only if it is closed.*

Proof. Let A be a closed BCC-ideal of a group-like weak BCC-algebra $\mathfrak{X}$. If it is not strong, then there are $a \in A$ and $x \notin A$ such that $z = ax \in A$. Then $0z \in A$, which, by Theorem 2.5.2(5), implies $xa = 0 \cdot ax \in A$, and consequently $x \in A$.

Obtained contradiction proves that the BCC-ideal A is strong.

Conversely, suppose that there exists a strong BCC-ideal A, which is not closed. Then $0x \notin A$ for some $x \in A$. But in this case for $y = 0x$ we have $0y = 0 \cdot 0x = x \in A$. Thus $0y \in A$ for $y \notin A$. This contradicts the assumption that A is strong.

So, the BCC-ideal A of $\mathfrak{X}$ is closed. □

Definition 4.3.11. A BCC-ideal A of a weak BCC-algebra $\mathfrak{X}$ is called *regular* if it satisfies the following condition:

$$xy \in A,\ x \in A \text{ imply } y \in A.$$

Theorem 4.3.12. *A BCC-ideal A of a weak BCC-algebra $\mathfrak{X}$ is regular if and only if it is closed and anti-grouped.*

Proof. Let A be a regular ideal of a weak BCC-algebra $\mathfrak{X}$. If $\varphi^2(x) \in A$, then, by Lemma 2.2.2(2), we have $\varphi^2(x)x = 0 \in A$, whence, by the regularity of A, we get $x \in A$. This proves that A is anti-grouped. Moreover, for $x \in A$, by Lemma 2.2.2(2) and Lemma 2.3.13, we have $\varphi^2(x) \in A$. But $0\varphi(x) = \varphi^2(x) \in A$, according to the regularity, gives $\varphi(x) \in A$. Thus A is closed.

On the other hand, for any closed and anti-grouped ideal A of a weak BCC-algebra $\mathfrak{X}$ for $xy \in A$, $x \in A$ it has to be $\varphi(xy), \varphi(x) \in A$. Whence, by Lemma 2.2.2(4) and Theorem 4.1.4, we obtain $\varphi^2(y) = \varphi(xy)\varphi(x) \in A$. So, $y \in A$, because A is anti-grouped. This proves that A is regular. □

As a summary of the above results, we have the following corollary:

Corollary 4.3.13. *For BCC-ideal A of a weak BCC-algebra $\mathfrak{X}$, the following statements are equivalent:*

(1) *A is a strong ideal,*

(2) *A is a regular ideal,*

(3) *A is a closed anti-grouped ideal.*

For any BCC-ideal A of a weak BCC-algebra $\mathfrak{X}$, we define the set

$$Z(A) = \{x \in X \mid \varphi^2(x) \in A\}.$$

From Lemma 2.2.2(2) it follows that $A \subset Z(A)$. Moreover, $A = Z(A)$ if and only if A is an anti-grouped BCC-ideal of $\mathfrak{X}$. In a weak BCC-algebra $\mathfrak{X}$ defined by Table 1.1.3, $A = \{0, 2\}$ is an example of a proper BCK-ideal for which $Z(A) = X$.

Corollary 4.3.14. *$Z(A)$ is a proper BCC-ideal of $\mathfrak{X}$ if and only if there exists at least one initial element $a \in I(X)$ such that $a \notin Z(A)$.*

Proof. The proof follows directly from Theorem 2.5.2(2). □

Theorem 4.3.15. *$Z(A)$ is an anti-grouped BCC-ideal of a weak BCC-algebra $\mathfrak{X}$.*

Proof. Obviously, $0 \in Z(A)$. Let y and $xy \cdot z$ belong to $Z(A)$. Then $\varphi^2(y) \in A$ and

$$\varphi^2(x)\varphi^2(y) \cdot \varphi^2(z) = \varphi^2(xy \cdot z) \in A,$$

because φ^2 is an endomorphism of $\mathfrak{X}$ (Corollary 2.2.3).

Thus $\varphi^2(xz) = \varphi^2(x)\varphi^2(z) \in A$, which implies $xz \in Z(A)$. So, $Z(A)$ is a BCC-ideal of $\mathfrak{X}$.

Assume that $\varphi^2(x) \in Z(A)$ for some $x \in X$. Then, by Lemma 2.2.2(5),

$$\varphi^2(x) = \varphi(\varphi(x)) = \varphi(\varphi^3(x)) = \varphi^4(x) \in A,$$

which implies $x \in Z(A)$. So, $Z(A)$ is anti-grouped. □

Corollary 4.3.16. *In a weak BCC-algebra any BCC-ideal is contained in some anti-grouped BCC-ideal.*

Theorem 4.3.17. *A BCC-ideal $Z(A)$ is closed if and only if A is closed.*

Proof. Let $\mathfrak{X}$ be a weak BCC-algebra. If a BCC-ideal $Z(A)$ is closed, then $\varphi(x) \in Z(A)$ for every $x \in Z(A)$. Thus, in particular, $\varphi(y) \in Z(A)$ for $y \in A$. So, $\varphi(y) = \varphi^3(y) \in A$.

Let us assume that A is closed and $x \in Z(A)$. Then $\varphi^2(x) \in A$, and consequently, $\varphi(\varphi^2(x)) \in A$. Since $\varphi(\varphi^2(x)) = \varphi^2(\varphi(x))$, the above implies $\varphi(x) \in Z(A)$. Thus $Z(A)$ is a closed BCC-ideal. □

Comparing this result with Theorem 4.3.10, we obtain:

Corollary 4.3.18. *$Z(A)$ is a strong BCC-ideal of weak BCC-algebra $\mathfrak{X}$ if and only if a BCC-ideal A of $\mathfrak{X}$ is strong, or equivalently, if and only if it is closed.*

4.4 Associative Ideals

Definition 4.4.1. Let $\mathfrak{X}$ be a weak BCC-algebra. A BCC-ideal A of $\mathfrak{X}$ is called associative ideal of $\mathfrak{X}$ if it satisfies

$$x\varphi(z) \in A \Longrightarrow zx \in A. \tag{4.16}$$

Theorem 4.4.2. *For an associative BCC-ideal A of a weak BCC-algebra $\mathfrak{X}$, the following statements are true:*

(1) *A is an anti-grouped BCC-ideal,*

(2) *A is a closed BCC-ideal.*

Proof. Let A be an associative BCC-ideal of $\mathfrak{X}$. By replacing x with 0 in (4.16), we get $\varphi^2(z) \Longrightarrow z$, i.e., A is an anti-grouped ideal of $\mathfrak{X}$, which proves (1).

Next, for any $x \in A$ we have $x\varphi(0) = x \in A$, which implies $\varphi(x) \in A$. So, A is closed and (2) holds. □

Theorem 4.4.3. *Let $\mathfrak{X}$ be a weak BCC-algebra and A an associative ideal of $\mathfrak{X}$. Then A is a T-ideal of $\mathfrak{X}$.*

Proof. Let us suppose that $x\varphi(z) \in A$. By the definition of associative ideal we have $zx \in A$. So $zx \cdot 00 \in A$, by (4.16) we get $0 \cdot zx \in A$. Using Lemma 2.2.2(8) we get

$$0(0 \cdot xz) = \varphi^2(xz) = \varphi(zx) = 0 \cdot zx \in A.$$

By Theorem 4.4.2(1) and the definition of anti-grouped ideal, we obtain $xz \in A$. By Theorem 4.2.14 we get that A is a T-ideal of $\mathfrak{X}$. □

Theorem 4.4.4. *Let $\mathfrak{X}$ be a weak BCC-algebra and A an anti-grouped T-ideal of $\mathfrak{X}$. Then A is an associative ideal of $\mathfrak{X}$.*

Proof. Assume $x\varphi(z) \in A$. Then $xz \in A$ by Corollary 4.2.13. By Theorem 4.2.25(3) we have $\varphi(xz) \cdot xz \in A$, so $\varphi(xz) \in A$, by Theorem 2.3.4. Since $\varphi^2(zx) = \varphi(xz)$, by Lemma 2.2.2(8), we have $\varphi^2(zx) \in A$. By the definition of anti-grouped ideal, we get $zx \in A$, this means that (4.16) holds for A, i.e., A is an associative ideal of $\mathfrak{X}$. □

Combining Theorems 4.4.2(1), 4.4.3, 4.4.4, we obtain:

Corollary 4.4.5. *Associative ideal and anti-grouped T-ideal coincide in weak BCC-algebra.*

Corollary 4.4.6. *Let $\mathfrak{X}$ be a weak BCC-algebra. Then $\mathfrak{X}$ is associative if and only if $\mathfrak{X}$ is a group-like T-type weak BCC-algebra.*

Proof. The proof is obvious. □

Combining Theorem 2.3.4, Corollary 4.2.21, Corollary 4.2.22, Theorem 4.2.24, Corollary 4.4.5 and Theorem 4.4.6, we obtain:

Corollary 4.4.7. *A weak BCC-algebra $\mathfrak{X}$ is associative if and only if every BCC-ideal A of $\mathfrak{X}$ is associative BCC-ideal or zero-ideal $\{0\}$ of $\mathfrak{X}$ is an associative BCC-ideal.*

Corollary 4.4.8. *If A is a BCC-ideal of a weak BCC-algebra $\mathfrak{X}$, then the quotient algebra $\mathfrak{X}/\mathfrak{A}$ is associative if and only if A is an associative BCC-ideal.*

Theorem 4.4.9. *A T-ideal of a weak BCC-algebra $\mathfrak{X}$ is anti-grouped if and only if it is associative.*

Proof. Let A be an anti-grouped T-ideal and let $x\varphi(z) \in A$.
Then $0 \in A$ and $xz \in A$, by Theorem 4.2.25(1). Whence, by the condition (3) of the same theorem, it follows $\varphi(xz) \cdot xz \in A$. So, we have shown that

$$\varphi(xz) \cdot (xz) = \varphi(xz) \cdot (xz)0 \in A \text{ and } xz \in A.$$

From Definition 4.2.10(T2) we get $\varphi(xz) \in A$, which, by Lemma 2.2.2(8), implies

$$\varphi^2(zx) = \varphi(xz) \in A, \text{ i.e., } \varphi^2(zx) \in A.$$

However, A is anti-grouped, so, $zx \in A$. This proves that A is associative.

The converse statement is a consequence of Theorem 4.4.2.
The proof is completed. □

4.5 p-Ideals

In this section we describe ideals of weak BCC-algebras, which like the anti-grouped ideals, together with an element x, contain the whole branch to which this element belongs.

We will present their characterization and connections with other objects such as BCC- and BCK-ideals, and group-like weak BCC-algebras. We will also indicate that each (m, n)-fold p-ideal is regular.

We end this section with the theorem that each BCC-ideal of a weak BCC-algebra is contained in an n-fold p-ideal.

The (m,n)-fold p-ideals of a weak BCC-algebra are described in, for example, [58].

Definition 4.5.1. Let $\mathfrak{X}$ be a weak BCC-algebra. A nonempty subset A of X is called an *(m,n)-fold p-ideal* of $\mathfrak{X}$ if it contains 0 and

$$xz^m \cdot yz^n,\ y \in A \Longrightarrow x \in A, \tag{4.17}$$

for $x, z \in X$ and nonnegative integers m, n.

An (n,n)-fold p-ideal is called an *n-fold p-ideal.*

Since $(0,0)$-fold p-ideals coincide with BCK-ideals we will consider (m,n)-fold p-ideals only for $m \geq 1$ and $n \geq 1$. Moreover, it will be assumed that $m \neq n+1$ because for $m = n+1$ we have

$$xx^{n+1} \cdot 0x^n = 0x^n \cdot 0x^n = 0 \in A,$$

which implies $x \in A$. So, $A = X$ for every $(n+1,n)$-fold p-ideal A of $\mathfrak{X}$.

Note that the concept of 1-fold p-ideals coincides with the concept of p-ideals studied in BCI-algebras (cf. [157]).

Example 4.5.2. It is easy to see that in a weak BCC-algebra defined by Table 1.1.2 the set $A = \{0,1\}$ is an n-fold p-ideal for every $n \geq 1$. It is not an (m,n)-fold p-ideal, where m is odd and n is even because in this case $(2 \cdot 2^m)(0 \cdot 2^n) \in A$ and $0 \in A$, but $2 \notin A$. □

Putting $z = 0$ in (4.17), we see that each (m,n)-fold p-ideal of a weak BCC-algebra is a BCK-ideal. The converse statement is not true since, as it follows from Theorem 4.5.3 proven below, each (m,n)-fold ideal contains the branch $B(0)$, which for BCK-ideals is not true.

Theorem 4.5.3. *Any (m,n)-fold p-ideal contains $B(0)$.*

Proof. Let A be an (m,n)-fold p-ideal of a weak BCC-algebra $\mathfrak{X}$. For every $x \in B(0)$ we have

$$xx^m \cdot 0x^n = 0x^{m-1} \cdot 0x^n = 0 \in A,$$

which, according to (4.17), gives $x \in A$. Thus $B(0) \subseteq A$. □

Corollary 4.5.4. *An (m,n)-fold ideal A together with an element $x \in A$ contains whole branch containing this element.*

Proof. Let $x \in A$ and y be an arbitrary element from the branch $B(a)$ containing x. Then, according to Theorem 2.4.20, we have $yx \in B(0) \subset A$. Since A is also a BCK-ideal, the last implies $y \in A$. Thus $B(a) \subset A$. □

Corollary 4.5.5. *For any n-fold p-ideal A from $x \leq y$ and $x \in A$, it follows $y \in A$. In other words, any n-fold p-ideal is regular.*

Corollary 4.5.6. *In a weak BCC-algebra $\mathfrak{X}$ for any chain in $B(a)$, $a \in I(X)$, if an element of a chain is in (m,n)-fold ideal A, then the whole chain belongs to A.*

Proof. All elements of a chain are comparable, i.e., they are all in one branch. The thesis follows from Corollary 4.5.4. □

Theorem 4.5.7. *A nonempty subset A of a solid weak BCC-algebra $\mathfrak{X}$ is its (m,n)-fold p-ideal if and only if*

(1) *$I(A)$ is an (m,n)-fold p-ideal of $\mathfrak{I}(\mathfrak{X})$,*

(2) *$A = \bigcup\{B(a) \mid a \in I(A)\}$.*

Proof. Let A be an (m,n)-fold p-ideal of $\mathfrak{X}$. Then clearly $I(A) = A \cap I(X) \neq \emptyset$ is an (m,n)-fold p-ideal of $\mathfrak{I}(\mathfrak{X})$. By Corollary 4.5.4, A is the set-theoretic sum of all branches $B(a)$ such that $a \in I(A)$. So, any (m,n)-fold p-ideal A satisfies the above two conditions.

Let us suppose that a nonempty subset A of X satisfies these two conditions. Let $x, y, z \in X$. If $x \in B(a)$, $y \in B(b)$, $z \in B(c)$ and y, $xz^m \cdot yz^n \in A$, then $xz^m \cdot yz^n \in B(ac^m \cdot bc^n)$, which, by (2), implies b, $ac^m \cdot bc^n \in I(A)$. This, by (1), gives $a \in I(A)$. So, $B(a) \subset A$. Hence $x \in A$. □

Note that in some situations, the converse of Theorem 4.5.3 is true.

Theorem 4.5.8. *A BCK-ideal A of a weak BCC-algebra $\mathfrak{X}$ is its n-fold p-ideal if and only if $B(0) \subset A$.*

Proof. By Theorem 4.5.3, any n-fold p-ideal contains $B(0)$.

On the other hand, if A is an ideal of $\mathfrak{X}$ and $B(0) \subset A$, then from $y \in A$ and $xz^n \cdot yz^n \in A$, by Lemma 1.1.32, it follows

$$xz^n \cdot yz^n \leq xy,$$

so $xz^n \cdot yz^n$ and xy, as comparable elements, are in the same branch (Corollary 2.4.21). Hence, $xy \cdot (xz^n \cdot yz^n) \in B(0) \subset A$, by Theorem 2.4.20.

Since $xz^n \cdot yz^n \in A$ and A is a BCC-ideal (and a BCK-ideal), $xy \cdot (xz^n \cdot yz^n) \in A$ implies $xy \in A$. Consequently, $x \in A$. So, A is an n-fold p-ideal. □

Corollary 4.5.9. *Any BCK-ideal containing an n-fold p-ideal is also an n-fold p-ideal.*

Proof. Suppose that a BCK-ideal B contains some n-fold p-ideal A. Then $B(0) \subset A \subset B$, which completes the proof. □

Corollary 4.5.10. *A BCC-ideal A of a weak BCC-algebra $\mathfrak{X}$ is its n-fold p-ideal if and only if the implication*

$$xz^n \cdot yz^n \in A \Longrightarrow xy \in A \tag{4.18}$$

is valid for all $x, y, z \in X$.

Proof. Let A be an n-fold p-ideal of $\mathfrak{X}$. Since $xz^n \cdot yz^n \leq xy$, from $xz^n \cdot yz^n \in A$ and Corollary 4.5.5 we obtain $xy \in A$. So, the condition (4.18) is satisfied.

The converse statement is obvious. □

Theorem 4.5.11. *An n-fold p-ideal is a k-fold p-ideal for any $k \leq n$.*

Proof. As in the proof of Lemma 1.1.32, we have

$$xz^n \cdot yz^n \leq xz^{n-1} \cdot yz^{n-1} \leq \ldots \leq xz^k \cdot yz^k$$

for every $1 \leq k \leq n$. Thus, $xz^n \cdot yz^n$ and $xz^k \cdot yz^k$ are in the same branch. Hence, if A is an n-fold p-ideal and $xz^k \cdot yz^k \in A$, then by Corollary 4.5.4 also $xz^n \cdot yz^n \in A$. This, together with $y \in A$, implies $x \in A$. Therefore, A is a k-fold ideal. □

Theorem 4.5.12. *$B(0)$ is the smallest n-fold p-ideal of each weak BCC-algebra.*

Proof. Obviously $0 \in B(0)$. If $y \in B(0)$, then $0 \leq y,\ 0z^n \leq yz^n$ and

$$xz^n \cdot yz^n \leq xz^n \cdot 0z^n \leq xz^{n-1} \cdot 0z^{n-1} \leq \ldots \leq x0 = x,$$

by (1.4) and (1.5). Thus $xz^n \cdot yz^n \leq x$.

Since $xz^n \cdot yz^n \in B(0)$ means $0 \leq xz^n \cdot yz^n$, from the above we obtain

$$0 \leq xz^n \cdot yz^n \leq x \text{ implies } 0 \leq x.$$

So, $x \in B(0)$. Hence $B(0)$ is an n-fold p-ideal.

By Theorem 4.5.3, it is the smallest n-fold p-ideal of each weak BCC-algebra. □

Theorem 4.5.13. *Let $\mathfrak{X}$ be a weak BCC-algebra. If $I(X)$ has k elements and k divides $|m-n|$, then $B(0)$ is an (m,n)-fold p-ideal of $\mathfrak{X}$.*

Proof. By Theorem 2.2.6 and Corollary 2.5.4, $\mathfrak{I}(\mathfrak{X})$ is a group-like subalgebra of $\mathfrak{X}$. Hence, if $I(X)$ has k elements, then in the group $\langle I(X); *, 0\rangle$ corresponding to $\mathfrak{I}(\mathfrak{X})$ (Theorem 2.5.7) we have $b^{ks} = 0$ for every $b \in I(X)$ and any integer s.

We have to consider two cases.

Case 1^0. Let $m \geq n$.
If $xz^m \cdot yz^n \in B(0)$ for some $x \in B(a)$, $y \in B(0)$, $z \in B(c)$, then, by (BCC1$'$), we have

$$xz^m \cdot yz^n \leq xz^{m-n} \cdot y.$$

Hence, $xz^{m-n} \cdot y$ and $xz^m \cdot yz^n$, as comparable elements, are in the same branch. Consequently,

$$(xz^{m-n} \cdot y)(xz^m \cdot yz^n) \in B(0).$$

Since, by Theorem 2.3.7, $B(0)$ is a BCC-ideal in each weak BCC-algebra, from the last we obtain $xz^{m-n} \cdot y \in B(0)$, and consequently, $xz^{m-n} \in B(0)$. But, by Theorem 2.4.18, $xz^{m-n} \in B(ac^{m-n})$, so $B(0) = B(ac^{m-n})$, i.e., $0 = ac^{m-n}$. This in the group $\langle I(X); *, 0\rangle$ corresponding to $\mathfrak{I}(\mathfrak{X})$ gives $0 = a * c^{n-m} = a$.

So, $x \in B(0)$.

Case 2^0. Now, let $m < n$. Then

$$xz^m \cdot yz^n \leq x \cdot yz^{n-m}.$$

This, as in the previous case, for $xz^m \cdot yz^n \in B(0)$ gives

$$(x \cdot yz^{n-m})(xz^m \cdot yz^n) \in B(0).$$

Consequently, $x \cdot yz^{n-m} \in B(0) \cap B(a \cdot 0c^{n-m})$. So, $0 = a \cdot 0c^{n-m}$. This in the group $\langle I(X); *, 0\rangle$ implies $0 = a * c^{m-n} = a$.

Hence, $x \in B(0)$.

The proof is completed. □

The assumption on the number of elements of the set $I(X)$ is essential if k is not a divisor of $|m-n|$, then $B(0)$ may not be an (m,n)-fold p-ideal.

Example 4.5.14. In a solid weak BCC-algebra $\mathfrak{X}$ with the operation $\cdot$ defined in Table 4.5.1:

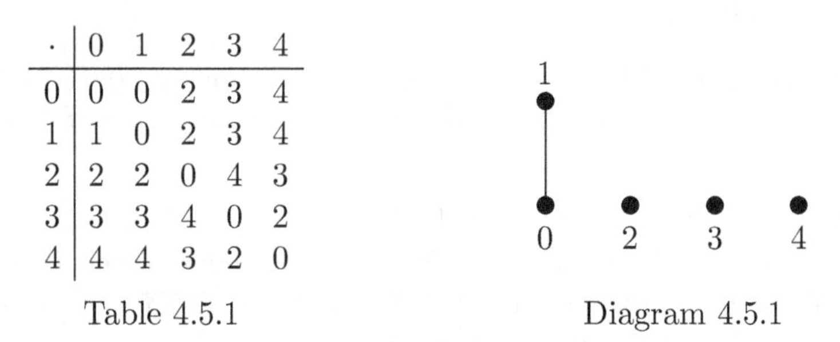

$\cdot$	0	1	2	3	4
0	0	0	2	3	4
1	1	0	2	3	4
2	2	2	0	4	3
3	3	3	4	0	2
4	4	4	3	2	0

Table 4.5.1

Diagram 4.5.1

the set $I(X)$ has four elements. The set $B(0) = \{0,1\}$ is an n-fold p-ideal for every natural n, but it is not a $(4,1)$-fold ideal because $(2 \cdot 2^4) \cdot (0 \cdot 2) \in B(0)$ and $2 \notin B(0)$. □

In the case when $B(0)$ has only one element, the equivalence relation induced by $B(0)$ has one-element equivalence classes. Since these equivalence classes are branches, a weak BCC-algebra with this property is group-like. Direct computations show that in this case $B(0)$ is an n-fold p-ideal for every natural n.

This observation together with just proven results suggests simple characterization of group-like weak BCC-algebras.

Theorem 4.5.15. *A weak BCC-algebra $\mathfrak{X}$ is group-like if and only if for some $n \geq 1$ and all $x, z \in X$*

$$xz^n \cdot 0z^n = 0 \Longrightarrow x = 0. \tag{4.19}$$

Proof. Let $\mathfrak{X}$ be a group-like weak BCC-algebra.

Since for $x, y, z \in X$ we have $xz^n \cdot yz^n \leq xy$ and, by Theorem 2.5.6, a group-like weak BCC-algebra satisfies the identity $xz^n \cdot yz^n = xy$. In particular, for $y = 0$ we have

$$xz^n \cdot 0z^n = x0 = x.$$

So,

$$xz^n \cdot 0z^n = 0 \text{ implies } x = 0,$$

i.e., condition (4.19) is satisfied.

Conversely, if the implication (4.19) is valid for all $x, z \in X$, then

$$0 = xz^n \cdot 0z^n \leq x0 = x \text{ implies } 0 \leq x.$$

This, according to the assumption, implies $x = 0$. Hence $B(0) = \{0\}$. Theorem 2.5.2(6) completes the proof. □

Theorem 4.5.16. *A closed BCK-ideal A of a group-like weak BCC-algebra $\mathfrak{X}$ is its (m, n)-fold p-ideal if $0z^k \in A$ for every $z \in X$, where $k = |m - n|$.*

Proof. Indeed, according to Theorem 2.5.2(3), in any group-like weak BCC-algebra $\mathfrak{X}$ we have

$$(xz^m)(yz^n) = (xz^{m-n})y,$$

if $m \geq n$, or

$$(xz^m)(yz^n) = x(yz^{n-m}),$$

if $m < n$.

Thus

$$(xz^m)(yz^n) \in A \text{ implies } (xz^{m-n})y \in A$$

or

$$x(yz^{n-m}) \in A,$$

which in the corresponding group (Theorem 2.5.7) means that in both cases

$$x * z^{-(m-n)} * y^{-1} \in A.$$

But, by Corollary 4.1.9, A is a subgroup of the corresponding group and $0z^k = z^{-k}$. Therefore,

$$x * z^{-(m-n)} * y^{-1} \in A \text{ implies } x \in A.$$

So, A is an (m, n)-fold p-ideal of $\mathfrak{X}$. □

Corollary 4.5.17. *A nonempty subset A of a group-like weak BCC-algebra $\mathfrak{X}$ is an (m, n)-fold p-ideal if and only if it is a subalgebra of $\mathfrak{X}$ and $0z^k \in A$ for every $z \in X$, where $k = |m - n|$.*

Theorem 4.5.18. *For an (m, n)-fold p-ideal A of a solid weak BCC-algebra $\mathfrak{X}$, the following statements are equivalent:*

(1) *A is a closed (m,n)-fold p-ideal of $\mathfrak{X}$,*

(2) *$I(A)$ is a closed (m,n)-fold p-ideal of $\mathfrak{I}(\mathfrak{X})$,*

(3) *$\mathfrak{I}(\mathfrak{A})$ is a subalgebra of $\mathfrak{I}(\mathfrak{X})$,*

(4) *$\mathfrak{A}$ is a subalgebra of $\mathfrak{X}$.*

Proof. The implication (1) $\Longrightarrow$ (2) follows from Theorem 4.5.7.

(2) $\Longrightarrow$ (3) Observe first that $I(A)$ is a closed BCK-ideal of $\mathfrak{I}(\mathfrak{X})$. By Theorem 2.2.12, for any $a, b \in I(A)$ we have $ab = c \in I(X)$. Since $\mathfrak{I}(\mathfrak{X})$ is a group-like subalgebra of $\mathfrak{X}$ (Theorem 2.2.6 and Corollary 2.5.4), in the group $\langle I(X); *, 0\rangle$ we have $c = a * b^{-1}$ (Theorem 2.5.7), which means that $c * b = a \in I(A)$. Thus

$$c \cdot 0b = c * (0b)^{-1} = a = c * (b^{-1})^{-1} = c * b \in I(A).$$

Hence $c \cdot 0b \in I(A)$.

But $0b \in I(A)$ and $I(A)$ is a BCK-ideal of $\mathfrak{I}(\mathfrak{X})$. Therefore, $c \in I(A)$.

Consequently, $ab \in I(A)$ for every $a, b \in I(A)$.

So, $\mathfrak{I}(\mathfrak{A})$ is a subalgebra of $\mathfrak{I}(\mathfrak{X})$.

(3) $\Longrightarrow$ (4) $I(A) \subset A$, so $0 \in A$. Let $x \in B(a)$, $y \in B(b)$. If $x, y \in A$, then $a, b \in I(A)$, and, by the assumption, $ab \in I(A)$. Hence applying Theorem 2.4.18, we obtain $xy \in B(a)B(b) = B(ab)$, which together with Theorem 4.5.7 proves $xy \in A$. Hence $\mathfrak{A}$ is a subalgebra of $\mathfrak{X}$.

The implication (4) $\Longrightarrow$ (1) is obvious. □

Let us look at n-fold p-ideals again.

In group-like weak BCC-algebras BCK-ideals coincide with n-fold p-ideals since, by Theorem 2.5.6, in these BCC-algebras $xz^n \cdot yz^n = xy$. We will generalize the statement for BCC-ideals.

Lemma 4.5.19. *If A is an n-fold p-ideal of a weak BCC-algebra $\mathfrak{X}$, then*

$$\varphi^2(x) \in A \Longleftrightarrow x \in A.$$

Proof. From Lemma 2.2.2(2) it follows that $x \in B(\varphi^2(x))$ and from Corollary 4.5.4 we get $x \in A$. □

Corollary 4.5.20. *Every n-fold p-ideal is an anti-grouped BCK-ideal.*

Corollary 4.5.21. *$Z(A) = A$ for every n-fold p-ideal A.*

Theorem 4.5.22. *A weak BCC-algebra $\mathfrak{X}$ is group-like if and only if every its BCC-ideal is an n-fold p-ideal.*

Proof. Assume that $\mathfrak{X}$ is a weak group-like BCC-algebra. Then, by Theorem 2.5.6, for all $x, y, z \in X$ we have $xz^n \cdot yz^n = xy$. Thus for every BCC-ideal A of $\mathfrak{X}$ from $xz^n \cdot yz^n \in A$ and $y \in A$ it follows $xy \in A$, whence we obtain $x \in A$. So, every BCC-ideal of $\mathfrak{X}$ is an n-fold p-ideal.

Conversely, let $\mathfrak{X}$ be a weak BCC-algebra in which every BCC-ideal is an n-fold p-ideal. Since in a weak BCC-algebra $\mathfrak{X}$ for $x, y \in X$, by Lemma 2.2.2(8), we have $\varphi^2(xy) = \varphi(yx)$, for $y = \varphi^2(x)$ the following identity:

$$\varphi^2(x\varphi^2(x)) = \varphi(\varphi^2(x)x)$$

is valid. By Lemma 2.2.2(2), it can be written in the form

$$\varphi^2(x\varphi^2(x)) = 0.$$

This, in view of Lemma 4.5.19 and the fact that $B(0) = \{0\}$ is an n-fold p-ideal, implies $x\varphi^2(x) = 0$. Thus $x \leq \varphi^2(x)$, which, together with Lemma 2.2.2(2), gives $x = \varphi^2(x)$ for every $x \in X$.

Therefore, by Theorem 2.5.2(2), $\mathfrak{X}$ is a weak group-like BCC-algebra. □

Corollary 4.5.23. *A weak BCC-algebra is group-like if and only if $\{0\}$ is its p-ideal.*

Corollary 4.5.24. *A weak BCC-algebra $\mathfrak{X}$ is group-like if and only if for all $x, z \in X$ the following implication:*

$$xz \cdot 0z = 0 \Longrightarrow x = 0$$

is valid.

The last corollary follows from Theorem 4.5.15 for $n = 1$, too.

Theorem 4.5.25. *Let A be a BCC-ideal of a weak BCC-algebra $\mathfrak{X}$. Then a quotient weak BCC-algebra $\mathfrak{X}/\mathfrak{A}$ is group-like if and only if A is a 1-fold p-ideal of $\mathfrak{X}$.*

Proof. Let A be a BCC-ideal of a weak BCC-algebra $\mathfrak{X}$. Then, as we already know, the relation θ defined by (2.8) is a congruence and the set $X/A = \{C_x \mid x \in X\}$, where $C_x = \{y \in X \mid (y, x) \in \theta\}$, is the base set of the weak BCC-algebra $\mathfrak{X}/\mathfrak{A}$.

Assume that $\mathfrak{X}/\mathfrak{A}$ is group-like. Then

$$C_xC_z \cdot C_yC_z = C_xC_y, \text{ i.e., } C_{xz\cdot yz} = C_{xy}.$$

Hence $xz \cdot yz = xy$ and, of course,

$$xy \cdot (xz \cdot yz) = 0 \in A.$$

If y and $xz \cdot yz$ are in A, then also $xy \in A$. Whence $x \in A$. Thus A is a 1-fold p-ideal of $\mathfrak{X}$.

To prove the converse statement, let us assume that A is a 1-fold p-ideal of $\mathfrak{X}$. Observe that, as in the proof of the previous theorem, we have

$$\varphi^2(x\varphi^2(x)) = \varphi(\varphi^2(x)x) \tag{4.20}$$

for $x \in X$.

Since, by Lemma 2.2.2(2),

$$\varphi^2(x)x = 0 \in A, \tag{4.21}$$

for $x \in X$ the equation (4.20) has the form

$$\varphi^2(x\varphi^2(x)) = 0 \in A,$$

which, by Lemma 4.5.19, implies

$$x\varphi^2(x) \in A. \tag{4.22}$$

So, by (4.21), (4.22) and (2.8), $(x, \varphi^2(x)) \in \theta$. Therefore, for every $x \in X$

$$C_x = C_{\varphi^2(x)},$$

i.e., by Theorem 2.5.2(2), $\mathfrak{X}/\mathfrak{A}$ is group-like BCC-algebra.

The proof is completed. □

Theorem 4.5.26. *Any BCC-ideal of a weak BCC-algebra $\mathfrak{X}$ is contained in some n-fold p-ideal.*

Proof. Obviously $A \subset Z(A)$ for any BCC-ideal A of $\mathfrak{X}$. By Theorem 4.3.15, $Z(A)$ is an anti-grouped BCC-ideal. To prove that $Z(A)$ is an n-fold p-ideal, consider an arbitrary element $xz^n \cdot yz^n \in Z(A)$, where $x, z \in X$ and $y \in Z(A)$. Since $xz^n \cdot yz^n \leqslant xy$, elements $xz^n \cdot yz^n$ and xy are in the same branch. So, according to Theorem 4.3.2, $xy \in Z(A)$, whence it follows $x \in A$.

By Definition 4.5.1, $Z(A)$ is an n-fold p-ideal.

This completes the proof. □

Corollary 4.5.27. *$Z(A)$ is an n-fold p-ideal for every $n \geq 1$.*

4.6 k-Nil Radicals of Solid Weak BCC-Algebras

The theory of radicals in BCI-algebras was considered by many mathematicians from China (cf. [146]). Obtained results show that this theory is almost parallel to the theory of radicals in rings. But the results proven for radicals in BCI-algebras cannot be transformed to weak BCC-algebras (cf. [131]).

In this section we characterize one analogue of nil radicals in weak BCC-algebras. Further, this characterization will be used to describe some ideals of solid weak BCC-algebras.

We begin with the following definition

Definition 4.6.1. Let $\mathfrak{X}$ be a weak BCC-algebra and let A be a subset of X. For any positive integer k by a *k-nil radical of A*, denoted by $[A;k]$, we mean the set of all elements of X such that $0x^k \in A$, i.e.,

$$[A;k] = \{x \in X \mid 0x^k \in A\}.$$

Example 4.6.2. In a weak BCC-algebra $\mathfrak{X}$ defined in Example 2.2.15 for $A = \{0,1\}$ and any natural k we have $[A;3k+1] = [A;3k+2] = B(0)$, $[A;3k] = X$. But for $B = \{1,5\}$ we get $[B;3k+1] = \{4\}$, $[B;3k+2] = B(3)$. The set $[B;3k]$ is empty. □

Example 4.6.3. In a weak BCC-algebra $\mathfrak{X}$ defined in Example 2.6.5 (Table 2.6.4) each k-nil radical of $A = \{0,3\}$ is equal to X; each k-nil radical of $B = \{1,5\}$ is empty. □

Let A be nonempty subset of X. Is a k-nil radical of A also nonempty? The answer is given by the following theorem:

Theorem 4.6.4. *Let $\mathfrak{X}$ be a solid weak BCC-algebra and let A be a nonempty subset of X. A k-nil radical $[A;k]$ of A is nonempty if and only if A contains at least one element $a \in I(X)$.*

Proof. From the proof of Theorem 2.6.19, it follows that $0x^k = 0a^k$ for every $x \in B(a)$ and any positive k. So, $x \in [A;k]$ if and only if $0a^k \in A$. Hence $0a^k \in A \cap I(X)$ because $\mathfrak{I}(\mathfrak{X})$ is a subalgebra of $\mathfrak{X}$. □

Corollary 4.6.5. *Let $\mathfrak{X}$ be a solid weak BCC-algebra. Then $[I(X);k] = X$ for every k.*

Proof. Indeed, $0x^k = 0a^k \in I(X)$ for every $x \in X$. Thus $x \in [I(X); k]$. This means $X \subseteq [I(X); k]$.

The converse is obvious. □

Corollary 4.6.6. *Let $\mathfrak{X}$ be a solid weak BCC-algebra. If $I(X)$ has n elements, then $[A; n] = X$ for any subset A of X containing* 0, *and $[A; n] = \emptyset$ if $0 \notin A$.*

Proof. As in previous proofs, we have $0x^k = 0a^k$ for every $x \in B(a)$ and any k. Since $0a^k \in I(X)$ and $\mathfrak{I}(\mathfrak{X})$ is a group-like subalgebra of $\mathfrak{X}$, $0x^k = a^{-k}$ in the group $\langle I(X); *, 0\rangle$. If $I(X)$ has n elements, then obviously $0x^n = a^{-n} = 0 \in A$. Hence $x \in [A; n]$.

This completes the proof. □

Corollary 4.6.7. *Let $\mathfrak{X}$ be a solid weak BCC-algebra and let $x \in B(a)$. Then $x \in [A; k]$ if and only if $B(a) \subset [A; k]$.*

Proof. Since $0x^k = 0a^k$, we have $x \in [A; k] \Longleftrightarrow a \in [A; k]$. □

Corollary 4.6.8. *Let $\mathfrak{X}$ be a solid weak BCC-algebra and let $A \subset X$. Then $[A; k] = \bigcup\{B(a) \mid 0a^k \in A\}$.*

Theorem 4.6.9. *Let $\mathfrak{X}$ be a solid weak BCC-algebra. Then for every positive integer k and any subalgebra $\mathfrak{A}$ of $\mathfrak{X}$ the algebra $\langle [A; k]; \cdot, 0\rangle$ is a subalgebra of $\mathfrak{X}$ such that $A \subset [A; k]$.*

Proof. Let $x, y \in [A; k]$. Then $0x^k, 0y^k \in A$ and $0(xy)^k = 0x^k \cdot 0y^k \in A$, by Theorem 2.6.19. Hence $xy \in [A; k]$. Clearly $A \subseteq [A; k]$. □

Theorem 4.6.10. *In a solid weak BCC-algebra $\mathfrak{X}$, a k-nil radical of a BCC-ideal is also a BCC-ideal.*

Proof. Let A be a BCC-ideal of $\mathfrak{X}$. If $y \in [A; k]$ and $xy \cdot z \in [A; k]$, then $0y^k \in A$ and

$$A \ni 0(xy \cdot z)^k = (0x^k \cdot 0y^k) \cdot 0z^k,$$

by Theorem 2.6.19. Hence

$$A \ni 0x^k \cdot 0z^k = 0(xz)^k.$$

Thus $xy \in [A; k]$. □

Let us note that for nonsolid weak BCC-algebras, the last two theorems are not true.

Example 4.6.11. Consider a weak BCC-algebra $\mathfrak{X}$ with the operation $\cdot$ defined by Table 2.3.3.

By Theorem 2.5.2(6), $\mathfrak{X}$ is group-like. So, by Theorem 2.5.7, $\mathfrak{X}$ is induced by the symmetric group S_3. Hence it is not solid. Routine calculations show that $A = \{0, 3\}$ is a subalgebra and a BCC-ideal of $\mathfrak{X}$, but $[A; 3] = \{0, 1, 2, 3\}$ is neither BCC-ideal nor subalgebra of $\mathfrak{X}$. □

Theorem 4.6.12. *In a solid weak BCC-algebra a k-nil radical of an (m, n)-fold p-ideal is also an (m, n)-fold p-ideal.*

Proof. By Theorem 4.6.10, a k-nil radical of an (m, n)-fold p-ideal A of $\mathfrak{X}$ is a BCC-ideal of $\mathfrak{X}$.

If y, $xz^m \cdot yz^n \in [A; k]$, then $0y^k$, $0(xz^m \cdot yz^n)^k \in A$. Hence, applying Theorem 2.6.19, we obtain

$$(0x^k \cdot (0z^k)^m)(0y^k \cdot (0z^k)^n) = 0(xz^m \cdot yz^n)^k \in A.$$

Thus $0x^k \in A$. So, $x \in [A; k]$. □

Let us note that in general a k-nil radical $[A; k]$ of a BCC-ideal A does not save all properties of a BCC-ideal A. For example, if an BCC-ideal A is a horizontal ideal, then a k-nil radical $[A; k]$ may not be a horizontal ideal. Such situation takes place in a weak BCC-algebra defined by Table 2.2.1. In this algebra we have $0x^3 = 0$ for all elements. Hence $x \in [A; 3] \cap B(0)$ means that $0x^3 \in A$ and $x \in B(0)$ which is true also for $x \neq 0$.

Nevertheless, properties of many main types of ideals are saved by their k-nil radicals. We have shown two examples in the following corollaries:

Corollary 4.6.13. *In a solid weak BCC-algebra, a k-nil radical of a closed ideal is also a closed ideal.*

Proof. Let $x \in [A; k]$. Then $0x^k \in A$. Thus from Corollary 2.6.27

$$0 \cdot (0x)^k = 0 \cdot 0x^k \in A.$$

So, $0x \in [A; k]$. □

Corollary 4.6.14. *In a solid weak BCC-algebra, a k-nil radical of an anti-grouped ideal is also an anti-grouped ideal.*

Proof. Let $\varphi^2(x) \in [A; k]$. Then $0(\varphi^2(x))^k \in A$. Since, by Corollary 2.2.3, φ^2 is an endomorphism of each weak BCC-algebra, we have

$$\varphi^2(0x^k) = \varphi^2(0)(\varphi^2(x))^k = 0(\varphi^2(x))^k \in A.$$

Thus $\varphi^2(0x^k) \in A$, which according to the definition implies $0x^k \in A$. Hence $x \in [A; k]$. □

4.7 Fuzzy BCC-Ideals

Fuzzy ideals for BCK-algebras were defined in the 1990s. Since then, they have been studied by many mathematicians. At the turn of the 20th and 21st centuries, the idea of fuzzy ideals was generalized to BCC-algebras. Many of the results for BCK-algebras have been generalized by producing partially new proofs fitted to the axiomatics of BCC-algebras. There were also many interesting new results for BCC-algebras.

Jun and Kim [87] introduced the notion of fuzzy ideals of BCK-algebras with respect to a given t-norm and obtained some of their properties. These ideas have been generalized for BCC-algebras and supplemented with further results by W.A. Dudek and Y.B. Jun in a series of their joint publications (see Bibliography).

In this chapter, we will introduce the fuzzy ideals of BCC-algebras and show the most important results on this topic.

We start by giving the fuzzy version of Definitions 2.3.1 and 2.3.3 of BCK- and BCC-ideals.

Definition 4.7.1. A fuzzy set μ is called *a fuzzy BCK-ideal of a BCC-algebra* $\mathfrak{X}$ if it satisfies the following conditions:

$$(\text{FI1})_{\text{BCK}} \quad \mu(0) \geq \mu(x),$$

$$(\text{FI2})_{\text{BCK}} \quad \mu(x) \geq \min\{\mu(xy), \mu(y)\}$$

for all $x, y \in X$.

Definition 4.7.2. A fuzzy set μ is called *a fuzzy BCC-ideal of a BCC-algebra* $\mathfrak{X}$ if it satisfies the following conditions:

(FI1) $\mu(0) \geq \mu(x)$,

(FI2) $\mu(xz) \geq \min\{\mu((xy)z), \mu(y)\}$

for all $x, y, z \in X$.

If the level subset μ_t of μ is a BCC-ideal, then it is called *a level BCC-ideal.*

Example 4.7.3. Let $\langle\{0, 1, 2, 3, 4, 5\}; \cdot, 0\rangle$ be a proper BCC-algebra with the operation $\cdot$ defined by Table 2.1.1.

We define a fuzzy set μ in $\mathfrak{X}$ by

$$\mu(x) = \begin{cases} 0.3 & \text{if} \quad x = 5, \\ 0.5 & \text{if} \quad x \neq 5. \end{cases}$$

Then μ is a fuzzy BCC-ideal of $\mathfrak{X}$. □

The next theorem is Theorem 2.3.4 for fuzzy BCC- and BCK-ideals of a BCC-algebra.

Theorem 4.7.4. *In a BCC-algebra, every fuzzy BCC-ideal is a fuzzy BCK-ideal.*

Proof. Let $\mathfrak{X}$ be a BCC-algebra and let μ be a fuzzy BCC-ideal of $\mathfrak{X}$. Clearly, μ satisfies the condition $(\text{FI1})_{BCK}$.

When we put $z = 0$ in (FI2), then by (BCC3) we get

$$\mu(x) \geq \min\{\mu(xy), \mu(y)\}$$

for all $x, y \in X$. So, the condition $(\text{FI2})_{BCK}$ is satisfied, too.

This means that fuzzy BCC-ideal μ of $\mathfrak{X}$ is a fuzzy BCK-ideal of $\mathfrak{X}$. □

In the following example, we show that the converse of Theorem 4.7.4 is in general not true.

Example 4.7.5. Let $\langle\{0, 1, 2, 3, 4\}; \cdot, 0\rangle$ be a proper BCC-algebra with the operation $\cdot$ defined by Table 2.3.1. Let $\mu : X \longrightarrow [0, 1]$ be a fuzzy set defined in the following way:

$$\mu(x) = \begin{cases} 0.9 & \text{if} \quad x \in \{0, 1\}, \\ 0.3 & \text{if} \quad x \in \{2, 3, 4\}. \end{cases}$$

It is easy to verify that μ is a fuzzy BCK-ideal of $\mathfrak{X}$.
Since

$$\mu(4 \cdot 2) < \min\{\mu((4 \cdot 1) \cdot 2), \mu(1)\}$$

it is not a fuzzy BCC-ideal of $\mathfrak{X}$. □

The next theorem is Theorem 2.3.5 for fuzzy BCC- and BCK-ideals of a BCC-algebra.

Theorem 4.7.6. *In a BCK-algebra, every fuzzy BCK-ideal is a fuzzy BCC-ideal.*

Proof. Let μ be a fuzzy BCK-ideal of a BCK-algebra $\mathfrak{X}$.

Condition (FI1) is satisfied. Then for $x, y, z \in X$ we have

$$\mu(xz) \geq \min\{\mu((xz)y), \mu(y)\} = \min\{\mu((xy)z), \mu(y)\}.$$

Condition (FI2) is also satisfied and so μ is a fuzzy BCC-ideal of $\mathfrak{X}$. □

The next result is Theorem 2.3.6 for fuzzy BCK-ideals of fuzzy BCC-algebra $\mathfrak{X}$.

Theorem 4.7.7. *Every fuzzy BCK-ideal of a BCC-algebra $\mathfrak{X}$ is a fuzzy BCC-subalgebra of $\mathfrak{X}$.*

Proof. Let μ be a fuzzy BCK-ideal of a BCC-algebra $\mathfrak{X}$. Then by $(\text{FI1})_{BCK}$ $\mu(0) \geq \mu(x)$ for all $x \in X$. From Theorem 1.1.30(4) it follows

$$\mu((xy)x) = \mu(0) \geq \mu(x)$$

for all $x, y \in X$.

Thus

$$\mu(xy) \geq \min\{\mu((xy)x), \mu(x)\} = \mu(x) \geq \min\{\mu(x), \mu(y)\}$$

for all $x, y \in X$.

So, by Definition 3.6.2, μ is a BCC-subalgebra of $\mathfrak{X}$. □

By combining Theorems 4.7.4 and 4.7.7, we get the following corollary:

Corollary 4.7.8. *Every fuzzy BCC-ideal of a BCC-algebra $\mathfrak{X}$ is a fuzzy BCC-subalgebra.*

Theorem 4.7.9. *A fuzzy set μ in a BCC-algebra $\mathfrak{X}$ is a fuzzy BCC-ideal of $\mathfrak{X}$ if and only if for every $t \in [0, 1]$, μ_t is either empty or a BCC-ideal of $\mathfrak{X}$.*

Proof. Let, for every $t \in [0,1]$, μ_t be empty or a BCC-ideal of $\mathfrak{X}$. If $\mu(0) \geq \mu(x)$ for all $x \in X$ is not true, then there is $x_0 \in X$ such that $\mu(0) < \mu(x_0)$. But in this case for $a_0 = (\mu(0) + \mu(x_0))/2$ we have $\mu(0) < a_0 < \mu(x_0)$, which implies $x_0 \in \mu_{a_0}$, i.e., $\mu_{a_0} \neq \emptyset$. We know that μ_{a_0} is a BCC-ideal. So, $\mu(0) \geq a_0$, which is impossible.

Hence $\mu(0) \geq \mu(x)$ for all $x \in X$.

Now, if (FI2) is not true, then

$$\mu(x_0 z_0) < \min\{\mu((x_0 y_0)z_0), \mu(y_0)\}$$

for some $x_0, y_0, z_0 \in X$.

Let

$$b_0 = (\mu(x_0 z_0) + \min\{\mu((x_0 y_0)z_0), \mu(y_0)\})/2.$$

Then clearly $b_0 \in [0,1]$ and

$$\mu(x_0 z_0) < b_0 < \min\{\mu((x_0 y_0)z_0), \mu(y_0)\},$$

which shows that $(x_0 y_0)z_0 \in \mu_{b_0}$ and $y_0 \in \mu_{b_0}$. Hence, since μ_{b_0} is a BCC-ideal, we have $x_0 z_0 \in \mu_{b_0}$. Thus $\mu(x_0 z_0) \geq b_0$, which is a contradiction.

Therefore, (FI2) is true and μ has to be a fuzzy BCC-ideal.

The converse follows from Theorem 3.6.5, because any fuzzy BCC-ideal is a BCC-subalgebra. □

Corollary 4.7.10. *A fuzzy set μ in a BCK-algebra $\mathfrak{X}$ is a fuzzy BCC-ideal of $\mathfrak{X}$ if and only if for every $t \in [0,1]$, μ_t is either empty or a BCC-ideal of $\mathfrak{X}$.*

The next result can be proven in the same way as Theorem 4.7.9:

Theorem 4.7.11. *A fuzzy set μ in a BCC-algebra $\mathfrak{X}$ is a fuzzy BCK-ideal if and only if for every $t \in [0,1]$, μ_t is either empty or a BCK-ideal of $\mathfrak{X}$.*

The proof of the next result is similar to that of Theorem 3.6.6.

Theorem 4.7.12. *Any BCC-ideal of a BCC-algebra $\mathfrak{X}$ can be realized as a level BCC-ideal of some fuzzy BCC-ideal of $\mathfrak{X}$.*

Theorem 4.7.13. *Two level BCC-ideals μ_s, μ_t $(s < t)$ of a fuzzy BCC-ideal are equal if and only if there is no $x \in X$ such that $s \leq \mu(x) < t$.*

Proof. The statement of the theorem follows from Theorem 4.7.11, Theorem 4.7.12 and Theorem 3.6.7. □

Corollary 4.7.14. *Let μ be a fuzzy BCC-ideal of $\mathfrak{X}$. If $Im(\mu) = \{t_1, t_2, ...t_n\}$, where $t_1 < t_2 < ... < t_n$, then the family of levels μ_{t_i}, for $1 \leq i \leq n$, constitutes all level BCC-ideals of μ.*

As further extensions of Definitions 2.3.1 and 2.3.3, we have the definitions of T-fuzzy BCK-ideals and BCC-ideals.

Definition 4.7.15. A fuzzy set μ in X is called a *T-fuzzy BCC-ideal* of a BCC-algebra $\mathfrak{X}$, if it satisfies the following conditions:

(TI1) $\mu(0) \geq \mu(x)$,

(TI2) $\mu(xz) \geq T(\mu(xy \cdot z), \mu(y))$

for all $x, y, z \in X$.

The full name of such ideal is *a fuzzy ideal* of a BCC-algebra $\mathfrak{X}$ with respect to a t-norm T.

A fuzzy set μ satisfying (TI1) and the condition

(TI3) $\mu(x) \geq T(\mu(xy), \mu(y))$

is called a *T-fuzzy BCK-ideal* of $\mathfrak{X}$.

This means that any T-fuzzy BCC-ideal is a T-fuzzy BCK-ideal, but not conversely, as it is shown in the following example.

Example 4.7.16. Let us consider the BCC-algebra defined by Table 2.3.1. By routine calculations we know that a fuzzy set μ in X defined by

$$\mu(x) = \begin{cases} 0.9 & \text{if} \quad x = 0, \\ 0.9 & \text{if} \quad x = 1, \\ 0.3 & \text{if} \quad x = 2, \\ 0.3 & \text{if} \quad x = 3, \\ 0.3 & \text{if} \quad x = 4 \end{cases}$$

is a T_m-fuzzy BCK-ideal of $\mathfrak{X}$, which is not a T_m-fuzzy BCC-ideal because the condition (FI2) is not satisfied. Indeed,

$$\mu(4 \cdot 2) = \mu(4) = 0.3.$$

On the other hand,

$$T_m(\mu((4 \cdot 1) \cdot 2), \mu(1)) = T_m(\mu(1), \mu(1)) = 0.9.$$

Hence $\mu(4 \cdot 2) < T_m(\mu((4 \cdot 1) \cdot 2), \mu(1))$. □

The next lemma is a version of Theorem 2.3.5 for T-fuzzy BCK-ideals of a BCK-algebra $\mathfrak{X}$.

Lemma 4.7.17. *In a BCK-algebra every T-fuzzy BCK-ideal is a T-fuzzy BCC-ideal.*

Proof. Let $\mathfrak{X}$ be a BCK-algebra and let μ be a T-fuzzy BCK-ideal of $\mathfrak{X}$ with T as a t-norm.

Since $\mathfrak{X}$ is a BCK-algebra, then it satisfies (1.6). So, we have

$$\mu(xz) \geq T(\mu(xz \cdot y), \mu(y)) = T(\mu(xy \cdot z), \mu(y))$$

for all $x, y, z \in X$. Hence μ is a T-fuzzy BCC-ideal of $\mathfrak{X}$. □

4.8 Cubic Bipolar BCC-Ideals

Zadeh [149] made an extension of the concept of a fuzzy set by an interval-valued fuzzy set, i.e., a fuzzy set with an interval-valued membership function. The idea of an intuitionistic fuzzy set was first introduced by Atanassov [9] as a generalization of the notion of fuzzy sets. In 1991, Ougen (cf. [123]) applied the concept of fuzzy sets to BCI- and BCK-algebras. In [110], the notion of bipolar-valued fuzzy sets, which is an extension of fuzzy sets, was introduced, and in [109] their notion of BCK-algebra was presented.

Using a fuzzy set and an interval-valued fuzzy set, Jun et al. (cf. [85]) introduced a new notion, called a cubic set, and investigated its several properties.

In this section we present the concepts that have been translated into BCC-algebras and BCC-ideals (cf. [108]) and we generalize the results presented in [109] for BCI-, BCK-algebras to the case of (weak) BCC-algebras. We begin with the following definition:

Definition 4.8.1. A *bipolar-valued fuzzy set* or simply *bipolar fuzzy set* Φ in a weak BCC-algebra $\mathfrak{X}$ is the set of triples

$$\Phi = \{(x, \mu_\Phi^P(x), \mu_\Phi^N(x)) \mid x \in X\}, \tag{4.23}$$

where

$$\mu_\Phi^P : X \Longrightarrow [0,1]$$

and

$$\mu_\Phi^N : X \Longrightarrow [-1,0]$$

are mappings.

The term $\mu_\Phi^P(x)$ denotes the satisfaction degree of an element x to the property corresponding to a bipolar fuzzy set given in (4.23). Analogously, $\mu_\Phi^N(x)$ denotes the satisfaction degree of x to some implicit counter-property of (4.23).

In the case when $\mu_\Phi^P \neq 0$ and $\mu_\Phi^N = 0$ we say that x is regarded as having only positive satisfaction of (4.23). On the other hand, when $\mu_\Phi^P = 0$ and $\mu_\Phi^N \neq 0$ we say that x does not satisfy the property (4.23), but it somewhat satisfies the counter-property of (4.23).

In the case when the property and counter-property are not disjoint, it is possible that an element x is $\mu_\Phi^P \neq 0$ and $\mu_\Phi^N \neq 0$.

In the sequel, we will use for simplicity the notation $\Phi = (\mu_\Phi^P, \mu_\Phi^N)$.

Definition 4.8.2. A bipolar fuzzy set $\Phi = (\mu_\Phi^P, \mu_\Phi^N)$ in a weak BCC-algebra $\mathfrak{X}$ is called a *bipolar fuzzy subalgebra* of $\mathfrak{X}$, if it satisfies the following conditions:

$$\mu_\Phi^P(xy) \geq \min\{\mu_\Phi^P(x), \mu_\Phi^P(y)\} \tag{4.24}$$

$$\mu_\Phi^N(xy) \leq \max\{\mu_\Phi^N(x), \mu_\Phi^N(y)\} \tag{4.25}$$

for all $x, y \in X$.

Strictly speaking, when we say "the subalgebra Φ" we mean the simplified version of the algebra $\langle \Phi; \cdot, 0 \rangle$.

Example 4.8.3. Let $\mathfrak{X}$ be a weak BCC-algebra with the operation $\cdot$ defined by Table 3.6.1. Let $\Phi = (\mu_\Phi^P, \mu_\Phi^N)$ be a bipolar fuzzy set in $\mathfrak{X}$ defined by the following table:

	0	1	2	3
μ_Φ^P	0.6	0.6	0.3	0.6
μ_Φ^N	−0.7	−0.7	−0.2	−0.7

Table 4.8.1

Then $\Phi = (\mu_\Phi^P, \mu_\Phi^N)$ is a bipolar fuzzy subalgebra of $\mathfrak{X}$. □

Lemma 4.8.4. *If $\Phi = (\mu_\Phi^P, \mu_\Phi^N)$ is a bipolar fuzzy subalgebra of a weak BCC-algebra $\mathfrak{X}$, then*

$$\mu_\Phi^P(0) \geq \mu_\Phi^P(x) \text{ and } \mu_\Phi^N(0) \leq \mu_\Phi^N(x)$$

for all $x \in X$.

Proof. For $x \in X$ we have

$$\mu_\Phi^P(0) = \mu_\Phi^P(xx) \geq \min\{\mu_\Phi^P(x), \mu_\Phi^P(x)\} = \mu_\Phi^P(x).$$

Similarly,

$$\mu_\Phi^N(0) = \mu_\Phi^N(xx) \leq \max\{\mu_\Phi^N(x), \mu_\Phi^N(x)\} = \mu_\Phi^N(x).$$

The proof is completed. □

Definition 4.8.5. For a bipolar fuzzy set $\Phi = (\mu_\Phi^P, \mu_\Phi^N)$ and for a couple $(s,t) \in [-1,0] \times [0,1]$, we define the so-called *positive t-cut* of $\Phi = (\mu_\Phi^P, \mu_\Phi^N)$ in the following way:

$$\Phi_t^P = \{x \in X \mid \mu_\Phi^P(x) \geq t\}. \tag{4.26}$$

Similarly, we define the *negative s-cut* of $\Phi = (\mu_\Phi^P, \mu_\Phi^N)$, as

$$\Phi_s^N = \{x \in X \mid \mu_\Phi^N(x) \leq s\}. \tag{4.27}$$

We call the set

$$\Phi_k = \Phi_k^P \cap \Phi_{-k}^N$$

the *k-cut* of $\Phi = (\mu_\Phi^P, \mu_\Phi^N)$.

Theorem 4.8.6. *Let $\Phi = (\mu_\Phi^P, \mu_\Phi^N)$ be a bipolar fuzzy subalgebra of a weak BCC-algebra $\mathfrak{X}$. Then the following statements are true:*

(1) $\Phi_t^P \neq \emptyset \Longrightarrow \Phi_t^P$ *is a subalgebra of $\mathfrak{X}$ for any $t \in [0,1]$,*

(2) $\Phi_s^N \neq \emptyset \Longrightarrow \Phi_s^N$ *is a subalgebra of $\mathfrak{X}$ for any $s \in [-1,0]$.*

Proof. (1) Let $t \in [0,1]$ and let $\Phi_t^P \neq \emptyset$.
For $x, y \in \Phi_t^P$ we have $\mu_\Phi^P(x) \geq t$ and $\mu_\Phi^P(y) \geq t$. It follows that

$$\mu_\Phi^P(xy) \geq \min\{\mu_\Phi^P(x), \mu_\Phi^P(y)\} \geq t,$$

i.e., $xy \in \Phi_t^P$. So, Φ_t^P is a subalgebra of $\mathfrak{X}$.

(2) Let us assume that $s \in [-1,0]$ and $\Phi_s^N \neq \emptyset$. In this case the proof is analogous. Indeed, if $x, y \in \Phi_s^N$, then

$$\mu_\Phi^N(xy) \leq \max\{\mu_\Phi^N(x), \mu_\Phi^N(y)\} \leq s,$$

i.e., $xy \in \Phi_s^N$. So, Φ_s^N is a subalgebra of $\mathfrak{X}$. □

From the above theorem the following corollary immediately follows:

Corollary 4.8.7. *Let $\Phi = (\mu_\Phi^P, \mu_\Phi^N)$ be a bipolar fuzzy subalgebra of a weak BCC-algebra $\mathfrak{X}$. Then the sets $\Phi^P_{\mu_\Phi^P(0)}$ and $\Phi^N_{\mu_\Phi^N(0)}$ are subalgebras of $\mathfrak{X}$.*

Now we define the bipolar fuzzy BCK- and BCC-ideals as a generalization of both Definitions 2.3.1 and 2.3.3 and their equivalents for fuzzy BCC-subalgebras.

Definition 4.8.8. A bipolar fuzzy set $\Phi = (\mu_\Phi^P, \mu_\Phi^N)$ in a weak BCC-algebra $\mathfrak{X}$ is called a *bipolar fuzzy BCK-ideal* of $\mathfrak{X}$ if it satisfies:

$(\mathrm{BFI1_P})_{\mathrm{BCK}}$ $\mu_\Phi^P(0) \geq \mu_\Phi^P(x)$,

$(\mathrm{BFI1_N})_{\mathrm{BCK}}$ $\mu_\Phi^N(0) \leq \mu_\Phi^N(x)$,

$(\mathrm{BFI2_P})_{\mathrm{BCK}}$ $\mu_\Phi^P(x) \geq \min\{\mu_\Phi^P(xy), \mu_\Phi^P(y)\}$,

$(\mathrm{BFI2_N})_{\mathrm{BCK}}$ $\mu_\Phi^N(x) \leq \max\{\mu_\Phi^N(xy), \mu_\Phi^N(y)\}$

for all $x, y \in X$.

Definition 4.8.9. A bipolar fuzzy set $\Phi = (\mu_\Phi^P, \mu_\Phi^N)$ in a weak BCC-algebra $\mathfrak{X}$ is called a *bipolar fuzzy BCC-ideal* of $\mathfrak{X}$ if it satisfies:

$(\mathrm{BFI1_P})$ $\mu_\Phi^P(0) \geq \mu_\Phi^P(x)$,

$(\mathrm{BFI1_N})$ $\mu_\Phi^N(0) \leq \mu_\Phi^N(x)$,

$(\mathrm{BFI2_P})$ $\mu_\Phi^P(xz) \geq \min\{\mu_\Phi^P(xy \cdot z), \mu_\Phi^P(y)\}$,

$(\mathrm{BFI2_N})$ $\mu_\Phi^N(xz) \leq \max\{\mu_\Phi^N(xy \cdot z), \mu_\Phi^N(y)\}$

for all $x, y, z \in X$.

Example 4.8.10. Let $\mathfrak{X} = \langle\{0,1,2,3,4\}; \cdot, 0\rangle$ be a nonproper BCC-algebra with the operation $\cdot$ defined by the following table:

$\cdot$	0	1	2	3	4
0	0	0	0	0	0
1	1	0	1	0	0
2	2	2	0	0	0
3	3	3	3	0	0
4	4	3	4	1	0

Table 4.8.2

Let $\Phi = (\mu_\Phi^P, \mu_\Phi^N)$ be a bipolar fuzzy set in $\mathfrak{X}$ defined by the following table:

	0	1	2	3	4
μ_Φ^P	0.7	0.2	0.7	0.2	0.2
μ_Φ^N	−0.8	−0.7	−0.8	−0.7	−0.7

Table 4.8.3

Then $\Phi = (\mu_\Phi^P, \mu_\Phi^N)$ is a bipolar fuzzy BCK-ideal of $\mathfrak{X}$. □

Theorem 4.8.11. *Every bipolar fuzzy BCK-ideal of a BCK-algebra $\mathfrak{X}$ is a bipolar fuzzy BCC-ideal of $\mathfrak{X}$.*

Proof. Let $\Phi = (\mu_\Phi^P, \mu_\Phi^N)$ be a bipolar fuzzy BCK-ideal of a BCK-algebra $\mathfrak{X}$. Conditions ($\mathrm{BFI1}_P$) and ($\mathrm{BFI1}_N$) are satisfied. Then for $x, y, z \in X$, by $(\mathrm{BFI2}_P)_{BCK}$ and (1.6), we have

$$\mu_\Phi^P(xz) \geq \min\{\mu_\Phi^P(xz \cdot y), \mu_\Phi^P(y)\} = \min\{\mu_\Phi^P(xy \cdot z), \mu_\Phi^P(y)\},$$

which shows that ($\mathrm{BFI2}_P$) is satisfied.

Similarly, by $(\mathrm{BFI2}_N)_{BCK}$ and (1.6), we have

$$\mu_\Phi^N(xz) \leq \max\{\mu_\Phi^N(xz \cdot y), \mu_\Phi^N(y)\} = \max\{\mu_\Phi^N(xy \cdot z), \mu_\Phi^N(y)\}.$$

Hence ($\mathrm{BFI2}_N$) is also satisfied. □

Theorem 4.8.12. *Every bipolar fuzzy BCC-ideal of a weak BCC-algebra $\mathfrak{X}$ is a bipolar fuzzy BCK-ideal of $\mathfrak{X}$.*

Proof. Let $\mathfrak{X}$ be a weak BCC-algebra and let $\Phi = (\mu_\Phi^P, \mu_\Phi^N)$ be a bipolar fuzzy BCC-ideal of $\mathfrak{X}$. Then conditions $(\mathrm{BFI1}_P)_{BCK}$ and $(\mathrm{BFI1}_N)_{BCK}$ are satisfied.

When we put $z = 0$ in $(\mathrm{BFI2}_P)$, we get $(\mathrm{BFI2}_P)_{BCK}$. Similarly, by putting $z = 0$ in $(\mathrm{BFI2}_N)$ we get condition $(\mathrm{BFI2}_N)_{BCK}$. □

Lemma 4.8.13. *Let $\Phi = (\mu_\Phi^P, \mu_\Phi^N)$ be a bipolar fuzzy BCK-ideal of weak BCC-algebra $\mathfrak{X}$. If the inequality $xy \leq z$ holds for $x, y, z \in X$, then*

$$\mu_\Phi^P(x) \geq \min\{\mu_\Phi^P(y), \mu_\Phi^P(z)\}, \tag{4.28}$$

$$\mu_\Phi^N(x) \leq \max\{\mu_\Phi^N(y), \mu_\Phi^N(z)\}. \tag{4.29}$$

Proof. Let $x, y, z \in X$ be such that $xy \leq z \iff xy \cdot z = 0$. Then we have

$$\begin{aligned}
\mu_\Phi^P(x) &\geq \min\{\mu_\Phi^P(xy), \mu_\Phi^P(y)\} \\
&\geq \min\{\min\{\mu_\Phi^P(xy \cdot z), \mu_\Phi^P(z)\}, \mu_\Phi^P(y)\} \\
&= \min\{\min\{(\mu_\Phi^P(0), \mu_\Phi^P(z)\}, \mu_\Phi^P(y)\} \\
&= \min\{\mu_\Phi^P(y), \mu_\Phi^P(z)\}.
\end{aligned}$$

Similarly,

$$\begin{aligned}
\mu_\Phi^N(x) &\leq \max\{\mu_\Phi^N(xy), \mu_\Phi^N(y)\} \\
&\leq \max\{\max\{\mu_\Phi^N(xy \cdot z), \mu_\Phi^N(z)\}, \mu_\Phi^N(y)\} \\
&= \max\{\max\{(\mu_\Phi^N(0), \mu_\Phi^N(z)\}, \mu_\Phi^N(y)\} \\
&= \max\{\mu_\Phi^N(y), \mu_\Phi^N(z)\}.
\end{aligned}$$

The proof is completed. □

Lemma 4.8.14. *Let $\Phi = (\mu_\Phi^P, \mu_\Phi^N)$ be a bipolar fuzzy BCK-ideal of a weak BCC-algebra $\mathfrak{X}$. If the inequality $x \leq y$ holds for $x, y \in X$, then*

$$\mu_\Phi^P(x) \geq \mu_\Phi^P(y) \text{ and } \mu_\Phi^N(x) \leq \mu_\Phi^N(y).$$

Proof. Let $x, y \in X$ be such that $x \leq y$. Then

$$\mu_\Phi^P(x) \geq \min\{\mu_\Phi^P(xy), \mu_\Phi^P(y)\} = \min\{\mu_\Phi^P(0), \mu_\Phi^P(y)\} = \mu_\Phi^P(y),$$

i.e., $\mu_\Phi^P(x) \geq \mu_\Phi^P(y)$.
Similarly,

$$\mu_\Phi^N(x) \leq \max\{\mu_\Phi^N(xy), \mu_\Phi^N(y)\} = \max\{\mu_\Phi^N(0), \mu_\Phi^N(y)\} = \mu_\Phi^N(y),$$

i.e., $\mu_\Phi^N(x) \leq \mu_\Phi^N(y)$. □

Theorem 4.8.15. *Every bipolar fuzzy BCK-ideal of a weak BCC-algebra $\mathfrak{X}$ is a bipolar fuzzy subalgebra of $\mathfrak{X}$.*

Proof. Let $\Phi = (\mu_\Phi^P, \mu_\Phi^N)$ be a bipolar fuzzy BCK-ideal of a weak BCC-algebra $\mathfrak{X}$. Since for all $x, y \in X$, by Theorem 1.1.30(4), $xy \leq x$, from Lemma 4.8.14 it follows that

$$\mu_\Phi^P(xy) \geq \mu_\Phi^P(x)$$

and

$$\mu_\Phi^N(xy) \leq \mu_\Phi^N(x).$$

Now, by $(\mathrm{BFI2}_P)_{BCK}$, we get

$$\begin{aligned} \mu_\Phi^P(xy) &\geq \mu_\Phi^P(x) \\ &\geq \min\{\mu_\Phi^P(xy), \mu_\Phi^P(y)\} \\ &\geq \min\{\mu_\Phi^P(x), \mu_\Phi^P(y)\}. \end{aligned}$$

Similarly,

$$\begin{aligned} \mu_\Phi^N(xy) &\leq \mu_\Phi^N(x) \\ &\leq \max\{\mu_\Phi^N(xy), \mu_\Phi^N(y)\} \\ &\leq \max\{\mu_\Phi^N(x), \mu_\Phi^N(y)\}. \end{aligned}$$

From Definition 4.8.2 we know that $\Phi = (\mu_\Phi^P, \mu_\Phi^N)$ is a bipolar fuzzy subalgebra of $\mathfrak{X}$. □

By combining Theorems 4.8.15 and 4.8.12, we get the following corollary:

Corollary 4.8.16. *Every bipolar fuzzy BCC-ideal of a weak BCC-algebra $\mathfrak{X}$ is a bipolar fuzzy BCC-subalgebra of $\mathfrak{X}$.*

The converse of Theorem 4.8.15 is not true. For example, the bipolar fuzzy subalgebra $\Phi = (\mu_\Phi^P, \mu_\Phi^N)$ in Example 4.8.3 is not a bipolar fuzzy BCK-ideal of $\mathfrak{X}$. Indeed, we have

$$\mu_\Phi^N(2) = -0.2 > -0.7 = \max\{\mu_\Phi^N(2 \cdot 1), \mu_\Phi^N(1)\} = \max\{\mu_\Phi^N(1), \mu_\Phi^N(1)\}.$$

This case does not satisfy condition $(\text{BFI2}_N)_{BCK}$.

Theorem 4.8.17. *A bipolar fuzzy set $\Phi = (\mu_\Phi^P, \mu_\Phi^N)$ in a weak BCC-algebra $\mathfrak{X}$ is a bipolar fuzzy BCK-ideal of $\mathfrak{X}$ if and only if it satisfies the following two conditions:*

(1) $\Phi_t^P \neq \emptyset \Longrightarrow \Phi_t^P$ *is a BCK-ideal of $\mathfrak{X}$ for any $t \in [0,1]$,*

(2) $\Phi_s^N \neq \emptyset \Longrightarrow \Phi_s^N$ *is a BCK-ideal of $\mathfrak{X}$ for any $s \in [-1,0]$.*

Proof. Let $\Phi = (\mu_\Phi^P, \mu_\Phi^N)$ be a bipolar fuzzy BCK-ideal of a weak BCC-algebra $\mathfrak{X}$ and let $(s,t) \in [-1,0] \times [0,1]$ be such that $\Phi_t^P \neq \emptyset$ and $\Phi_s^N \neq \emptyset$. Clearly, $0 \in \Phi_t^P \cap \Phi_s^N$.

Let now $x_t y_t \in \Phi_t^P$ and $y_t \in \Phi_t^P$ for $x_t, y_t \in X$ and let $x_s y_s \in \Phi_s^N$ and $y_s \in \Phi_s^N$ for $x_s, y_s \in X$. Then we have

$$\mu_\Phi^P(x_t y_t) \geq t,\ \mu_\Phi^P(y_t) \geq t,\ \mu_\Phi^N(x_s y_s) \leq s,\ \mu_\Phi^N(y_s) \leq s.$$

It follows from $(\text{BFI2}_P)_{BCK}$

$$\mu_\Phi^P(x_t) \geq \min\{\mu_\Phi^P(x_t y_t), \mu_\Phi^P(y_t)\} \geq t$$

and similarly, from $(\text{BFI2}_N)_{BCK}$

$$\mu_\Phi^N(x_s) \leq \max\{\mu_\Phi^N(x_s y_s), \mu_\Phi^N(y_s)\} \leq s$$

so that $x_t \in \Phi_t^P$ and $x_s \in \Phi_s^N$. Hence Φ_t^P and Φ_s^N are BCK-ideals of $\mathfrak{X}$.

Conversely, let us suppose that the implications (1) and (2) are satisfied. Let $\mu_\Phi^P(x) = t$ and $\mu_\Phi^N(x) = s$ for any $x \in X$. Then, of course, $x \in \Phi_t^P \cap \Phi_s^N$, i.e., Φ_t^P and Φ_s^N are nonempty.

From assumption Φ_t^P and Φ_s^N are BCK-ideals of $\mathfrak{X}$ and hence $0 \in \Phi_t^P \cap \Phi_s^N$.

That implies that $\mu_\Phi^P(0) \geq t = \mu_\Phi^P(x)$ and $\mu_\Phi^N(0) \leq s = \mu_\Phi^N(x)$ for all $x \in X$. If there exist the elements $x_t', y_t', x_s', y_s' \in X$ satisfying the conditions

$$\mu_\Phi^P(x_t') < \min\{\mu_\Phi^P(x_t' y_t'), \mu_\Phi^P(y_t')\}$$

and

$$\mu_\Phi^N(x_s^{'}) > \max\{\mu_\Phi^N(x_s^{'}y_s^{'}), \mu_\Phi^N(y_s^{'})\},$$

then by putting

$$t_0 = \tfrac{1}{2}(\mu_\Phi^P(x_t^{'}) + \min\{\mu_\Phi^P(x_t^{'}y_t^{'}), \mu_\Phi^P(y_t^{'})\}),$$

$$s_0 = \tfrac{1}{2}(\mu_\Phi^N(x_s^{'}) + \max\{\mu_\Phi^N(x_s^{'}y_s^{'}), \mu_\Phi^N(y_s^{'})\}),$$

we have

$$\mu_\Phi^P(x_t^{'}) < t_0 < \min\{\mu_\Phi^P(x_t^{'}y_t^{'}), \mu_\Phi^P(y_t^{'})\},$$

$$\mu_\Phi^N(x_s^{'}) > s_0 > \max\{\mu_\Phi^N(x_s^{'}y_s^{'}), \mu_\Phi^N(y_s^{'})\}.$$

This gives $x_t^{'} \notin \Phi_{t_0}^P, x_t^{'}y_t^{'} \in \Phi_{t_0}^P, y_t^{'} \in \Phi_{t_0}^P$ and $x_s^{'} \notin \Phi_{s_0}^N, x_s^{'}y_s^{'} \in \Phi_{s_0}^N$, $y_s^{'} \in \Phi_{s_0}^N$. As it can be easily seen, this is a contradiction.

Thus $\Phi = (\mu_\Phi^P, \mu_\Phi^N)$ is a bipolar fuzzy BCK-ideal of $\mathfrak{X}$. □

Corollary 4.8.18. *If $\Phi = (\mu_\Phi^P, \mu_\Phi^N)$ is a bipolar fuzzy BCK-ideal of a weak BCC-algebra $\mathfrak{X}$, then the intersection of a nonempty positive t-cut and a nonempty negative s-cut of $\Phi = (\mu_\Phi^P, \mu_\Phi^N)$ is a BCK-ideal of $\mathfrak{X}$ for all $(s,t) \in [-1,0] \times [0,1]$.*

In particular, the nonempty k-cut of $\Phi = (\mu_\Phi^P, \mu_\Phi^N)$ is a BCK-ideal of $\mathfrak{X}$ for all $k \in [0,1]$.

In the following example we show such couple $(s,t) \in [-1,0] \times [0,1]$ that the set-theoretic union of a nonempty positive t-cut and nonempty negative s-cut of a bipolar fuzzy BCK-ideal $\Phi = (\mu_\Phi^P, \mu_\Phi^N)$ of a weak BCC-algebra $\mathfrak{X}$ is not a BCK-ideal of $\mathfrak{X}$.

Example 4.8.19. Let $\mathfrak{X} = \langle\{0,1,2,3\}; \cdot, 0\rangle$ be a nonproper weak BCC-algebra with the operation $\cdot$ defined by the following table (cf. the algebra I_{4-2-1} on p. 337 in [146]):

$\cdot$	0	1	2	3
0	0	1	2	3
1	1	0	3	2
2	2	3	0	1
3	3	2	1	0

Table 4.8.4

Let $\Phi = (\mu_\Phi^P, \mu_\Phi^N)$ be a bipolar fuzzy set in $\mathfrak{X}$ defined by the following table:

	0	1	2	3
μ_Φ^P	0.7	0.6	0.4	0.4
μ_Φ^N	−0.8	−0.3	−0.7	−0.3

Table 4.8.5

Then

$$\Phi_t^P = \begin{cases} \emptyset & \text{if } \ 0.7 < t, \\ \{0\} & \text{if } \ 0.6 < t \leq 0.7, \\ \{0,1\} & \text{if } \ 0.4 < t \leq 0.6, \\ X & \text{if } \ 0 \leq t \leq 0.4 \end{cases}$$

and

$$\Phi_s^N = \begin{cases} \emptyset & \text{if } \ -1 \leq s < -0.8, \\ \{0\} & \text{if } \ -0.8 \leq s < -0.7, \\ \{0,2\} & \text{if } \ -0.7 \leq s < -0.3, \\ X & \text{if } \ -0.3 \leq s < 0. \end{cases}$$

From Theorem 4.8.17, it follows that $\Phi = (\mu_\Phi^P, \mu_\Phi^N)$ is a bipolar fuzzy BCK-ideal of $\mathfrak{X}$. But

$$\Phi_{0.5}^P \cup \Phi_{-0.6}^N = \{0,1\} \cup \{0,2\} = \{0,1,2\}$$

is not a BCK-ideal of $\mathfrak{X}$. Indeed, we have $3 \cdot 1 = 2 \in \{0,1,2\}$ and $1 \in \{0,1,2\}$, but $3 \notin \{0,1,2\}$, i.e., the condition $(I2)_{BCK}$ is not satisfied. □

In the next example, we show that there exists such number $k \in [0,1]$ that the set-theoretic union of a nonempty positive k-cut and nonempty negative $(-k)$-cut of a bipolar fuzzy BCK-ideal $\Phi = (\mu_\Phi^P, \mu_\Phi^N)$ of a weak BCC-algebra $\mathfrak{X}$ is not a BCK-ideal of $\mathfrak{X}$.

Example 4.8.20. Let us consider the weak BCC-algebra defined in Example 4.5.14. Let $\Phi = (\mu_\Phi^P, \mu_\Phi^N)$ be a bipolar fuzzy set in $\mathfrak{X}$ defined by the following table:

	0	1	2	3	4
μ_Φ^P	0.8	0.6	0.5	0.3	0.3
μ_Φ^N	−0.7	−0.7	−0.2	−0.5	−0.2

Table 4.8.6

Then

$$\Phi_t^P = \begin{cases} \emptyset & \text{if } \ 0.8 < t, \\ \{0\} & \text{if } \ 0.6 < t \le 0.8, \\ \{0,1\} & \text{if } \ 0.5 < t \le 0.6, \\ \{0,1,2\} & \text{if } \ 0.3 < t \le 0.5, \\ X & \text{if } \ 0 \le t \le 0.3 \end{cases}$$

and

$$\Phi_s^N = \begin{cases} \emptyset & \text{if } \ -1 \le s < -0.7, \\ \{0,1\} & \text{if } \ -0.7 \le s < -0.5, \\ \{0,1,3\} & \text{if } \ -0.5 \le s < -0.2, \\ X & \text{if } \ -0.2 \le s < 0. \end{cases}$$

From Theorem 4.8.17 it follows that $\Phi = (\mu_\Phi^P, \mu_\Phi^N)$ is a bipolar fuzzy BCK-ideal of $\mathfrak{X}$. But

$$\Phi_{0.4}^P \cup \Phi_{-0.4}^N = \{0,1,2\} \cup \{0,1,3\} = \{0,1,2,3\}$$

is not a BCK-ideal of $\mathfrak{X}$. Indeed, we have $4 \cdot 3 = 2 \in \{0,1,2,3\}$ and $3 \in \{0,1,2,3\}$, but $4 \notin \{0,1,2,3\}$, i.e., the condition $(I2)_{BCK}$ is not satisfied. □

In the sequel we show the condition that has to be satisfied in order for the set-theoretic union of a nonempty positive k-cut and a nonempty negative $(-k)$-cut of $\Phi = (\mu_\Phi^P, \mu_\Phi^N)$ to be a BCK-ideal of a weak BCC-algebra $\mathfrak{X}$.

Theorem 4.8.21. *If $\Phi = (\mu_\Phi^P, \mu_\Phi^N)$ is a bipolar fuzzy BCK-ideal of a weak BCC-algebra $\mathfrak{X}$ satisfying for all $x \in X$ the following condition:*

$$\mu_\Phi^P(x) + \mu_\Phi^N(x) \ge 0, \tag{4.30}$$

then the set-theoretic union of a nonempty positive k-cut and a nonempty negative $(-k)$-cut of $\Phi = (\mu_\Phi^P, \mu_\Phi^N)$ is a BCK-ideal of $\mathfrak{X}$ for all $k \in [0,1]$.

Proof. Let $k \in [0,1]$. Since $\Phi_k^P \neq \emptyset$ and $\Phi_{-k}^N \neq \emptyset$, by Theorem 4.8.17, they are BCK-ideals of $\mathfrak{X}$. Hence $0 \in \Phi_k^P \cup \Phi_{-k}^N$.

Let now $xy \in \Phi_k^P \cup \Phi_{-k}^N$ and $y \in \Phi_k^P \cup \Phi_{-k}^N$ hold for $x, y \in X$. We have to consider the following four cases:

1^0. $xy \in \Phi_k^P$ and $y \in \Phi_k^P$.
It follows that $x \in \Phi_k^P \subseteq \Phi_k^P \cup \Phi_{-k}^N$.

2^0. $xy \in \Phi_k^P$ and $y \in \Phi_{-k}^N$.
We have $\mu_\Phi^P(xy) \geq k$ and $\mu_\Phi^N(y) \leq -k$. From $(\mathrm{BFI2}_P)_{BCK}$ and (4.30) it follows that

$$\mu_\Phi^P(x) \geq \min\{\mu_\Phi^P(xy), \mu_\Phi^P(y)\} \geq \min\{\mu_\Phi^P(xy), -\mu_\Phi^N(y)\} \geq k.$$

This means that $x \in \Phi_k^P \subseteq \Phi_k^P \cup \Phi_{-k}^N$.

3^0. $xy \in \Phi_{-k}^N$ and $y \in \Phi_k^P$.
We have $\mu_\Phi^N(xy) \leq -k$ and $\mu_\Phi^P(y) \geq k$. From $(\mathrm{BFI2}_P)_{BCK}$ and (4.30) we get

$$\mu_\Phi^P(x) \geq \min\{\mu_\Phi^P(xy), \mu_\Phi^P(y)\} \geq \min\{-\mu_\Phi^N(xy), \mu_\Phi^P(y)\} \geq k,$$

i.e., $x \in \Phi_k^P \subseteq \Phi_k^P \cup \Phi_{-k}^N$.

4^0. $xy \in \Phi_{-k}^N$ and $y \in \Phi_{-k}^N$.
It follows that $x \in \Phi_{-k}^N \subseteq \Phi_k^P \cup \Phi_{-k}^N$. So, $\Phi_k^P \cup \Phi_{-k}^N$ is a BCK-ideal of $\mathfrak{X}$. □

For the next theorem we need the following two subsets:

$$\Lambda^P \subset [0,1], \ \Lambda^P \neq \emptyset$$

and

$$\Lambda^N \subset [-1,0], \ \Lambda^N \neq \emptyset.$$

So, we can provide the following result:

Theorem 4.8.22. *Let $\{A_k \,|\, k \in \Lambda^P \cup \Lambda^N\}$ be a finite family of BCK-ideals of a weak BCC-algebra $\mathfrak{X}$ satisfying the following two conditions:*

$$(1) \ \ X = \left(\bigcup\{A_t \,|\, t \in \Lambda^P\}\right) \cup \left(\bigcup\{A_s \,|\, s \in \Lambda^N\}\right),$$

$$(2) \ \ m > n \iff A_m \subset A_n \text{ for any } m, n \in \Lambda^P \cup \Lambda^N.$$

Then the bipolar fuzzy set $\Phi = (\mu_\Phi^P, \mu_\Phi^N)$ *in* $\mathfrak{X}$ *defined by*

$$\mu_\Phi^P(x) = \sup\{t \in \Lambda^P \mid x \in A_t\},$$

$$\mu_\Phi^N(x) = \inf\{s \in \Lambda^N \mid x \in A_s\}$$

for all $x \in X$ *is a bipolar fuzzy BCK-ideal of* $\mathfrak{X}$.

Proof. Let $(s,t) \in [-1,0] \times [0,1]$ be such that $\Phi_t^P \neq \emptyset$ and $\Phi_s^N \neq \emptyset$. It is sufficient to prove that Φ_t^P and Φ_s^N are BCK-ideals of $\mathfrak{X}$. We begin by considering the following two cases:

1^0. $t = \sup\{r \in \Lambda^P \mid r < t\}$ and

2^0. $t \neq \sup\{r \in \Lambda^P \mid r < t\}$.

In case 1^0, we have

$$x \in \Phi_t^P \Longleftrightarrow x \in A_r \text{ for all } r < t \Longleftrightarrow x \in \bigcap\{A_r \mid r < t\},$$

which gives

$$\Phi_t^P = \bigcap\{A_r \mid r < t\},$$

i.e., Φ_t^P is a BCK-ideal of $\mathfrak{X}$.

Now, let us consider case 2^0.

We will show that $\Phi_t^P = \bigcup\{A_r \mid r \geq t\}$.

($\supseteq$) If $x \in \bigcup\{A_r \mid r \geq t\}$, then $x \in A_r$ for some $r \geq t$. But it follows that $\mu_\Phi^P(x) \geq r \geq t$. Hence $x \in \Phi_t^P$ and $\Phi_t^P \supseteq \bigcup\{A_r \mid r \geq t\}$.

($\subseteq$) If $x \notin \bigcup\{A_r \mid r \geq t\}$, then $x \notin A_r$ for all $r \geq t$. Since from the assumption $t \neq \sup\{r \in \Lambda^P \mid r < t\}$, there exists $\varepsilon > 0$ such that $(t-\varepsilon, t) \cap \Lambda^P = \emptyset$. Hence $x \notin A_r$ for all $r > t - \varepsilon$, which means that if $x \in A_r$, then $r \leq t - \varepsilon$. Thus $\mu_\Phi^P(x) \leq t - \varepsilon < t$, and so $x \notin \Phi_t^P$, which means that $\Phi_t^P \subseteq \bigcup\{A_r \mid r \geq t\}$.

Therefore, $\Phi_t^P = \bigcup\{A_r \mid r \geq t\}$.

It is a BCK-ideal of $\mathfrak{X}$ because, by (2), $\{A_k\}$ forms a chain.

Next, we show that Φ_s^N is a BCK-ideal of $\mathfrak{X}$. Here we have also two cases to consider:

3^0. $s = \inf\{q \in \Lambda^N \mid s < q\}$ and

4^0. $s \neq \inf\{q \in \Lambda^N \mid s < q\}$.

From case 3^0, it follows that

$$x \in \Phi_s^N \Longleftrightarrow x \in A_q \text{ for all } s < q \Longleftrightarrow x \in \bigcap\{A_q \,|\, s < q\}.$$

Hence $\Phi_s^N = \bigcap\{A_s \,|\, s < q\}$, i.e., Φ_s^N is a BCK-ideal of $\mathfrak{X}$.

Let us consider case 4^0. Similarly to case 2^0, we show that $\Phi_s^N = \bigcup\{A_q \,|\, q \leq s\}$.

($\supseteq$) If $x \in \bigcup\{A_q \,|\, q \leq s\}$, then $x \in A_q$ for some $q \leq s$. It follows that $\mu_\Phi^N(x) \leq q \leq s$, so that $x \in \Phi_s^N$. Hence $\Phi_s^N \supseteq \bigcup\{A_q \,|\, q \leq s\}$.

($\subseteq$) Conversely, if $x \notin \bigcup\{A_q \,|\, q \leq s\}$, then $x \notin A_q$ for all $q \leq s$. The assumption $s \neq \inf\{q \in \Lambda^N \mid s < q\}$ implies that there exists $\varepsilon > 0$ such that $(s, s+\varepsilon) \cap \Lambda^N = \emptyset$, which means that $x \neq A_q$ for all $q < s+\varepsilon$. This says that if $x \in A_q$, then $q \geq s + \varepsilon$. Thus $\mu_\Phi^N(x) \geq s + \varepsilon > s$, and so $x \notin \Phi_s^N$. Therefore, $\Phi_s^N \subseteq \bigcup\{A_q \,|\, q \leq s\}$, which together with the previous inclusion gives $\Phi_s^N = \bigcup\{A_q \,|\, q \leq s\}$ and Φ_s^N is a BCK-ideal of $\mathfrak{X}$.

The proof is completed. □

Further in this section we will look at so-called cubic bipolar subalgebras of BCC-algebras. First, we give the definitions of the used objects. We begin with the concept of interval-valued fuzzy sets.

Definition 4.8.23. We denote the set of all closed subintervals of [0,1] by $D[0,1]$ and we have

$$D[0,1] = \{\tilde{a} = [a^-, a^+] \,|\, \tilde{a} \subset [0,1] \text{ and } 0 \leq a^- \leq a^+ \leq 1\}.$$

Each closed subinterval $\tilde{a}$ of $D[0,1]$ is called an *interval valued number.*

Definition 4.8.24. We define for $\tilde{a} = [a^-, a^+]$ and $\tilde{b} = [b^-, b^+]$ the following five binary operations on $D[0,1]$ as follows:

(1) $\tilde{a} \preceq \tilde{b} \Longleftrightarrow a^- \leq b^-$ and $a^+ \leq b^+$,
(2) $\tilde{a} \succeq \tilde{b} \Longleftrightarrow a^- \geq b^-$ and $a^+ \geq b^+$,
(3) $\tilde{a} = \tilde{b} \Longleftrightarrow a^- = b^-$ and $a^+ = b^+$,
(4) $\operatorname{rmin}\{\tilde{a}, \tilde{b}\} = [\min\{a^-, b^-\}, \min\{a^+, b^+\}]$,
(5) $\operatorname{rmax}\{\tilde{a}, \tilde{b}\} = [\max\{a^-, b^-\}, \max\{a^+, b^+\}]$.

The operations rmin and rmax are known as *refined minimum* and *refined maximum* of two elements $\tilde{a}, \tilde{b}$ in $D[0,1]$, respectively.

Obviously $\tilde{0} = [0,0]$ and $\tilde{1} = [1,1]$ are the least and the greatest elements, respectively.

We define so-called *refined infimum*, or briefly rinf, and *refined supremum*, or briefly rsup, of two elements in $D[0,1]$.

Definition 4.8.25. Let $\{\tilde{a}_\xi \,|\, \xi \in \Lambda\} = \{[a_\xi^-, a_\xi^+] \,|\, \xi \in \Lambda\} \subset D[0,1]$. Then its rinf and rsup are as follows:

$$\operatorname*{rinf}_{\xi\in\Lambda} \tilde{a}_\xi = \left[\inf_{\xi\in\Lambda} a_\xi^-, \inf_{\xi\in\Lambda} a_\xi^+\right]$$

and

$$\operatorname*{rsup}_{\xi\in\Lambda} \tilde{a}_\xi = \left[\sup_{\xi\in\Lambda} a_\xi^-, \sup_{\xi\in\Lambda} a_\xi^+\right].$$

Let $D[-1,0]$ denote the set of all closed subintervals of $[-1,0]$. Then the above operations are defined similarly for elements of $D[-1,0]$.

Definition 4.8.26. A mapping $\mu : X \longrightarrow D[0,1]$, for a nonempty set X, is called an *interval-valued fuzzy set* in X (briefly denoted by *IVF*) $\tilde{\mu}$ defined on a nonempty set X is given by

$$\tilde{\mu} = \{(x, [\mu^-(x), \mu^+(x)]) \,|\, x \in X\}, \tag{4.31}$$

where μ^- and μ^+ are two fuzzy sets in X such that $\mu^-(x) \leq \mu^+(x)$ for all $x \in X$. They are called a *lower* and an *upper* fuzzy set in X, respectively.

The set $\tilde{\mu}$ in (4.31) is briefly denoted by $\tilde{\mu} = [\mu^-, \mu^+]$.

For any IVF set $\tilde{\mu}$ on X and $x \in X$

$$\tilde{\mu}(x) = [\mu(x)^-, \mu^+(x)]$$

is called the *degree of membership* of an element x to $\tilde{\mu}$, in which $\mu^-(x)$ and $\mu^+(x)$ are referred to as the *lower degree* and *upper degree* of membership of x to μ.

We have already all definitions of the objects we need to describe cubic sets, so we continue with the following definition:

Definition 4.8.27. A *cubic set* $\mathcal{C}$ in X is a structure

$$\mathcal{C} = \{\langle x, \tilde{\mu}(x), \mu(x)\rangle \mid x \in X\},$$

which is briefly denoted by $\mathcal{C} = \langle \tilde{\mu}, \mu\rangle$, where $\tilde{\mu} = [\mu^-, \mu^+]$ is an IVF set in X and μ is a fuzzy set in X.

Definition 4.8.28. For a nonempty set X, a pair $A = (A^N, A^P)$ is called an *interval-valued bipolar fuzzy set* (briefly, *IVBFS*) in X if $A^N : X \longrightarrow D[-1, 0]$ and $A^P : X \longrightarrow D[0, 1]$.

In this case for any $x \in X$ we have $A^N(x) = [A^{N,-}(x), A^{N,+}(x)]$ and $A^P(x) = [A^{P,-}(x), A^{P,+}(x)]$.

Next, we will provide the definition of cubic bipolar BCC-ideal in BCC-algebra.

Definition 4.8.29. Let us assume that the set X is nonempty. Then a pair $\mathcal{C} = \langle A, \Phi\rangle$ is called a *cubic bipolar fuzzy set* in X if

(1) $A = (A^N, A^P)$ is an interval-valued bipolar fuzzy set,

(2) $\Phi = (\mu_\Phi^P, \mu_\Phi^N)$ is a bipolar fuzzy set in X.

For a weak BCC-algebra $\mathfrak{X}$ the set of all cubic bipolar fuzzy sets in X we will denote as $CB(X)$.

Definition 4.8.30. Let $\mathfrak{X}$ be a weak BCC-algebra and let $\mathcal{C} \in CB(X)$. We call $\mathcal{C} = \langle A, \Phi\rangle$ a *cubic bipolar subalgebra* of $\mathfrak{X}$ if it satisfies the following conditions for all $x, y \in X$:

(CBS1) $A^P(xy) \geq \min\{A^P(x), A^P(y)\}$,

(CBS2) $A^N(xy) \leq \max\{A^N(x), A^N(y)\}$,

(CBS3) $\mu_\Phi^P(xy) \geq \min\{\mu_\Phi^P(x), \mu_\Phi^P(y)\}$,

(CBS4) $\mu_\Phi^N(xy) \leq \max\{\mu_\Phi^N(x), \mu_\Phi^N(y)\}$.

Example 4.8.31. The algebra $\mathfrak{X} = \langle\{0, 1, 2, 3\}; \cdot, 0\rangle$ with the operation $\cdot$ defined by Table 1.2.13 is a BCC-algebra. Let us consider the cubic bipolar fuzzy set $\mathcal{C} = \langle A, \Phi\rangle$ defined as follows:

$$A(x) = \begin{cases} ([-0.9, -0.5], [0.3, 0.9]) & \text{if} x \in \{0, 1\}, \\ ([-0.8, -0.2], [0.1, 0.6]) & \text{otherwise} \end{cases}$$

and

$$\Phi(x) = \begin{cases} (-0.5, 0.7) & \text{for} \quad x = 0, \\ (-0.3, 0.6) & \text{for} \quad x = 1, \\ (-0.2, 0.5) & \text{for} \quad x = 2, \\ (-0.1, 0.4) & \text{for} \quad x = 3. \end{cases}$$

It is not difficult to check that $\mathcal{C}$ is a cubic bipolar subalgebra of $\mathfrak{X}$. □

Lemma 4.8.32. *Let $\mathfrak{X}$ be a BCC-algebra and let $\mathcal{C} \in CB(X)$. If $\mathcal{C}$ is a cubic bipolar subalgebra of $\mathfrak{X}$, then it satisfies the following conditions:*

(1) $A^P(0) \geq A^P(x)$,

(2) $A^N(0) \leq A^N(x)$,

(3) $\mu_\Phi^P(0) \geq \mu_\Phi^P(x)$,

(4) $\mu_\Phi^N(0) \leq \mu_\Phi^N(x)$.

Proof. The conditions follow directly from Definition 4.8.30. Indeed, for the proof of (1), for $x \in X$ we have

$$A^P(0) = A^P(xx) \geq \min\{A^P(x), A^P(x)\} = A^P(x).$$

In a similar way, we prove condition (2).

Now we prove (4):

$$\mu_\Phi^N(0) = \mu_\Phi^N(xx) \leq \max\{\mu_\Phi^N(x), \mu_\Phi^N(x)\} = \mu_\Phi^N(x).$$

The proof of condition (3) is similar to the above. □

Let us now turn to a consideration of the ideals in the environment of cubic bipolar weak BCC-algebras. First, we define the ideals of interest to us.

Definition 4.8.33. Let $\mathfrak{X}$ be a weak BCC-algebra and let $\mathcal{C} \in CB(X)$. We call $\mathcal{C}$ a *cubic bipolar BCK-ideal* of $\mathfrak{X}$ if it satisfies the following conditions for all $x, y \in X$:

$(\mathrm{CBI1})_{\mathrm{BCK}}$ $\quad A^P(x) \geq \min\{A^P(xy), A^P(y)\}$,

$(\mathrm{CBI2})_{\mathrm{BCK}}$ $\quad A^N(x) \leq \max\{A^N(xy), A^N(y)\}$,

$(\mathrm{CBI3})_{\mathrm{BCK}}$ $\quad \mu_\Phi^P(x) \geq \min\{\mu_\Phi^P(xy), \mu_\Phi^P(y)\}$,

$(\mathrm{CBI4})_{\mathrm{BCK}}$ $\quad \mu_\Phi^N(x) \leq \max\{\mu_\Phi^N(xy), \mu_\Phi^N(y)\}$.

Example 4.8.34. Let $\mathfrak{X} = \langle\{0,1,2,3\};\ \cdot, 0\rangle$ be a BCC-algebra with the operation $\cdot$ defined by Table 1.2.13. Let us consider the cubic bipolar fuzzy set $\mathcal{C} = \langle A, \Phi\rangle$ defined as follows:

$$A(x) = \begin{cases} ([-0.9,-0.5],[0.5,0.9]) & \text{if} x \in \{0,2\}, \\ ([-0.6,-0.2],[0.2,0.6]) & \text{otherwise} \end{cases}$$

and

$$\Phi(x) = \begin{cases} (-0.4, 0.6) & \text{for} \quad x = 0, \\ (-0.1, 0.3) & \text{for} \quad x = 1, \\ (-0.4, 0.6) & \text{for} \quad x = 2, \\ (-0.1, 0.3) & \text{for} \quad x = 3. \end{cases}$$

It is not difficult to check that $\mathcal{C}$ is a cubic bipolar BCK-ideal of $\mathfrak{X}$. □

Definition 4.8.35. Let $\mathfrak{X}$ be a weak BCC-algebra and let $\mathcal{C} \in CB(X)$. We call $\mathcal{C}$ a *cubic bipolar BCC-ideal* of $\mathfrak{X}$ if it satisfies the following conditions for all $x, y \in X$:

(CBI1) $A^P(xz) \geq \min\{A^P(xy \cdot z), A^P(y)\}$,

(CBI2) $A^N(xz) \leq \max\{A^N(xy \cdot z), A^N(y)\}$,

(CBI3) $\mu_\Phi^P(xz) \geq \min\{\mu_\Phi^P(xy \cdot z), \mu_\Phi^P(y)\}$,

(CBI4) $\mu_\Phi^N(xz) \leq \max\{\mu_\Phi^N(xy \cdot z), \mu_\Phi^N(y)\}$.

Example 4.8.36. Let algebra $\mathfrak{X} = \langle\{0,1,2,3\};\ \cdot, 0\rangle$ with the operation $\cdot$ defined by Table 1.2.13 being a BCC-algebra. Let us consider the cubic bipolar fuzzy set $\mathcal{C} = \langle A, \Phi\rangle$ defined as follows:

$$A(x) = \begin{cases} ([-0.9,-0.5],[0.2,0.8]) & \text{if} x \in \{0,1\}, \\ ([-0.8,-0.2],[0.1,0.5]) & \text{otherwise} \end{cases}$$

and

$$\Phi(x) = \begin{cases} (-0.3, 0.5) & \text{for} \quad x = 0 \text{ and } x = 1, \\ (-0.2, 0.3) & \text{for} \quad x = 2 \text{ and } x = 3. \end{cases}$$

Then it can be easily checked that $\mathcal{C}$ is a cubic bipolar BCC-ideal of $\mathfrak{X}$. □

Theorem 4.8.37. *In a weak BCC-algebra $\mathfrak{X}$ every cubic bipolar BCC-ideal is a cubic bipolar BCK-ideal of $\mathfrak{X}$.*

Proof. Putting $z = 0$ in (CBI1) and (CBI2), we receive $(\text{CBI1})_{BCK}$ and $(\text{CBI2})_{BCK}$, respectively. □

The converse of the above theorem is not true. We show it in the next example.

Example 4.8.38. Let $\mathcal{C} = \langle A, \Phi \rangle$ be the cubic bipolar BCK-ideal of $\mathfrak{X}$ given in Example 4.8.34. Then we have

$$A^P(1 \cdot 3) = [0.2, 0.6] \not\geq [0.5, 0.9] = \min\{A^P((1 \cdot 2) \cdot 3), A^P(2)\}.$$

As we can see, $\mathcal{C}$ is not a cubic bipolar BCC-ideal of $\mathfrak{X}$. □

Lemma 4.8.39. *Let $\mathfrak{X}$ be a BCC-algebra and let $\mathcal{C} \in CB(X)$. If $\mathcal{C}$ is a cubic bipolar BCC-ideal of $\mathfrak{X}$, then it satisfies the following conditions:*

(1) $A^P(0) \geq A^P(x)$,
(2) $A^N(0) \leq A^N(x)$,
(3) $\mu_\Phi^P(0) \geq \mu_\Phi^P(x)$,
(4) $\mu_\Phi^N(0) \leq \mu_\Phi^N(x)$

for any $x \in X$.

Proof. It is straightforward. □

Lemma 4.8.40. *Let $\mathfrak{X}$ be a BCC-algebra and let $\mathcal{C}$ be a BCC-ideal of $\mathfrak{X}$. Then the following conditions are satisfied for any $x, y \in X$ such that $x \leq y$:*

(1) $A^P(x) \geq A^P(y)$,
(2) $A^N(x) \leq A^N(y)$,
(3) $\mu_\Phi^P(x) \geq \mu_\Phi^P(y)$,
(4) $\mu_\Phi^N(x) \leq \mu_\Phi^N(y)$.

Proof. Let $x, y \in X$ and let $x \leq y$. Then we have

$$\begin{aligned}
A^P(x) &= A^P(x \cdot 0) && \text{by (BCC3)} \\
&\geq \min\{A^P(xy \cdot 0), A^P(y)\} && \text{by (CBI1)} \\
&= \min\{A^P(0 \cdot 0), A^P(y)\} && \text{from assumption} \\
&= \min\{A^P(0), A^P(y)\} && \text{by (BCC2)} \\
&= A^P(y) && \text{by Lemma 4.8.39.}
\end{aligned}$$

In a similar way, we show the condition (2).

Now we prove the condition (3). We have

$$
\begin{aligned}
\mu^P(x) &= \mu_\Phi^P(x \cdot 0) && \text{by (BCC3)} \\
&\geq \min\{\mu_\Phi^P(xy \cdot 0), \mu_\Phi^P(y)\} && \text{by (CBI3)} \\
&= \min\{\mu_\Phi^P(0 \cdot 0), \mu_\Phi^P(y)\} && \text{from assumption} \\
&= \min\{\mu_\Phi^P(0), \mu_\Phi^P(y)\} && \text{by (BCC2)} \\
&= \mu_\Phi^P(y) && \text{by Lemma 4.8.39.}
\end{aligned}
$$

Similarly, we prove the condition (4).

The proof is completed. □

Lemma 4.8.41. *Let $\mathfrak{X}$ be a BCC-algebra and let $\mathcal{C}$ be a BCC-ideal of $\mathfrak{X}$. Then for any $x, y, z \in X$ such that $xy \leq z$ the following conditions are satisfied:*

(1) $A^P(x) \geq \min\{A^P(y), A^P(z)\}$,

(2) $A^N(x) \leq \max\{A^N(y), \mu^N(z)\}$,

(3) $\mu_\Phi^P(x) \geq \min\{\mu_\Phi^P(y), \mu_\Phi^P(z)\}$,

(4) $\mu_\Phi^N(x) \leq \max\{\mu_\Phi^N(y), \mu_\Phi^N(z)\}$.

Proof. Let for $x, y, z \in X$ the condition $xy \leq z$ be satisfied. First, observe that from Lemma 4.8.40 it follows

$$A^P(xy) \geq A^P(z) \tag{4.32}$$

and

$$\mu_\Phi^N(xy) \leq \mu_\Phi^N(z). \tag{4.33}$$

Then we have

$$
\begin{aligned}
A^P(x) &= A^P(x \cdot 0) && \text{by (BCC3)} \\
&\geq \min\{A^P(xy \cdot 0), A^P(y)\} && \text{by (CBI1)} \\
&= \min\{A^P(xy), A^P(y)\} \\
&\geq \min\{A^P(z), A^P(y)\} && \text{by (4.32)} \\
&= \min\{A^P(y), A^P(z)\}.
\end{aligned}
$$

We have proven condition (1).

The condition (2) can be proven similarly.

Next, we have

$$\begin{aligned} \mu_\Phi^N(x) &= \mu_\Phi^N(x \cdot 0) && \text{by (BCC3)} \\ &\leq \max\{\mu_\Phi^N(xy \cdot 0), \mu_\Phi^N(y)\} && \text{by (CBI4)} \\ &= \max\{\mu_\Phi^N(xy), \mu_\Phi^N(y)\} \\ &\leq \max\{\mu_\Phi^N(z), \mu_\Phi^N(y)\} && \text{by (4.33)} \\ &= \max\{\mu_\Phi^N(y), \mu_\Phi^N(z)\}. \end{aligned}$$

So, we have proven (4). The proof of condition (3) is similar. □

Theorem 4.8.42. *Every cubic bipolar BCK-ideal of a BCC-algebra $\mathfrak{X}$ is a cubic bipolar subalgebra of $\mathfrak{X}$.*

Proof. Let $\mathcal{C}$ be a cubic bipolar BCK-ideal of a BCC-algebra $\mathfrak{X}$ and let $x, y \in X$. From Theorem 1.1.30(4) we get $xy \leq x$, which implies

$$\begin{aligned} A^P(xy) &\geq \min\{A^P(xy \cdot x), A^P(x)\} && \text{by (CBI1)}_{\text{BCK}} \\ &= \min\{A^P(0), A^P(x)\} \\ &\geq \min\{A^P(y), A^P(x)\} && \text{by Lemma 4.8.40} \\ &= \min\{A^P(x), A^P(y)\}. \end{aligned}$$

We have shown that condition (CBSI1) holds. Similarly we show that condition (CBS2) holds. Now,

$$\begin{aligned} \mu_\Phi^N(xy) &\leq \max\{\mu_\Phi^N(xy \cdot x), \mu_\Phi^N(x)\} && \text{by (CBI4)}_{\text{BCK}} \\ &= \max\{\mu_\Phi^N(0), \mu_\Phi^N(x)\} \\ &\leq \max\{\mu_\Phi^N(y), \mu_\Phi^N(x)\} && \text{by Lemma 4.8.40} \\ &= \max\{\mu_\Phi^N(x), \mu_\Phi^N(y)\}. \end{aligned}$$

We have shown that condition (CBS4) holds. Condition (CBS3) can be shown in the same way.

Thus the BCK-ideal $\mathcal{C}$ is a subalgebra of $\mathfrak{X}$. □

The following corollary is known from the section about BCC-ideals, as well as from a few others, in this case it follows from Theorem 4.8.37 and Theorem 4.8.42.

Corollary 4.8.43. *Every cubic bipolar BCC-ideal of a BCC-algebra $\mathfrak{X}$ is a cubic bipolar subalgebra of $\mathfrak{X}$.*

As with the other types of ideals, the converse of the result is not true in this case as well. We will show it in an example:

Example 4.8.44. Let $\mathfrak{X} = \langle\{0,1,2,3\};\cdot,0\rangle$ with the operation $\cdot$ defined by Table 1.2.13 be a BCC-algebra and $\mathcal{C}$ be the cubic bipolar subalgebra of $\mathfrak{X}$ defined in Example 4.8.31. Then

$$\min\{\mu_\Phi^P(3\cdot 2), \mu_\Phi^P(2)\} = 0.5 \not\leq 0.4 = \mu_\Phi^P(3),$$

i.e., $\mathcal{C}$ is not a BCK-ideal $\mathfrak{X}$.
Next, we have

$$\min\{\mu_\Phi^P((3\cdot 2)\cdot 1), \mu_\Phi^P(2)\} = 0.5 \not\leq 0.4 = \mu_\Phi^P(3\cdot 1),$$

i.e., $\mathcal{C}$ is not a BCC-ideal of $\mathfrak{X}$. □

The next result shows that not all results that are true for cubic bipolar BCK-algebras can be generalized to BCC-algebras.

Theorem 4.8.45. *Let $\mathcal{C}$ be a cubic bipolar subalgebra of a BCK-algebra $\mathfrak{X}$. Then $\mathcal{C}$ is a cubic bipolar BCC-ideal of $\mathfrak{X}$ if the following conditions are satisfied for any $x, y, z \in X$ such that $xy \leq z$:*

(1) $A^P(x) \geq \min\{A^P(y), A^P(z)\}$,
(2) $A^N(x) \leq \max\{A^N(y), A^N(z)\}$,
(3) $\mu_\Phi^P(x) \geq \min\{\mu_\Phi^P(y), \mu_\Phi^P(z)\}$,
(4) $\mu_\Phi^N(x) \leq \max\{\mu_\Phi^N(y), \mu_\Phi^N(z)\}$.

Proof. Let $x, y, z \in X$ be such that $xy \leq z$. Then, by Theorem 1.1.21, condition (1.3) is satisfied and by (BCC1) we have

$$xz\cdot(xy\cdot z) \leq x\cdot xy \leq y.$$

Hence, by Lemma 4.8.41, we get

$$A^P(xz) \geq \min\{A^P(xz\cdot y), \mu^P(y)\}$$

and

$$\mu_\Phi^N(xz) \leq \max\{\mu_\Phi^N(xz\cdot y), \mu_\Phi^N(y)\}.$$

We have shown that conditions (CBI1) and (CBI4) are satisfied. Similarly it can be shown that conditions (CBI2) and (CBI3) are also satisfied.

Thus $\mathcal{C}$ is a cubic bipolar BCC-ideal of $\mathfrak{X}$. □

Since a proper BCC-algebra does not satisfy condition (1.3), the above theorem is not true for BCC-algebras.

To present further results, we need some definitions of the used objects. We begin with the definition of cubic bipolar numbers.

Definition 4.8.46. The elements of the Cartesian products $(D[-1,0] \times D[0,1]) \times ([-1,0] \times [0,1])$ are called *cubic bipolar numbers* and denoted by $\tilde{\bar{a}}, \tilde{\bar{b}}, \tilde{\bar{c}}, \ldots$, where

$$\tilde{\bar{a}} = \langle \tilde{a}, \bar{a} \rangle = \left\langle ([a^{N,-}, a^{N,+}], [a^{P,-}, a^{P,+}]), (a^N, a^P) \right\rangle .$$

For cubic bipolar numbers we define the binary operations $\preceq$, $=$, min and max of arbitrary cubic bipolar numbers.

Let $\tilde{\bar{a}}, \tilde{\bar{b}} \in (D[-1,0] \times D[0,1]) \times ([-1,0] \times [0,1])$, where

$$\tilde{\bar{a}} = \langle \tilde{a}, \bar{a} \rangle = \left\langle ([a^{N,-}, a^{N,+}], [a^{P,-}, a^{P,+}]), (a^N, a^P) \right\rangle$$

and

$$\tilde{\bar{b}} = \langle \tilde{b}, \bar{b} \rangle = \left\langle ([b^{N,-}, b^{N,+}], [b^{P,-}, b^{P,+}]), (b^N, b^P) \right\rangle ,$$

then

(1) $\tilde{\bar{a}} \preceq \tilde{\bar{b}} \iff a^{N,-} \geq b^{N,-}\,,\ a^{P,+} \leq b^{P,+}\,,\ a^N \geq b^N\,,\ a^P \leq b^P,$

(2) $\tilde{\bar{a}} = \tilde{\bar{b}} \iff \tilde{a} = \tilde{b}\,,\ \bar{a} = \bar{b},$

(3) $\min\{\tilde{\bar{a}}, \tilde{\bar{b}}\} = \langle ([\max\{a^{N,-}, b^{N,-}\}, \max\{a^{N,+}, b^{N,+}\}],$
$[\min\{a^{P,-}, b^{P,-}\}, \min\{a^{P,+}, b^{P,+}\}]), (\max\{a^N, b^N\}, \min\{a^P, b^P\})\rangle,$

(4) $\max\{\tilde{\bar{a}}, \tilde{\bar{b}}\} = \langle ([\min\{a^{N,-}, b^{N,-}\}, \min\{a^{N,+}, b^{N,+}\}],$
$[\max\{a^{P,-}, b^{P,-}\}, \max\{a^{P,+}, b^{P,+}\}]), (\min\{a^N, b^N\}, \max\{a^P, b^P\})\rangle.$

Definition 4.8.47. Let X be a nonempty set, let $\mathcal{C} \in CB(X)$ and let $\tilde{\bar{a}}$ be a cubic bipolar number. Then the subset $[\mathcal{C}]_{\tilde{\bar{a}}}$ of X defined as

$$\begin{aligned}[\mathcal{C}]_{\tilde{\bar{a}}} = \{x \in X \mid \tilde{A}^P(x) \geq \tilde{a}^P,\ \tilde{A}^N(x) \leq \tilde{a}^N, \\ A^P(x) \geq a^P,\ A^N(x) \leq a^N\}\end{aligned}$$

is called the $\tilde{\bar{a}}$-level set of $\mathcal{C}$, where

$$\tilde{a}^P = [a^{P,-}, a^{P,+}] \in D[0,1] \text{ and } \tilde{a}^N = [a^{N,-}, a^{N,+}] \in D[-1,0].$$

Theorem 4.8.48. *Let $\mathfrak{X}$ be a BCC-algebra, let $\tilde{\tilde{a}}$ be a cubic bipolar number and let $\mathcal{C} \in CB(X)$. Then $\mathcal{C}$ is a cubic bipolar BCC-ideal of $\mathfrak{X}$ if and only if $[\mathcal{C}]_{\tilde{\tilde{a}}}$ is a nonempty BCC-ideal of $\mathfrak{X}$.*

Proof. Let us suppose that $\mathcal{C}$ is a nonempty cubic bipolar BCC-ideal of $\mathfrak{X}$. For $x, y, z \in X$, let $xy \cdot z, y \in [\mathcal{C}]_{\tilde{\tilde{a}}}$ and let $\tilde{a}^P = [a^{P,-}, a^{P,+}], \tilde{a}^N = [a^{N,-}, a^{N,+}]$.
Then we have

$$A^P(xy \cdot z) \geq \tilde{a}^P, \tag{4.34}$$

hence

$$A^P(y) \geq \tilde{a}^P. \tag{4.35}$$

And

$$A^N(xy \cdot z) \leq \tilde{a}^N, \tag{4.36}$$

which gives

$$A^N(y) \leq \tilde{a}^N. \tag{4.37}$$

Similarly,

$$\mu_\Phi^P(xy \cdot z) \geq a^P, \tag{4.38}$$

which gives

$$\mu_\Phi^P(y) \geq a^P. \tag{4.39}$$

And

$$\mu_\Phi^N(xy \cdot z) \leq a^N, \tag{4.40}$$

i.e.,

$$\mu_\Phi^N(y) \leq a^N. \tag{4.41}$$

This implies

$$\begin{aligned} A^P(xz) &\geq \min\{A^P(xy \cdot z), A^P(y)\} && \text{from assumption} \\ &\geq \min\{\tilde{a}^P, \tilde{a}^P\} && \text{by (4.34) and by (4.35)} \\ &= \tilde{a}^P \end{aligned}$$

and

$$\begin{aligned} \mu_\Phi^P(xz) &\geq \min\{\mu_\Phi^P(xy \cdot z), \mu_\Phi^P(y)\} && \text{from assumption} \\ &\geq \min\{\tilde{a}^P, \tilde{a}^P\} && \text{by (4.34) and by (4.35)} \\ &= \tilde{a}^P. \end{aligned}$$

In the same way we can show that

$$A^N(xz) \leq \tilde{a}^N$$

and

$$\mu_\Phi^N(xz) \leq a^N.$$

So, we have proven that $xz \in [\mathcal{C}]_{\tilde{\bar{a}}}$. Thus $[\mathcal{C}]_{\tilde{\bar{a}}}$ is a BCC-ideal of $\mathfrak{X}$.

Conversely, let us suppose that a nonempty $[\mathcal{C}]_{\tilde{\bar{a}}}$ is a BCC-ideal of $\mathfrak{X}$. We will show that for $\mathcal{C}$ the conditions of Definition 4.8.35 hold. We will consider four cases.

Case 1^0. We assume that the condition (CBI1) does not hold. Then there are $x_0, y_0, z_0 \in X$ such that

$$A^P(x_0z_0) < \min\{A^P(x_0y_0 \cdot z_0), A^P(y_0)\}.$$

Now we put

$$A^P(x_0y_0 \cdot z_0) = [b^{P,-}, b^{P,+}],\ A^P(y_0) = [c^{P,-}, c^{P,+}] \text{ and}$$
$$A^P(x_0z_0) = [a^{P,-}, a^{P,+}],$$

where $[a^{P,-}, a^{P,+}]$, $[b^{P,-}, b^{P,+}]$, $[c^{P,-}, c^{P,+}] \in D[0,1]$.

Then, as it can be easily seen,

$$\begin{aligned}[a^{P,-}, a^{P,+}] &< \min\{[b^{P,-}, b^{P,+}], [c^{P,-}, c^{P,+}]\} \\ &= [\min\{b^{P,-}, c^{P,-}\}, \min\{b^{P,+}, c^{P,+}\}].\end{aligned}$$

Next, let

$$[d^{P,-}, d^{P,+}] = [\frac{1}{2}\min\{a^{P,-} + b^{P,-}, c^{P,-}\}, \frac{1}{2}\min\{a^{P,+} + b^{P,+}, c^{P,+}\}].$$

Then we obtain the following inequalities:

$$\min\{b^{P,-}, c^{P,-}\} > d^{P,-} = \frac{1}{2}\min\{a^{P,-} + b^{P,-}, c^{P,-}\} > a^{P,-}$$

and

$$\min\{b^{P,+}, c^{P,+}\} > d^{P,+} = \frac{1}{2}\min\{a^{P,+} + b^{P,+}, c^{P,+}\} > a^{P,+}.$$

Thus

$$\begin{aligned}[\min\{b^{P,-}, c^{P,-}\}, \min\{b^{P,+}, c^{P,+}\}] &> [d^{P,-}, d^{P,+}] \\ &> [a^{P,-}, a^{P,+}] \\ &= A^P(x_0z_0).\end{aligned}$$

So, we have $x_0y_0 \cdot z_0, y_0 \in [\mathcal{C}]_{\tilde{\tilde{a}}}$ but $x_0z_0 \notin [\mathcal{C}]_{\tilde{\tilde{a}}}$. This means that the condition (CBI1) is satisfied.

Case 2^0. We assume that condition (CBI2) is not satisfied. Then there are $x_0, y_0, z_0 \in X$ such that

$$A^N(x_0z_0) > \max\{A^N(x_0y_0 \cdot z_0), A^N(y_0)\}.$$

Let

$$A^N(x_0y_0 \cdot z_0) = [b^{N,-}, b^{N,+}],\ A^N(y_0) = [c^{N,-}, c^{N,+}] \text{ and}$$
$$A^N(x_0z_0) = [a^{N,-}, a^{N,+}],$$

where $[a^{N,-}, a^{N,+}]$, $[b^{N,-}, b^{N,+}]$, $[c^{N,-}, c^{N,+}] \in D[-1, 0]$. Then clearly,

$$\begin{aligned}[a^{N,-}, a^{N,+}] &> \max\{[b^{N,-}, b^{N,+}], [c^{N,-}, c^{N,+}]\} \\ &= [\max\{b^{N,-}, c^{N,-}\}, \max\{b^{N,+}, c^{N,+}\}].\end{aligned}$$

Let us put

$$[d^{N,-}, d^{N,+}] = [\frac{1}{2}\max\{a^{N,-} + b^{N,-}, c^{N,-}\}, \frac{1}{2}\max\{a^{N,+} + b^{N,+}, c^{N,+}\}].$$

Then it implies the following two inequalities:

$$\max\{b^{N,-}, c^{N,-}\} < d^{N,-} = \frac{1}{2}\max\{a^{N,-} + b^{N,-}, c^{N,-}\} < a^{N,-}$$

and

$$\max\{b^{N,+}, c^{N,+}\} < d^{N,+} = \frac{1}{2}\max\{a^{N,+} + b^{N,+}, c^{N,+}\} < a^{N,+}.$$

Thus

$$\begin{aligned}[\max\{b^{N,-}, c^{N,-}\}, \max\{b^{N,+}, c^{N,+}\}] &< [d^{N,-}, d^{N,+}] \\ &< [a^{N,-}, a^{N,+}] \\ &= A^N(x_0z_0).\end{aligned}$$

This means that $x_0y_0 \cdot z_0, y_0 \in [\mathcal{C}]_{\tilde{\tilde{a}}}$ but $x_0z_0 \notin [\mathcal{C}]_{\tilde{\tilde{a}}}$. Thus condition (CBI2) is satisfied.

Case 3^0. Let us assume that (CBI3) is not true. Then there are $x_0, y_0, z_0 \in X$ such that

$$\mu^P_\Phi(x_0z_0) < \min\{\mu^P_\Phi(x_0y_0 \cdot z_0), \mu^P_\Phi(y_0)\}.$$

Let
$$mu_{\Phi}^{P}(x_0y_0 \cdot z_0) = b^P,\ \mu_{\Phi}^{P}(y_0) = c^P,\ \mu_{\Phi}^{P}(x_0z_0) = a^P,$$
where $a^P, b^P, c^P \in [0,1]$. Then clearly, $a^P < \min\{b^P, c^P\}$.
When we put
$$d^P = \frac{1}{2}\min\{a^P + b^P, c^P\},$$
then we get the inequalities
$$\min\{b^P, c^P\} > d^P = \frac{1}{2}\min\{a^P + b^P, c^P\} > a^P.$$
Hence
$$\min\{b^P, c^P\} > d^P > a^P = \mu_{\Phi}^{P}(x_0z_0)$$
and
$$x_0y_0 \cdot z_0, y_0 \in [\mathcal{C}]_{\tilde{\tilde{a}}}$$
but $x_0z_0 \notin [\mathcal{C}]_{\tilde{\tilde{a}}}$. So, condition (CBI3) is satisfied.

Case 4^0. In this case we assume that condition (CBI4) does not hold.

Then there are $x_0, y_0, z_0 \in X$ such that
$$\mu_{\Phi}^{N}(x_0z_0) > \max\{\mu_{\Phi}^{N}(x_0y_0 \cdot z_0), \mu_{\Phi}^{N}(y_0)\}.$$
Let
$$\mu_{\Phi}^{N}(x_0y_0 \cdot z_0) = b^N,\ \mu_{\Phi}^{N}(y_0) = c^N,\ \mu_{\Phi}^{N}(x_0z_0) = a^N,$$
where $a^N, b^N, c^N \in [0,1]$.
Of course, $a^N > \max\{b^N, c^N\}$.

Let $d^N = \frac{1}{2}\max\{a^N + b^N, c^N\}$. Then we obtain the inequalities
$$\max\{b^N, c^N\} < d^N = \frac{1}{2}\{a^N + b^N, c^N\} > a^N.$$
Hence
$$\max\{b^N, c^N\} < d^N < a^N = \mu_{\Phi}^{N}(x_0z_0)$$
and
$$x_0y_0 \cdot z_0, y_0 \in [\mathcal{C}]_{\tilde{\tilde{a}}}$$
but $x_0z_0 \notin [\mathcal{C}]_{\tilde{\tilde{a}}}$. So, the condition (CBI4) is also satisfied.

We have proven that $[\mathcal{C}]_{\tilde{\tilde{a}}}$ is a cubic bipolar BCC-ideal.
The proof is completed. □

Theorem 4.8.49. *Let $\mathfrak{X}$ be a BCC-algebra and let $\mathcal{C} \in CB(X)$. Then $\mathcal{C}$ is a cubic bipolar BCC-ideal of $\mathfrak{X}$ if and only if A is a bipolar interval-valued fuzzy BCC-ideal and A is a bipolar fuzzy BCC-ideal of $\mathfrak{X}$.*

Proof. Let $\mathcal{C}$ be a cubic bipolar BCC-ideal of $\mathfrak{X}$. Then, by Lemma 4.8.39, for any $x \in X$, we have

$$\mu_\Phi^{P,-}(0) \geq \mu_\Phi^{P,-}(x),\ \mu_\Phi^{P,+}(0) \geq \mu_\Phi^{P,+}(x), \tag{4.42}$$

$$\mu_\Phi^{N,-}(0) \leq \mu_\Phi^{N,-}(x),\ \mu_\Phi^{N,+}(0) \leq \mu_\Phi^{N,+}(x). \tag{4.43}$$

This implies

$$\mu_\Phi^{P}(0) \geq \mu_\Phi^{P}(x),\ \mu_\Phi^{N}(0) \leq \mu_\Phi^{N}(x). \tag{4.44}$$

Let now $x, y, z \in X$. Then

$$\begin{aligned}
&[\mu_\Phi^{P,-}(xz), \mu_\Phi^{P,+}(xz)]\\
&= A^P(xz)\\
&\geq \min\{A^P(xy \cdot z), A^P(y)\}\\
&= \min\{[\mu_\Phi^{P,-}(xy \cdot z), \mu_\Phi^{P,+}(xy \cdot z)], [A^{P,-}(y), A^{P,+}(y)]\}\\
&= [\min\{A_\Phi^{P,-}(xy \cdot z), A^{P,-}(y)\}, \min\{A^{P,+}(xy \cdot z), A^{P,+}(y)\}].
\end{aligned}$$

Similarly,

$$\begin{aligned}
&[\mu_\Phi^{N,-}(xz), \mu_\Phi^{N,+}(xz)]\\
&= A^N(xz)\\
&\leq \max\{A^N(xy \cdot z), A^N(y)\}\\
&= \max\{[\mu_\Phi^{N,-}(xy \cdot z), \mu_\Phi^{N,+}(xy \cdot z)], [\mu_\Phi^{N,-}(y), \mu_\Phi^{N,+}(y)]\}\\
&= [\max\{\mu_\Phi^{N,-}(xy \cdot z), \mu_\Phi^{N,-}(y)\}, \max\{\mu_\Phi^{N,+}(xy \cdot z), \mu_\Phi^{N,+}(y)\}].
\end{aligned}$$

Thus we have

$$\mu_\Phi^{P,-}(xz) \geq \min\{\mu_\Phi^{P,-}(xy \cdot z), \mu_\Phi^{P,-}(y)\} \tag{4.45}$$

$$\mu_\Phi^{P,+}(xz) \geq \min\{\mu_\Phi^{P,+}(xy \cdot z), \mu_\Phi^{P,+}(y) \tag{4.46}$$

$$\mu_\Phi^{N,-}(xz) \leq \max\{\mu_\Phi^{N,-}(xy \cdot z), \mu_\Phi^{N,-}(y)\} \tag{4.47}$$

$$\mu_{\Phi}^{N,+}(xz) \leq \max\{\mu_{\Phi}^{N,+}(xy \cdot z), \mu_{\Phi}^{N,+}(y) \tag{4.48}$$

We can easily see that the following inequalities hold:

$$\mu^{P}(xz) \geq \min\{\mu^{P}(xy \cdot z), \mu^{P}(y)\}, \tag{4.49}$$

$$\mu^{N}(xz) \leq \max\{\mu^{N}(xy \cdot z), \mu^{N}(y)\}. \tag{4.50}$$

So by (4.42), (4.43), (4.45), (4.46), (4.47), (4.48), Lemma (4.8.39) and (4.44), (4.49), (4.50) A is a bipolar interval-valued fuzzy BCC-ideal of $\mathfrak{X}$ and A is a bipolar fuzzy BCC-ideal of $\mathfrak{X}$.

Conversely, let us suppose that A is a bipolar interval-valued fuzzy BCC-ideal of $\mathfrak{X}$ and μ is a bipolar fuzzy BCC-ideal of $\mathfrak{X}$.

It can be easily seen that the following inequalities hold for any $x \in X$:

$$A^{P}(0) \geq A^{P}(x)\,,\; A^{N}(0) \leq A^{N}(x), \tag{4.51}$$

$$\mu^{P}(0) \geq \mu^{P}(x)\,,\; \mu^{N}(0) \leq \mu^{N}(x), \tag{4.52}$$

Then for any $x, y, z \in X$ we have

$$\begin{aligned} A^{P}(xz) &= [\mu_{\Phi}^{P,-}(xz), \mu_{\Phi}^{P,+}(xz)] \\ &\geq [\min\{\mu_{\Phi}^{P,-}(xy \cdot z), \mu_{\Phi}^{P,-}(y)\}, \min\{\mu_{\Phi}^{P,+}(xy \cdot z), \mu_{\Phi}^{P,+}(y)\}] \\ &= \min\{[\mu_{\Phi}^{P,-}(xy \cdot z), \mu_{\Phi}^{P,+}(xy \cdot z)\}], [\mu_{\Phi}^{P,-}(y), \mu_{\Phi}^{P,+}(y)]\} \\ &= \min\{A^{P}(xy \cdot z), A^{P}(y)\}, \end{aligned}$$

i.e.,

$$A^{P}(xz) \geq \min\{A^{P}(xy \cdot z), A^{P}(y)\}.$$

In a similar way we get

$$A^{N}(xz) \leq \max\{A^{N}(xy \cdot z), A^{N}(y)\}.$$

Hence conditions (CBI1) and (CBI2) are satisfied.

It is not difficult to see that the following conditions also hold:

$$\mu_{\Phi}^{P}(xz) \geq \min\{\mu_{\Phi}^{P}(xz \cdot z), \mu_{\Phi}^{P}(y)\}$$

and

$$\mu^{N}(xz) \leq \max\{\mu^{N}(xz \cdot z), \mu^{N}(y)\}.$$

Thus C is a cubic bipolar BCC-ideal of $\mathfrak{X}$.

This completes the proof. □

Theorem 4.8.50. *Let $\{C_\xi\}_{\xi\in\Lambda}$ be a family of cubic bipolar BCC-ideals of a BCC-algebra $\mathfrak{X}$. Then $\bigcap_{\xi\in\Lambda} C_\xi$ is a cubic bipolar BCC-ideal of $\mathfrak{X}$.*

Proof. The proof is straightforward. □

4.9 Soft BCC-Ideals

Jun et al. [91] applied the notion of soft sets by Molodtsov to the ideal theory of d-algebras and provided their various properties. In this paper, the authors dealt with the ideal structure of BCC-algebras by applying soft set theory. They gave definitions of soft BCC-ideals in BCC-algebras and provided some properties. One of the important results was the relationship between a fuzzy BCC-ideal and an idealistic soft BCC-algebra. We will show these results in the sequel.

We begin as usual with the appropriate definition.

Definition 4.9.1. Let $\mathfrak{S}$ be a subalgebra of a BCC-algebra $\mathfrak{X}$. A subset I of X is called a BCC-ideal of $\mathfrak{X}$ related to $\mathfrak{S}$ (briefly, S-BCC-ideal of $\mathfrak{X}$), denoted by $I \lhd S$ if it satisfies the following conditions:

(1) $0 \in I$,

(2) $(\forall x, z \in S)(\forall y \in I)(xy \cdot z \in I \Longrightarrow xz \in I)$.

The next corollary is obvious.

Corollary 4.9.2. *Let $\mathfrak{S}$ be a subalgebra of a BCC-algebra $\mathfrak{X}$ and let I be a subset of X that contains S. Then I is an* S-BCC*-ideal of $\mathfrak{X}$.*

The next corollary follows from the above one.

Corollary 4.9.3. *Every BCC-ideal of $\mathfrak{X}$ is an* S-BCC*-ideal of $\mathfrak{X}$ for some subalgebra $\mathfrak{S}$ of $\mathfrak{X}$.*

The converse of the statement of the last corollary is not true. We show it in the following example:

Example 4.9.4. Let $\mathfrak{X}$ be BCC-algebra defined in Example 3.11.8. Then $\mathfrak{S} = \langle\{0,2\};\, \cdot, 0\rangle$ is a subalgebra of $\mathfrak{X}$ and $I = \{0,2,4\} \lhd S$, but I is not a BCC-ideal. Indeed, $4 \in I$ and $(4\cdot 4)\cdot 3 = 0 \in I$. But $4\cdot 3 = 1 \notin I$.

□

The next statement is obvious.

Corollary 4.9.5. *Let $\mathfrak{S}_1$ and $\mathfrak{S}_2$ be subalgebras of $\mathfrak{X}$ such that $S_1 \subset S_2$. Then every S_2-BCC-ideal of $\mathfrak{X}$ is an S_1-BCC-ideal of $\mathfrak{X}$.*

The converse of the last statement is not true. We will show it in the following example:

Example 4.9.6. Let $\mathfrak{X}$ be a BCC-algebra defined in Table 2.3.1. Let $S_1 = \{0,1,2\}$ and $S_2 = \{0,1,2,3\}$. Observe that $\mathfrak{S}_1 = \langle\{0,1,2\};\cdot,0\rangle$ and $\mathfrak{S}_2 = \langle\{0,1,2,3\};\cdot,0\rangle$ are both subalgebras of $\mathfrak{X}$. Indeed, $\mathfrak{S}_1$ is isomorphic to the BCC-algebra with the operation $\cdot$ defined by Table 1.2.6. $\mathfrak{S}_2$ is isomorphic to the BCC-algebra with the operation $\cdot$ defined by Table 1.2.13. Let $I = \{0,1,2\}$. Then I is an S_1-BCC-ideal of $\mathfrak{X}$. However, I is not an S_2-BCC-ideal of $\mathfrak{X}$ because $2 \in I$ and $(3\cdot 2)\cdot 1 = 1\cdot 1 = 0 \in I$ but $3\cdot 1 = 3 \notin I$. □

Definition 4.9.7. Let (δ, A) be a soft BCC-algebra over $\mathfrak{X}$. A soft set (γ, I) over $\mathfrak{X}$ is called a *soft BCC-ideal* of (δ, A), denoted by $(\gamma, I) \lhd (\delta, A)$ if it satisfies the following conditions:

(1) $I \subset A$,

(2) $\gamma(x) \lhd \delta(x)$ for all $x \in I$.

Let us illustrate this definition using an example.

Example 4.9.8. Let $\mathfrak{X} = \langle\{0,1,2,3,4\};\cdot,0\rangle$ be a BCC-algebra with the operation $\cdot$ defined by the following table:

$\cdot$	0	1	2	3	4
0	0	0	0	0	0
1	1	0	1	0	0
2	2	2	0	0	0
3	3	3	1	0	0
4	4	3	3	3	0

Table 4.9.1

For $A = X$, let $\delta : A \longrightarrow \mathscr{P}(X)$ be a set-valued function defined as follows:

$$\delta(x) = \Big\{y \in X \,|\, y\cdot yx \in \{0,1\}\Big\}$$

for all $x \in A$. Then

$$\delta(x) = \begin{cases} X & \text{for } x = 0, \\ \{0,1,2,3\} & \text{for } x = 1, \\ \{0,1\} & \text{for } x = 2, \\ \{0,1\} & \text{for } x = 3, \\ \{0,1\} & \text{for } x = 4. \end{cases}$$

Note that all three values of the function δ are subalgebras.

Indeed, the algebra $\langle\{0,1,2,3\};\cdot,0\rangle$ is isomorphic to the BCC-algebra defined by Table 1.2.13 and $\langle\{0,1\};\cdot,0\rangle$ is clearly a BCK-subalgebra of $\mathfrak{X}$.

Hence (δ, A) is a soft BCC-algebra over $\mathfrak{X}$.

We consider $I = \{0,1\} \subset A$ and we define a set-valued function $\gamma : I \longrightarrow \mathscr{P}(X)$ as follows:

$$\gamma(x) = \Big\{ y \in X \,|\, yx \in \{0\} \Big\}$$

for all $x \in I$.

It is easy to verify that

$$\gamma(0) = \{0\} \lhd \delta(0) = X \text{ and } \gamma(1) = \{0,1\} \lhd \delta(1) = \{0,1,2,3\}.$$

Hence $(\gamma, I)\tilde{\lhd}(\delta, A)$. □

As in Corollary 2.3.8 for BCC-ideals, also for the extended version we can say that every soft BCC-ideal is a soft BCC-subalgebra. And as in Corollary 2.3.8 the converse is not true. We show it in the following example:

Example 4.9.9. Consider the BCC-algebra $\mathfrak{X} = \langle\{0,1,2,3,4,5\};\cdot,0\rangle$ with the operation $\cdot$ defined by Table 2.1.1. Let $A = X$ and we define a set-valued function $\delta : A \longrightarrow \mathscr{P}(X)$ by

$$\delta(x) = \Big\{ y \in X \,|\, yx \cdot x \in \{0\} \Big\}$$

for all $x \in A$. Then we have

$$\delta(x) = \begin{cases} \{0\} & \text{for } x = 0, \\ \{0,1\} & \text{for } x = 1, \\ \{0,1,2,3\} & \text{for } x = 2, \\ \{0,1,2,3\} & \text{for } x = 3, \\ \{0,1,2,3,4\} & \text{for } x = 4, \\ \{0,5\} & \text{for } x = 5. \end{cases}$$

All values of $\delta(x)$ for $x \in \{0,1,2,3,4,5\}$ are subalgebras of $\mathfrak{X}$.

Indeed, $\langle \delta(2); \cdot, 0\rangle$ and $\langle \delta(3); \cdot, 0\rangle$ are isomorphic to the BCK-algebra B_{4-2-1} on page 315 in [146].

The algebra $\langle \delta(4); \cdot, 0\rangle$ is isomorphic to the BCC-algebra with the operation $\cdot$ defined in Table 1.3.2.

The other cases are obvious.

Then (δ, A) is a soft BCC-algebra over $\mathfrak{X}$.

Let $I = \{2,4\} \subset A$ and let $\gamma : I \longrightarrow \mathscr{P}(X)$ be a set-valued function defined as follows:

$$\gamma(x) = \Big\{ y \in X \mid yx \in \{0\} \Big\}$$

for all $x \in I$. Then we have

$$\gamma(x) = \begin{cases} \{0,1,2\} & \text{for} \quad x = 2, \\ \{0,1,4\} & \text{for} \quad x = 4. \end{cases}$$

The algebras $\langle \{0,1,2\}; \cdot, 0\rangle$ and $\langle \{0,1,4\}; \cdot, 0\rangle$ are subalgebras of $\mathfrak{X}$. They are both isomorphic to the BCK-algebra with the operation $\cdot$ defined in Table 1.2.5.

Hence (γ, I) is also a soft BCC-algebra over $\mathfrak{X}$.

Obviously $\gamma(2)$ and $\gamma(4)$ are subalgebras due to the operation $\cdot$ of $\delta(2)$ and $\delta(4)$, respectively. Thus $(\gamma, I)\tilde{<}(\delta, A)$. But, since $4 \in \gamma(4)$ and $(2 \cdot 4) \cdot 1 = 1 \cdot 1 = 0 \in \gamma(4)$ and $2 \cdot 1 = 2 \notin \gamma(4)$, $\gamma(4)$ is not a $\delta(4)$-*BCC*-ideal of $\mathfrak{X}$.

This means that (γ, I) is not a soft *BCC*-ideal of (δ, A). □

Theorem 4.9.10. *Let (δ, A) be a soft BCC-algebra over $\mathfrak{X}$. For any soft sets (γ_1, I_1) and (γ_2, I_2) over $\mathfrak{X}$ where $I_1 \cap I_2 \neq \emptyset$, we have*

$$(\gamma_1, I_1)\tilde{\lhd}(\delta, A),\ (\gamma_2, I_2)\tilde{\lhd}(\delta, A) \Longrightarrow (\gamma_1, I_1)\tilde{\cap}(\gamma_2, I_2)\tilde{\lhd}(\delta, A).$$

Proof. By Definition 3.11.2, we can write

$$(\gamma_1, I_1)\tilde{\cap}(\gamma_2, I_2) = (\gamma, I),$$

where $I = I_1 \cap I_2$ and $\gamma(x) = \gamma_1(x)$ or $\gamma(x) = \gamma_2(x)$ for all $x \in I$. Obviously, $I \subset A$ and $\gamma : I \longrightarrow \mathscr{P}(X)$ is a mapping. Hence (γ, I) is a soft set over $\mathfrak{X}$. Since $(\gamma_1, I_1)\tilde{\lhd}(\delta, A)$ and $(\gamma_2, I_2)\tilde{\lhd}(\delta, A)$, we know that $\gamma(x) = \gamma_1(x) \lhd \delta(x)$ or $\gamma(x) = \gamma_2(x) \lhd \delta(x)$ for all $x \in I$. Hence

$$(\gamma_1, I_1)\tilde{\cap}(\gamma_2, I_2) = (\gamma, I)\tilde{\lhd}(\delta, A).$$

The proof is completed. □

From the above theorem the following corollary immediately follows:

Corollary 4.9.11. *Let (δ, A) be a soft BCC-algebra over $\mathfrak{X}$. For any soft sets (γ, I) and (ρ, I) over $\mathfrak{X}$ the following implication is satisfied:*

$$\big((\gamma, I)\tilde{\lhd}(\delta, A),\ (\rho, I)\tilde{\lhd}(\delta, A)\big) \Longrightarrow (\gamma, I)\tilde{\cap}(\rho, I)\tilde{\lhd}(\delta, A).$$

Theorem 4.9.12. *Let (δ, A) be a soft BCC-algebra over $\mathfrak{X}$. For any soft sets (γ, I) and (ρ, J) over $\mathfrak{X}$ in which $I \cap J = \emptyset$, we have*

$$\big((\gamma, I)\tilde{\lhd}(\delta, A),\ (\rho, J)\tilde{\lhd}(\delta, A)\big) \Longrightarrow (\gamma, I)\tilde{\cup}(\rho, J)\tilde{\lhd}(\delta, A). \tag{4.53}$$

Proof. Let us assume that $(\gamma, I)\tilde{\lhd}(\delta, A)$ and $(\rho, J)\tilde{\lhd}(\delta, A)$. By Definition 3.11.3, we can use the notation

$$(\gamma, I)\tilde{\cup}(\rho, J) = (\kappa, U), \tag{4.54}$$

where $U = I \cup J$ and for every $x \in U$ we have,

$$\kappa(x) = \begin{cases} \gamma(x) & \text{if } \ x \in I \setminus J, \\ \rho(x) & \text{if } \ x \in J \setminus I, \\ \gamma(x) \cup \rho(x) & \text{if } \ x \in I \cap J. \end{cases}$$

We have assumed that $I \cap J = \emptyset$. Then either $x \in I \setminus J$ or $x \in J \setminus I$ for all $x \in U$.

In the case, when $x \in I \setminus J$, then $\kappa(x) = \gamma(x) \lhd \delta(x)$ since $(\gamma, I)\tilde{\lhd}(\delta, A)$.

In the case, when $x \in J \setminus I$, then since $(\rho, J)\tilde{\lhd}(\delta, A)$, we have $\kappa(x) = \rho(x) \lhd \delta(x)$.

Thus $\kappa(x) \lhd \delta(x)$ for all $x \in U$, i.e.,

$$(\gamma, I)\tilde{\cup}(\rho, J) = (\kappa, U)\tilde{\lhd}(\delta, A),$$

which, by assumption and (4.54), gives (4.53). □

In the case $I \cap J \neq \emptyset$, as we will show in the next example, Theorem 4.9.12 is not true.

Example 4.9.13. Let us consider the BCC-algebra $\mathfrak{X}$ in Example 4.9.8. Let $A = X$ and $\delta : A \longrightarrow \mathscr{P}(X)$ be a set-valued function defined by

$$\delta(x) = \left\{ y \in X \,|\, y \cdot xy \in \{0, 3\} \right\}$$

for all $x \in A$. Then we have

$$\delta(x) = \begin{cases} \{0,3\} & \text{for} \quad x = 0, \\ \{0,3\} & \text{for} \quad x = 1, \\ \{0,3\} & \text{for} \quad x = 2, \\ \{0,1,3\} & \text{for} \quad x = 3, \\ X & \text{for} \quad x = 4. \end{cases}$$

The subsets $\{0, 3\}$ and $\{0, 1, 3\}$ of X are, equipped with the operation $\cdot$, subalgebras. If we take $I = \{2, 3\} \subset A$ and define a set-valued function $\gamma : I \longrightarrow \mathscr{P}(X)$ in the following way:

$$\gamma(x) = \left\{ y \in X \,|\, y \cdot yx \in \{0\} \right\}$$

for all $x \in I$, then we have

$$\gamma(x) = \begin{cases} \{0,1\} & \text{for} \quad x = 2, \\ \{0\} & \text{for} \quad x = 3. \end{cases}$$

It is not difficult to verify that $\gamma(2) \lhd \delta(2)$ and $\gamma(3) \lhd \delta(3)$, i.e., $(\gamma, A)\tilde{\lhd}(\delta, A)$.

Now, let (ρ, J) be a soft set over $\mathfrak{X}$, where $J = \{2\} \subset A$ and $\rho : J \longrightarrow \mathscr{P}(X)$ is a set-valued function defined as follows:

$$\rho(x) = \left\{ y \in X \,|\, yx \cdot x \in \{0\} \right\}$$

for all $x \in J$. Then $\rho(2) = \{0, 2\} \lhd \delta(2)$, which means that $(\rho, J)\tilde{\lhd}(\delta, A)$. But $\gamma(2) \cup \rho(2) = \{0, 1, 2\}$ is not a $\delta(2)$-BCC-ideal of $\mathfrak{X}$ since $2 \in \{0, 1, 2\}$ and $(3 \cdot 2) \cdot 0 = 1 \in \{0, 1, 2\}$ and $3 \cdot 0 = 3 \notin \{0, 1, 2\}$.

Hence $(\gamma, I)\tilde{\cup}(\rho, J)$ is not a soft BCC-ideal of $\{\delta, A\}$. □

In the sequel we will deal with idealistic soft BCC-algebras over $\mathfrak{X}$. We begin with their definition and then we give an example to illustrate the definition.

Definition 4.9.14. Let (δ, A) be a soft set over $\mathfrak{X}$. Then (δ, A) is called an *idealistic soft BCC-algebra* over $\mathfrak{X}$ if $\delta(x)$ is a BCC-ideal of $\mathfrak{X}$ for all $x \in A$.

Example 4.9.15. Let us consider the BCC-algebra $\mathfrak{X}$ in Example 3.11.33. Let $A = X$ and we define a set-valued function $\delta : A \longrightarrow \mathscr{P}(X)$ in the following way:

$$\delta(x) = \Big\{ y \in X \,|\, y \cdot yx \in \{0\} \Big\}$$

for all $x \in A$. Then we have

$$\delta(x) = \begin{cases} X & \text{for} \quad x = 0, \\ \{0, 2\} & \text{for} \quad x = 1, \\ \{0, 1, 4\} & \text{for} \quad x = 2, \\ \{0\} & \text{for} \quad x = 3, \\ \{0, 2\} & \text{for} \quad x = 4. \end{cases}$$

We can verify that $\delta(x) \lhd \mathfrak{X}$ for all $x \in A$, i.e., (δ, A) is an idealistic soft BCC-algebra over $\mathfrak{X}$. □

The next result is obvious.

Theorem 4.9.16. *Let (δ, A) and (δ, B) be soft sets over $\mathfrak{X}$ where $B \subseteq A \subseteq X$. If (δ, A) is an idealistic soft BCC-algebra over $\mathfrak{X}$, then so is (δ, B).*

The converse of the above theorem is not true as you can see in the example below.

Example 4.9.17. The algebra $\mathfrak{X} = \langle \{0, 1, 2, 3, 4, 5\}; \cdot, 0 \rangle$ with the operation $\cdot$ defined by Table 2.1.1 is a BCC-algebra. Let (δ, A) be a soft set over $\mathfrak{X}$ defined in Example 4.9.9. For $B = \{0, 1, 2, 3, 4\} \subseteq A$ it is easy to verify that (δ, B) is an idealistic soft BCC-algebra over $\mathfrak{X}$.

It is not difficult to see that $\delta(5) = \{0, 5\}$ is not a *BCC*-ideal of $\mathfrak{X}$. We have $5 \in \delta(5)$ and $(2 \cdot 5) \cdot 4 = 0 \in \delta(5)$ but $2 \cdot 4 = 1 \notin \delta(5)$. Therefore, (δ, A) is not an idealistic soft BCC-algebra over $\mathfrak{X}$. □

We know, by Corollary 2.3.8, that every *BCC*-ideal of a BCC-algebra is its subalgebra. We know also that every idealistic soft BCC-algebra over a BCC-algebra $\mathfrak{X}$ is a soft BCC-algebra over $\mathfrak{X}$. But the converse is not true as we show it in the next example.

Example 4.9.18. Let us consider a soft set (δ, A) over $\mathfrak{X}$ which is given in Example 4.9.8. We know already that (δ, A) is a soft BCC-algebra over $\mathfrak{X}$. But $\delta(2) = \{0, 1\}$ is not a *BCC*-ideal since $1 \in \delta(2)$ and $(4 \cdot 1) \cdot 2 = 1 \in \delta(2)$ but $4 \cdot 2 = 3 \notin \delta(2)$. Hence (δ, A) is not an idealistic soft BCC-algebra over $\mathfrak{X}$. □

Theorem 4.9.19. *Let (δ, A) and (γ, B) be two idealistic soft BCC-algebras over $\mathfrak{X}$ such that $A \cap B \neq \emptyset$. Then the intersection $(\delta, A)\tilde{\cap}(\gamma, B)$ is an idealistic soft BCC-algebra over $\mathfrak{X}$.*

Proof. Using Definition 3.11.2 we can write

$$(\delta, A)\tilde{\cap}(\gamma, B) = (\rho, C),$$

where $C = A \cap B$ and $\rho(x) = \delta(x)$ or $\rho(x) = \gamma(x)$ for all $x \in C$. Note that $\rho : C \longrightarrow \mathscr{P}(X)$ is a mapping which implies (ρ, C) is a soft set over $\mathfrak{X}$. But (δ, A) and (γ, B) are, by assumption, idealistic soft BCC-algebras over $\mathfrak{X}$. Hence $\rho(x) = \delta(x)$ is a BCC-ideal of $\mathfrak{X}$ or $\rho(x) = \gamma(x)$ is a BCC-ideal of $\mathfrak{X}$ for all $x \in C$. It follows that $(\rho, C) = (\delta, A)\tilde{\cap}(\gamma, B)$ is an idealistic soft BCC-algebra over $\rho(x) = \delta(x)$ is a BCC-ideal of $\mathfrak{X}$. □

From the above theorem the corollary immediately follows:

Corollary 4.9.20. *The intersection $(\delta, A)\tilde{\cap}(\gamma, B)$ of two idealistic soft BCC-algebras (δ, A) and (γ, B) over $\mathfrak{X}$ is an idealistic soft BCC-algebra over $\mathfrak{X}$.*

Theorem 4.9.21. *Let (δ, A) and (γ, B) be two idealistic soft BCC-algebras over $\mathfrak{X}$ such that $A \cap B \neq \emptyset$. Then the union $(\delta, A)\tilde{\cup}(\gamma, B)$ is an idealistic soft BCC-algebra over $\mathfrak{X}$.*

Proof. Using Definition 3.11.3 we can write

$$(\delta, A)\tilde{\cup}(\gamma, B) = (\rho, C),$$

where $C = A \cup B$ and for every $x \in C$ we have

$$\rho(x) = \begin{cases} \delta(x) & \text{if } \ x \in A \setminus B, \\ \gamma(x) & \text{if } \ x \in B \setminus A, \\ \delta(x) \cup \gamma(x) & \text{if } \ x \in A \cap B. \end{cases}$$

Since A and B are disjoint, either $x \in A \setminus B$ or $x \in B \setminus A$ for all $x \in C$.

If $x \in A \setminus B$, then $\rho(x) = \delta(x)$ is a BCC-ideal of $\mathfrak{X}$ since (δ, A) is an idealistic soft BCC-algebra over $\mathfrak{X}$.

If $x \in B \setminus A$, then $\rho(x) = \gamma(x)$ is a BCC-ideal of $\mathfrak{X}$ since (γ, B) is an idealistic soft BCC-algebra over $\mathfrak{X}$.

Hence $(\rho, C) = (\delta, A)\tilde{\cup}(\gamma, B)$ is an idealistic soft BCC-algebra over $\mathfrak{X}$. □

Without the assumption that A and B are disjoint, the above theorem is not true. We show it in the following example:

Example 4.9.22. Let $\mathfrak{X} = \langle\{0,1,2,3,4\};\cdot,0\rangle$ be a BCC-algebra defined in Example 3.11.33. Let us consider the idealistic soft BCC-algebra (δ, A) over $\mathfrak{X}$ which is described in Example 4.9.15. If we take $B = \{2,4\}$, then for $A = X$ we have $B \cap A = B$. Now we define a set-valued function $\gamma : B \longrightarrow \mathscr{P}(X)$ as follows:

$$\gamma(x) = \Big\{ y \in X \mid yx \in \{0\} \Big\}$$

for all $x \in B$. We obtain

$$\gamma(x) = \begin{cases} \{0,2\} & \text{if} \quad x = 2, \\ \{0,1,4\} & \text{if} \quad x = 4. \end{cases}$$

This means that $\gamma(b) \lhd X$ and $\gamma(d) \lhd X$, too. Therefore (γ, B) is an idealistic soft BCC-algebra over $\mathfrak{X}$.

Let now $(\delta, A)\tilde{\cup}(\gamma, B) = (\rho, C)$. Then we get

$$\rho(2) = \delta(2) \cup \gamma(2) = \{0,1,4\} \cup \{0,2\} = \{0,1,2,4\}$$

and

$$\rho(4) = \delta(4) \cup \gamma(4) = \{0,2\} \cup \{0,1,4\} = \{0,1,2,4\}.$$

But we have $1 \in \{0,1,2,4\}$ and $(3 \cdot 1) \cdot 0 = 2 \in \{0,1,2,4\}$ and $3 \cdot 0 = 3 \notin \{0,1,2,4\}$, which means that $\rho(2)$ and $\rho(4)$ are not BCC-ideals of $\mathfrak{X}$. Therefore, $(\delta, A)\tilde{\cup}(\gamma, B)$ is not an idealistic soft BCC-algebra over $\mathfrak{X}$. □

Theorem 4.9.23. *If (δ, A) and (γ, B) are idealistic soft BCC-algebras over $\mathfrak{X}$, then $(\delta, A)\tilde{\wedge}(\gamma, B)$ is an idealistic soft BCC-algebra over $\mathfrak{X}$.*

Proof. By Definition 3.11.4 we know that

$$(\delta, A)\tilde{\wedge}(\gamma, B) = (\rho, A \times B),$$

where $\rho(x,y) = \delta(x) \cap \gamma(y)$ for all $(x,y) \in A \times B$. Since $\delta(x)$ and $\gamma(y)$ are BCC-ideals of $\mathfrak{X}$ their intersection $\delta(x) \cap \gamma(y)$ is also a BCC-ideal of $\mathfrak{X}$. Hence $\rho(x,y)$ is a BCC-ideal of $\mathfrak{X}$ for all $(x,y) \in A \times B$ and therefore, $(\delta, A)\tilde{\wedge}(\gamma, B) = (\rho, A \times B)$ is an idealistic soft BCC-algebra over $\mathfrak{X}$. □

Let us come back for a moment to "trivial" and "whole" concepts in the environment of idealistic soft BCC-algebras.

Definition 4.9.24. An idealistic soft BCC-algebra (δ, A) over $\mathfrak{X}$ is said to be *trivial* (resp., *whole*) if $\delta(x) = \{0\}$ (resp., $\delta(x) = X$) for all $x \in A$.

Example 4.9.25. The algebra $\mathfrak{X} = \langle\{0,1,2,3,4\}; \cdot, 0\rangle$ with the operation $\cdot$ defined by Table 2.3.1 is a BCC-algebra. Let us consider $A = \{0,1,2\} \subset X$ and a set-valued function $\delta : A \longrightarrow \mathscr{P}(X)$ defined by

$$\delta(x) = \Big\{y \in X \,|\, y \cdot xy \in \{0\}\Big\}$$

for all $x \in A$. Then $\delta(0) = \delta(1) = \delta(2) = \{0\}$ and so (δ, A) is a trivial idealistic soft BCC-algebra over $\mathfrak{X}$.

Let $\gamma : A \longrightarrow \mathscr{P}(X)$ be a set-valued function defined by

$$\gamma(x) = \Big\{y \in X \,|\, xy \in \{0, x\}\Big\}$$

for all $x \in A$. Then $\gamma(0) = \gamma(1) = \gamma(2) = X$. Hence (γ, A) is a whole idealistic soft BCC-algebra over $\mathfrak{X}$. □

Theorem 4.9.26. *Let $f : X \longrightarrow Y$ be an onto homomorphism of BCC-algebras and let (δ, A) be an idealistic soft BCC-algebra over $\mathfrak{X}$. Then the following conditions are satisfied:*

(1) *If $\delta(x) \subseteq kerf$ for all $x \in A$, then $(f(\delta), A)$ is a trivial idealistic soft BCC-algebra over $\mathfrak{Y}$,*

(2) *If (δ, A) is whole, then $(f(\delta), A)$ is a whole idealistic soft BCC-algebra over $\mathfrak{Y}$.*

Proof. (1) We assume that $\delta \subseteq kerf$ for all $x \in A$. Then $f(\delta)(x) = f(\delta(x)) = \{0y\}$ for all $x \in A$. Hence, from Definition 4.9.24, $(f(\delta), A)$ is a trivial idealistic soft BCC-algebra over $\mathfrak{Y}$.

(2) In this case we suppose that (δ, A) is whole. Then $\delta(x) = X$ for all $x \in A$. This implies $f(\delta)(x) = f(\delta(x)) = f(X) = Y$ for all $x \in A$. From Definition 4.9.24 it follows that $(f(\delta), A)$ is a whole idealistic soft BCC-algebra over $\mathfrak{Y}$. □

In the sequel we will investigate the idealistic soft BCC-algebras in environment of fuzzy BCC-sets and fuzzy BCC-ideals.

Theorem 4.9.27. *For every fuzzy BCC-ideal μ of $\mathfrak{X}$ there exists an idealistic soft BCC-algebra (δ, A) over $\mathfrak{X}$.*

Proof. Let μ be a fuzzy BCC-ideal of $\mathfrak{X}$. Then

$$U(\mu;t) = \{x \in X \mid \mu(x) \geq t\}$$

is a BCC-ideal of $\mathfrak{X}$ for all $t \in \operatorname{Im}(\mu)$. If we take $A = \operatorname{Im}(\mu)$ and consider a set-valued function $\delta : A \longrightarrow \mathscr{P}(X)$ given by $\delta(t) = U(\mu;t)$ for all $t \in A$, then (δ, A) is an idealistic soft BCC-algebra over $\mathfrak{X}$. □

The converse is obvious. We have namely the following theorem:

Theorem 4.9.28. *For any fuzzy set μ in $\mathfrak{X}$, if an idealistic soft BCC-algebra (δ, A) over $\mathfrak{X}$ is given by $A = \operatorname{Im}(\mu)$ and $\delta(t) = U(\mu;t)$ for all $t \in A$, then μ is a fuzzy BCC-ideal of $\mathfrak{X}$.*

Let us recall Equation (3.70). We will give an example that there is $t \in A$ such that $\delta(t)$ is not a BCC-ideal of $\mathfrak{X}$.

Example 4.9.29. We consider the fuzzy set μ and the soft set (δ, A) in Example 3.11.33.

Then $\delta(0.6) = \{0, 3, 4\}$ is not a BCC-ideal of $\mathfrak{X}$, because $3 \in \delta(0.6)$ and $(1 \cdot 3) \cdot 2 = 0 \cdot 2 = 0 \in \delta(0.6)$ but $1 \cdot 2 = 1 \notin \delta(0.6)$. □

Theorem 4.9.30. *Let μ be a fuzzy set in $\mathfrak{X}$ and let (δ, A) be a soft set over $\mathfrak{X}$ in which $A = [0,1]$ and $\delta : A \longrightarrow \mathscr{P}(X)$ is given by (3.70). Then the following statements are equivalent:*

(1) *μ is a fuzzy BCC-ideal of $\mathfrak{X}$,*

(2) *$(\forall t \in A)(\delta(t) \neq \emptyset \Longrightarrow (\delta, A)$ is idealistic soft BCC-algebra over $\mathfrak{X})$.*

Proof. Let μ be a fuzzy BCC-ideal of $\mathfrak{X}$. Let $t \in A$ and let $x, y \in \delta(t)$ be such that $\delta(t) \neq \emptyset$.

Let us take $x \in \delta(t)$, then $\mu(0) + t \geq \mu(x) + t > 1$, i.e., $0 \in \delta(t)$. Let now $t \in A$ and $x, y, z \in X$ be such that $y \in \delta(t)$ and $xy \cdot z \in \delta(t)$. Then $\mu(y) + t > 1$ and $\mu(xy \cdot z) + t > 1$. From (FI2) (Definition 4.7.2) it follows that

$$\begin{aligned}\mu(xz) + t &\geq \min\{\mu(xy \cdot z), \mu(y)\} + t \\ &= \min\{\mu(xy \cdot z) + t, \mu(y) + t\} \\ &> 1.\end{aligned}$$

Hence $xz \in \delta(t)$ and therefore $\delta(t)$ is a BCC-ideal of $\mathfrak{X}$ for all $t \in A$. Consequently, (δ, A) is an idealistic soft BCC-algebra over $\mathfrak{X}$.

Conversely, let us suppose that condition (2) holds. If there exists $x_0 \in X$ such that $\mu(0) < \mu(x_0)$, then we can select $t_0 \in A$ such that $\mu(0) + t_0 < 1 < \mu(x_0) + t_0$.
This implies that $0 \notin \delta(t_0)$, which contradicts our assumption. Hence $\mu(0) \geq \mu(x)$ for all $x \in X$.
Now, we assume that $\mu(ac) < \min\{\mu(ab \cdot c), \mu(b)\}$ for some $a, b, c \in X$. We take $s_0 \in A$ such that

$$\mu(ac) + s_0 < 1 < \min\{\mu(ab \cdot c), \mu(b)\} + s_0.$$

Then $ab \cdot c \in \delta(s_0)$ and $b \in \delta(s_0)$ but $ac \notin \delta(s_0)$. This is a contradiction.
Therefore, $\mu(xz) \geq \min\{\mu(xy \cdot z), \mu(y)\}$ for all $x, y, z \in X$. □

Corollary 4.9.31. *Let μ be a fuzzy set in $\mathfrak{X}$ such that $\mu(x) > 0.5$ for some $x \in X$ and let (δ, A) be a soft set over $\mathfrak{X}$ in which*

$$A = \{t \in \mathrm{Im}(\mu) \mid t > 0.5\}$$

and $\delta : A \longrightarrow \mathscr{P}(X)$ is given by (3.70). If μ is a fuzzy BCC-ideal of $\mathfrak{X}$, then (δ, A) is an idealistic soft BCC-algebra over $\mathfrak{X}$.

Proof. Straightforward. □

Theorem 4.9.32. *Let μ be a fuzzy set in $\mathfrak{X}$ and let (δ, A) be a soft set over $\mathfrak{X}$ in which $A = (0.5, 1]$ and $\delta : A \longrightarrow \mathscr{P}(X)$ is defined by*

$$\delta(t) = U(\mu; t)$$

for all $t \in A$.

Then (δ, A) is an idealistic soft BCC-algebra over $\mathfrak{X}$ if and only if the following statements are true:

(1) $\max\{\mu(0), 0.5\} \geq \mu(x)$,

(2) $\max\{\mu(xy), 0.5\} \geq \min\{\mu(xy \cdot z), \mu(y)\}$

for all $x, y, z \in X$.

Proof. Let us assume that (δ, A) is an idealistic soft BCC-algebra over $\mathfrak{X}$. If there exists $x_0 \in X$ such that $\max\{\mu(0), 0.5\} < \mu(x_0)$, then we can select $t_0 \in A$ such that

$$\max\{\mu(0), 0.5\} < t_0 < \mu(x_0).$$

This implies that $\mu(0) < t_0$, so that $0 \notin \delta(t_0)$. This is a contradiction, i.e., condition (1) is true.

Suppose that condition (2) is not true.

Then there exist $a, b, c \in X$ such that

$$\max\{\mu(ac), 0.5\} < \min\{\mu(ab \cdot c), \mu(b)\}.$$

Let us take $u_0 \in A$ such that

$$\max\{\mu(ac), 0.5\} < u_0 < \min\{\mu(ab \cdot c), \mu(b)\}.$$

Then $ab \cdot c \in \delta(u_0)$ and $b \in \delta(u_0)$, but $ac \notin \delta(u_0)$, which is a contradiction. Therefore, condition (2) is also true.

Conversely, let us suppose that both conditions (1) and (2) are true. Let $t \in A$. For any $x \in \delta(t)$, we have

$$\max\{\mu(0), 0.5\} \geq \mu(x) \geq t > 0.5$$

and so $\mu(0) \geq t$, i.e., $0 \in \delta(t)$. Let $x, y, z \in X$ be such that $y \in \delta(t)$ and $xy \cdot z \in \delta(t)$.

Then $\mu(xy \cdot z) \geq t$ and $\mu(y) \geq t$. From the second condition it follows that

$$\max\{\mu(xz), 0.5\} \geq \min\{\mu(xy \cdot z), \mu(y)\} \geq t > 0.5,$$

so that $\mu(xz) \geq t$, i.e., $xz \in \delta(t)$.

Therefore, (δ, A) is an idealistic soft BCC-algebra over $\mathfrak{X}$. □

Appendix

A. Proper Weak BCC-Algebras and BCC-Algebras of Order < 6

A.1. (Weak) BCC-Algebras of Order < 4

There are no proper BCC-algebras with order < 4 (cf. Theorem 1.2.6) and there are no proper weak BCC-algebras with order < 4 (cf. Corollary 1.2.7).

A.2. Weak BCC-Algebras of Order 4

There are 12 proper weak BCC-algebras of order 4. Among them are 2 nonisomorphic proper weak BCC-algebras. Their operations are given in Table 1.1.2 and Table 1.1.3.

A.3. BCC-Algebras of Order 4

There are 45 proper BCC-algebras of order 4. Among them are 8 nonisomorphic proper BCC-algebras. Their operations are given in Table 1.2.7 to Table 1.2.14.

A.4. Weak BCC-Algebras of Order 5

There are 960 proper weak BCC-algebras of order 5. Among them are 41 nonisomorphic proper weak BCC-algebras.

In Example 1.1.15 we have shown that there exist nonisomorphic weak BCC-algebras with the same partial order. So, we will present the diagrams of the partial order and the nonisomorphic proper weak BCC-algebras that share it.

We will use the notation:

$$\mathfrak{X}_{k-i} = \langle X; \cdot_{k-i}, 0\rangle,$$

where k is the cardinality of the base set X and i is the current number of the algebra.

1.

$\cdot_{5-i}$	0	1	2	3	4
0	0	0	0	0	4
1	1	0	0	0	4
2	2	a	0	0	4
3	3	b	c	0	4
4	4	4	4	4	0

i	1	2	3
abc	122	222	231

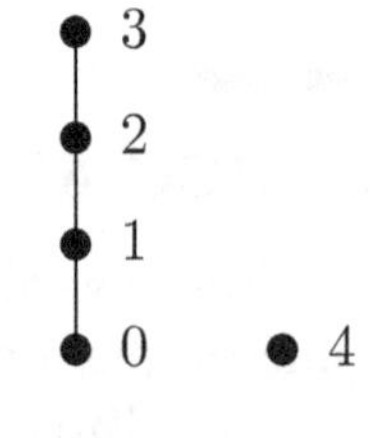

2.

$\cdot_{5-i}$	0	1	2	3	4
0	0	0	0	0	4
1	1	0	0	0	4
2	2	2	0	1	4
3	3	3	a	0	4
4	4	4	4	4	0

i	4	5
a	1	3

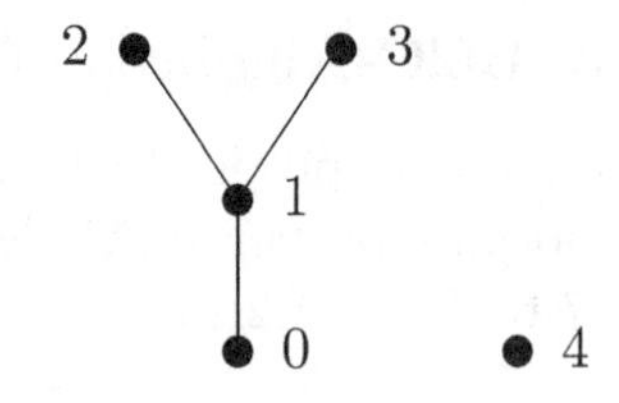

3.

$\cdot_{5-i}$	0	1	2	3	4
0	0	0	0	0	4
1	1	0	0	1	4
2	2	a	0	1	4
3	3	3	3	0	4
4	4	4	4	4	0

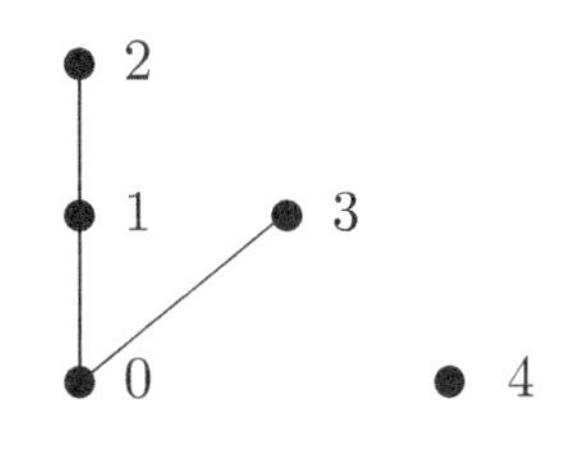

i	6	7
a	1	2

4.

$\cdot_{5-8}$	0	1	2	3	4
0	0	0	0	0	4
1	1	0	0	1	4
2	2	2	0	1	4
3	3	3	0	0	4
4	4	4	4	4	0

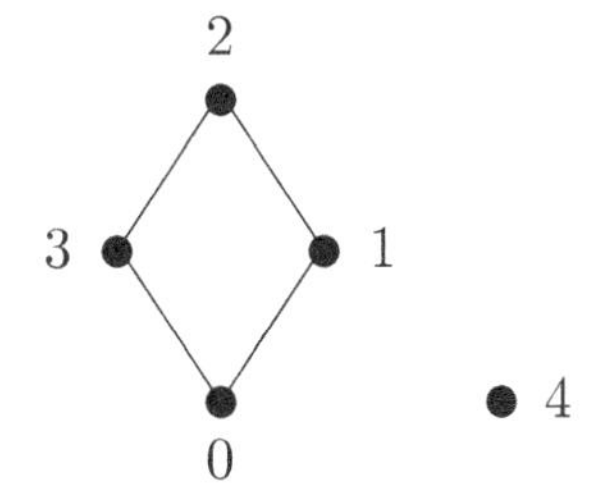

5.

$\cdot_{5-i}$	0	1	2	3	4
0	0	0	0	3	3
1	1	0	0	a	b
2	2	c	0	d	d
3	3	3	3	0	0
4	4	e	f	g	0

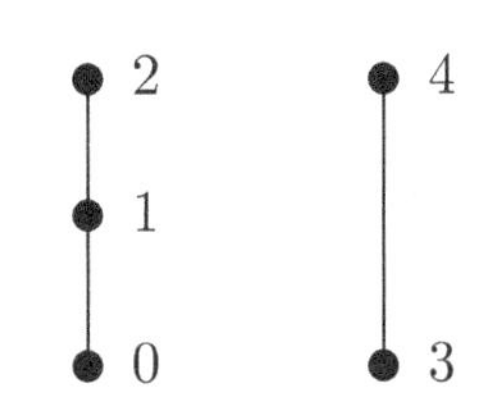

For $a = b = d = 3$

For $c = 1$

i	9	10	11
efg	332	441	442

For $c = 2$

i	12	13	14	15
efg	332	431	441	442

For $a = b = 3, d = 4$

For $c = 1$

i	16	17
efg	331	441

For $c = 2$

i	18	19	20	21
efg	331	431	441	442

For $a = 4, b = 3, d = 4$

For $c = 1$

i	22
efg	331

For $c = 2$

i	23
efg	331

For $a = b = d = 4$

For $c = 1$

i	24
efg	441

For $c = 2$

i	25
efg	441

6.

$\cdot_{5-i}$	0	1	2	3	4
0	0	0	0	3	3
1	1	0	1	3	3
2	2	2	0	a	a
3	3	3	3	0	0
4	4	b	c	d	0

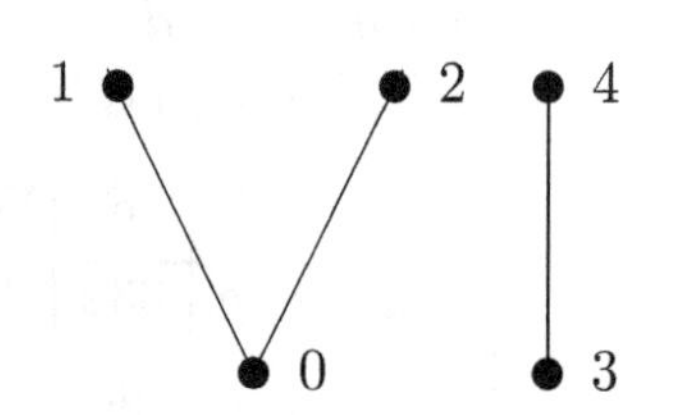

For $a = 3$

i	26	27	28
bcd	331	342	441

For $a = 4$

i	29
bcd	442

7.

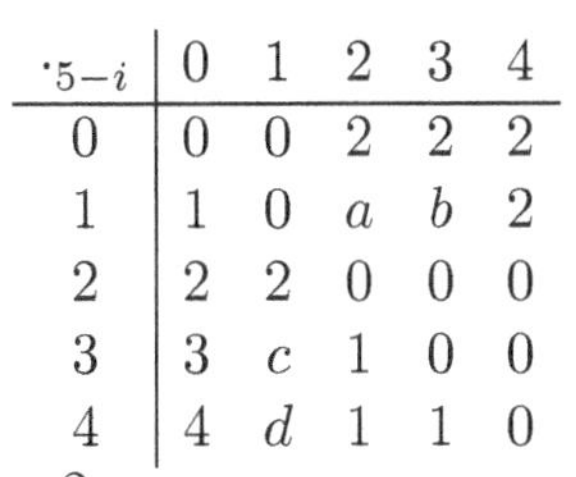

$\cdot_{5-i}$	0	1	2	3	4
0	0	0	2	2	2
1	1	0	a	b	2
2	2	2	0	0	0
3	3	c	1	0	0
4	4	d	1	1	0

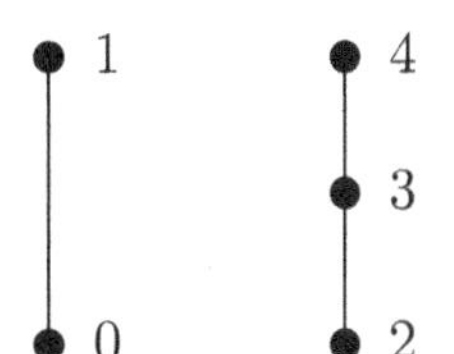

For $a = 2$

i	30	31	32	33
bcd	223	224	233	234

For $a = 3$

i	34	35	36	37
bcd	223	224	333	334

8.

$\cdot_{5-i}$	0	1	2	3	4
0	0	0	2	2	2
1	1	0	2	2	2
2	2	2	0	0	0
3	3	a	1	0	1
4	4	4	1	1	0

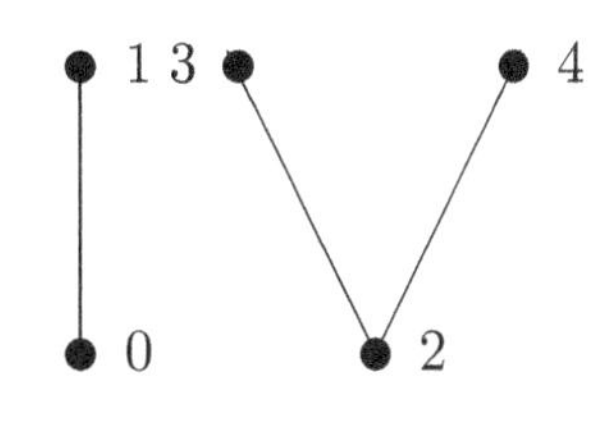

i	38	39
a	2	3

9.

$\cdot_{5-i}$	0	1	2	3	4
0	0	0	3	2	2
1	1	0	a	2	2
2	2	2	0	3	3
3	3	3	2	0	0
4	4	4	2	1	0

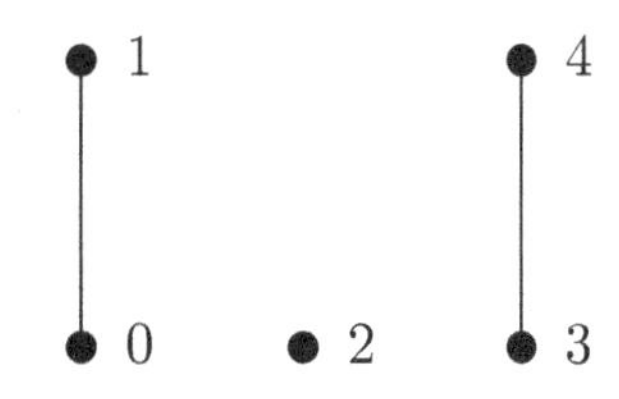

i	40	41
a	3	4

A.5. BCC-Algebras of Order 5

There are 6016 proper weak BCC-algebras of order 5. Among them are 268 nonisomorphic proper weak BCC-algebras.

We present them below.

1.

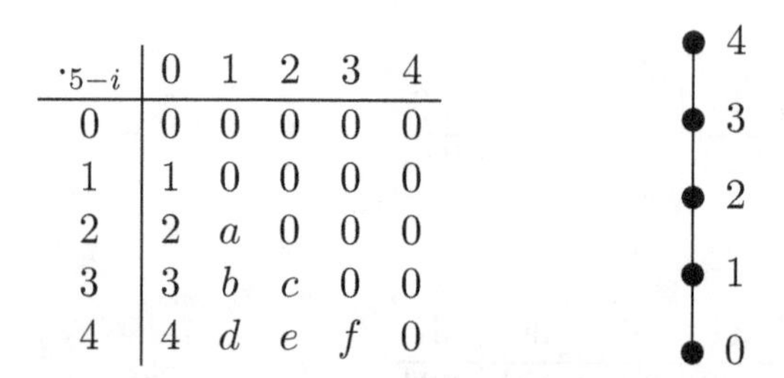

$\cdot_{5-i}$	0	1	2	3	4
0	0	0	0	0	0
1	1	0	0	0	0
2	2	a	0	0	0
3	3	b	c	0	0
4	4	d	e	f	0

For $a = 1$

For $b = 1$

For $c = 1$

i	1	2	3	4	5	6
def	221	222	311	322	332	333

For $a = 1$

For $b = 2$

For $c = 1$

i	7	8	9	10	11
def	221	222	322	332	333

For $c = 2$

i	12	13	14	15	16	17
def	221	222	322	332	333	444

For $b = 3$

For $c = 3$

i	18	19	20	21
def	332	333	441	442

For $a = 2$

For $b = 2$

For $c = 1$

i	22	23	24	25	26	27	28
def	222	311	322	332	333	411	433

For $c = 2$

i	29	30	31	32	33	34	35	36
def	221	222	322	332	333	422	433	444

For $a = 2$

For $b = 3$

For $c = 1$

i	37	38	39	40	41	42	43
def	311	322	333	411	422	433	444

For $c = 2$

i	44	45	46	47	48
def	322	332	333	421	433

For $c = 3$

i	49	50	51	52	53	54
def	332	333	431	433	441	442

2.

$\cdot_{5-i}$	0	1	2	3	4
0	0	0	0	0	0
1	1	0	0	0	0
2	2	a	0	0	0
3	3	b	c	0	d
4	4	e	f	g	0

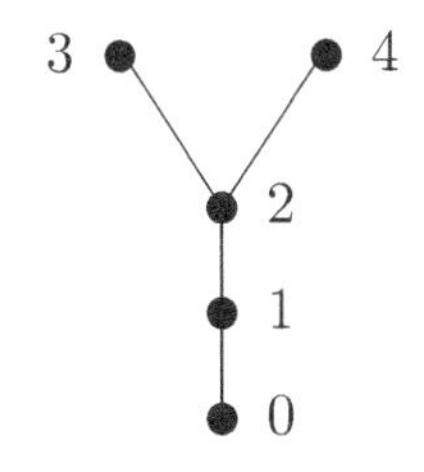

For $a = 1$

For $b = 1$

For $c = 1$

For $d = 1$

i	55	56
efg	221	222

For $b = 2$

For $c = 1$

For $d = 1$

i	57	58
efg	221	222

For $c = 2$

For $d = 1$

i	59	60	61
efg	221	222	444

For $d = 2$

i	62	63
efg	222	444

For $b = 3$

For $c = 3$

For $d = 1$

i	64	65	66
efg	441	442	444

For $d = 2$

i	67	68
efg	442	444

For $a = 2$

For $b = 2$

For $c = 1$

For $d = 1$

i	69	70
efg	222	411

For $c = 2$

For $d = 1$

i	71	72	73	74
efg	221	222	422	444

For $d = 2$

i	75	76	77	78
efg	222	411	422	444

For $b = 3$

For $c = 1$

For $d = 1$

i	79	80	81
efg	411	422	444

For $c = 2$

For $d = 1$

i	82	83	84
efg	421	422	444

For $c = 3$

For $d = 1$

i	85	86	87
efg	441	442	444

For $d = 2$

i	88	89
efg	442	444

3.

$\cdot_{5-i}$	0	1	2	3	4
0	0	0	0	0	0
1	1	0	0	0	0
2	2	d	0	0	a
3	3	e	f	0	g
4	4	b	c	0	0

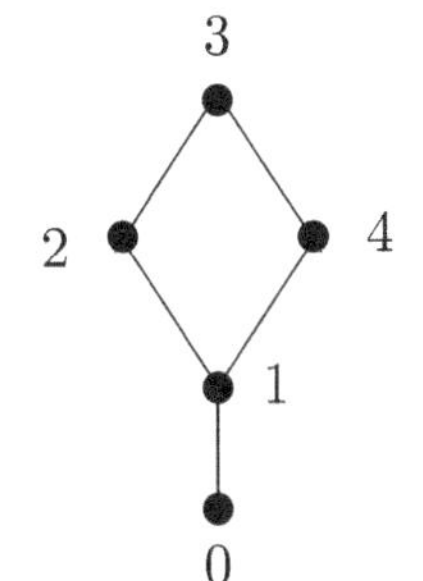

For $a = 1$

For $b = 1$

For $c = 1$

i	90	91	92
$defg$	1211	1221	1222

For $b = 4$

For $c = 1$

i	93	94	95
$defg$	2311	2322	2333

For $c = 4$

i	96	97	98	99	100	101
$defg$	1331	1332	1444	2331	2332	2333

i	102	103	104
$defg$	2334	2341	2344

For $a = 2$

For $b = c = 4$

i	105
$defg$	2332

4.

$\cdot_{5-i}$	0	1	2	3	4
0	0	0	0	0	0
1	1	0	0	0	0
2	2	e	0	0	a
3	3	f	g	0	h
4	4	b	c	d	0

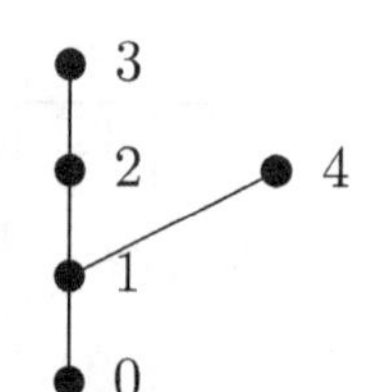

For $a = 1$

For $d = 1$

For $b = c = 1$

i	106	107	108	109	110
$efgh$	1211	1221	1222	2222	2313

For $b = 4$

For $c = 1$

i	111	112	113	114	115
$efgh$	2211	2222	2311	2322	2333

For $c = 4$

i	116	117	118	119	120	121
$efgh$	1331	1332	1333	2331	2332	2333

For $b = c = d = 4$

i	122	123	124	125	126	127	128
$efgh$	1211	1221	1222	1331	1332	2211	2221
i	129	130	131	132	133	134	135
$efgh$	2222	2311	2321	2322	2331	2332	2333

For $a = 2$

For $d = 1$

For $b = 4$

For $c = 1$

i	136	137	138	139	140	141	142
$efgh$	2212	2222	2312	2313	2322	2323	2333

For $c = 4$

i	143	144
$efgh$	2332	2333

For $b = c = d = 4$

i	145	146	147	148	149
$efgh$	2222	2312	2313	2322	2332

5.

$\cdot_{5-i}$	0	1	2	3	4
0	0	0	0	0	0
1	1	0	0	0	0
2	2	d	0	1	a
3	3	e	f	0	g
4	4	4	b	c	0

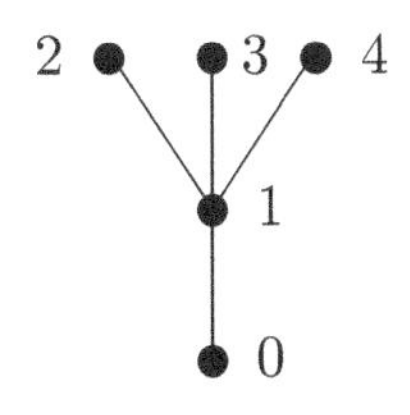

For $a = 1$

For $c = 1$

For $b = 1$

i	150
$defg$	2311

For $b = 4$

i	151	152	153
$defg$	1331	2331	2333

For $b = c = 4$

i	154	155	156	157	158
$defg$	1331	2311	2313	2331	2333

For $a = 2$

For $b = c = 4$

i	159
$defg$	2333

6.

$\cdot_{5-i}$	0	1	2	3	4
0	0	0	0	0	0
1	1	0	0	0	1
2	2	b	0	0	a
3	3	c	d	0	e
4	4	4	4	4	0

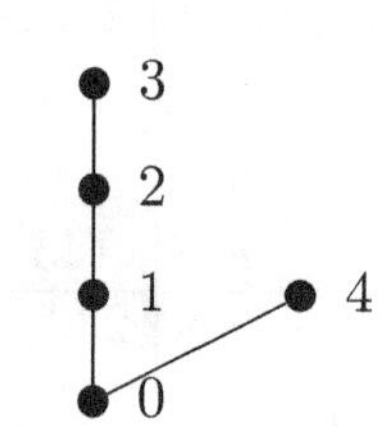

For $a = 1$

For $b = 1$

i	160	161	162	163	164	165	166	167	168	169	170
cde	111	112	113	211	212	221	222	223	331	332	333

For $b = 2$

i	171	172	173	174	175	176	177	178	179	180
cde	211	221	222	223	311	321	322	331	332	333

For $a = 2$

For $b = 1$

i	181	182	183	184	185
cde	112	212	222	223	332

For $b = 2$

i	186	187	188	189	190	191	192
cde	212	222	223	312	313	322	332

7.

$\cdot_{5-i}$	0	1	2	3	4
0	0	0	0	0	0
1	1	0	0	0	1
2	2	b	0	0	a
3	3	c	d	0	e
4	4	4	4	0	0

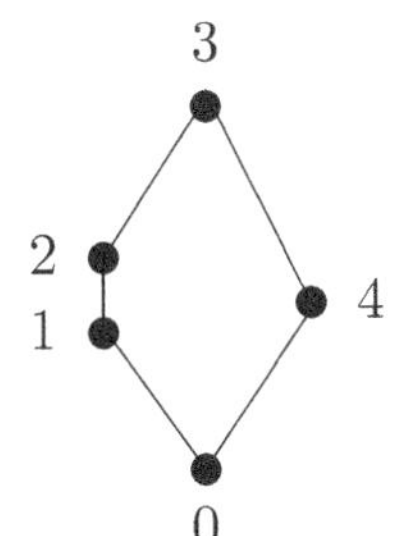

For $a = 1$

For $b = 1$

i	193	194	195
cde	331	332	333

For $b = 2$

i	196	197	198	199
cde	331	332	333	341

For $a = 2$

For $b = 1$

i	200
cde	332

For $b = 2$

i	201	202
cde	332	343

8.

$\cdot_{5-i}$	0	1	2	3	4
0	0	0	0	0	0
1	1	0	0	0	1
2	2	b	0	a	c
3	3	d	e	0	f
4	4	4	4	4	0

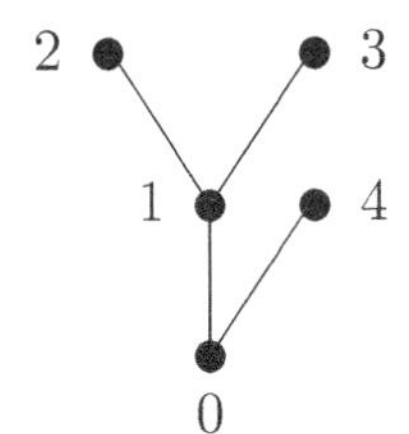

For $a = 1$

For $b = 1$

For $c = 1$

i	203	204	205	206
def	111	113	331	333

For $c = 2$

i	207
def	331

For $b = 2$

For $c = 1$

i	208	209	210
def	331	311	333

For $c = 2$

i	211	212
def	313	333

For $a = 2$

For $b = 2$

For $c = 1$

i	213	214
def	331	333

9.

$\cdot_{5-i}$	0	1	2	3	4
0	0	0	0	0	0
1	1	0	0	0	1
2	2	b	0	a	c
3	3	d	e	0	f
4	4	4	4	0	0

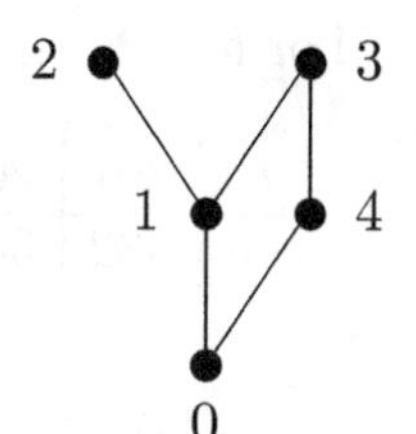

For $a = 1$

For $b = 1$

For $c = 1$

i	215	216	217	218
def	331	333	441	443

For $c = 2$

i	219	220
def	331	443

For $b = 2$

For $c = 1$

i	221	222	223	224	225
def	331	333	341	441	443

For $c = 2$

i	226	227
def	333	343

For $a = 2$

For $b = 2$

For $c = 2$

i	228	229	230	231
def	331	341	343	443

10.

$\cdot_{5-i}$	0	1	2	3	4
0	0	0	0	0	0
1	1	0	0	0	1
2	2	a	0	0	b
3	3	c	d	0	e
4	4	4	0	0	0

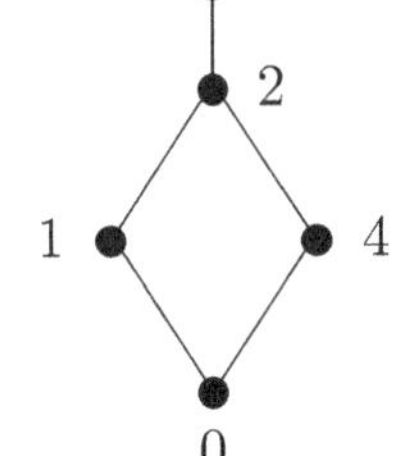

For $a = 2$

For $b = 1$

i	232	233	234	235	236	237
cde	211	222	223	311	322	333

For $b = 2$

i	238	239	240	241	242
cde	212	222	223	243	313

For $a = 4$

For $b = 1$

i	243	244	245	246
cde	211	222	311	322

11.

$\cdot_{5-i}$	0	1	2	3	4
0	0	0	0	0	0
1	1	0	0	0	1
2	2	2	0	1	a
3	3	b	c	0	d
4	4	4	0	0	0

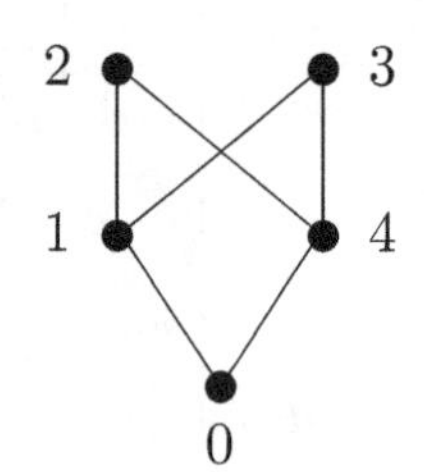

For $a = 1$

i	247	248	249
bcd	247	248	2493

For $a = 2$

i	250	251	252
bcd	313	333	343

12.

$\cdot_{5-i}$	0	1	2	3	4
0	0	0	0	0	0
1	1	0	0	1	1
2	2	a	0	b	c
3	3	3	3	0	0
4	4	d	e	f	0

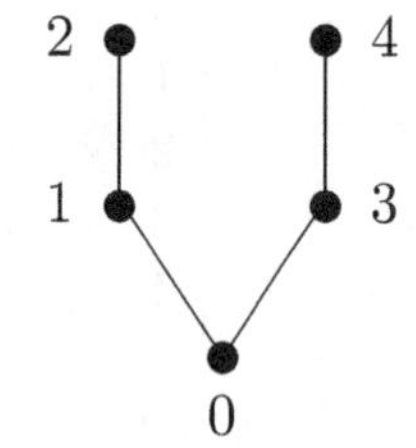

For $a = 1$

For $b = 1$

For $c = 1$

i	253	254	255	256
def	333	334	443	444

For $b = 2$

For $c = 1$

i	257	258
def	334	444

For $c = 2$

i	259
def	334

For $a = 2$

For $b = 1$

For $c = 1$

i	260	261
def	334	444

For $b = 2$

For $c = 1$

i	262	263
def	434	444

13.

$\cdot_{5-i}$	0	1	2	3	4
0	0	0	0	0	0
1	1	0	0	1	1
2	2	a	0	1	b
3	3	3	3	0	3
4	4	4	4	4	0

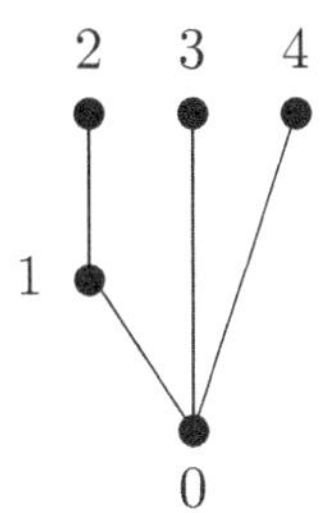

i	264	265	266	267
ab	11	12	21	22

14.

$\cdot_{5-268}$	0	1	2	3	4
0	0	0	0	0	0
1	1	0	0	1	1
2	2	2	0	1	2
3	3	3	0	0	3
4	4	4	4	4	0

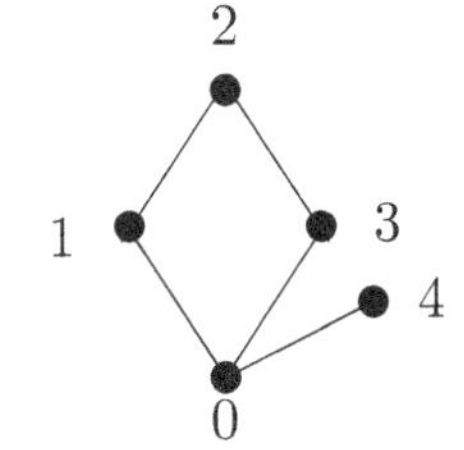

Bibliography

[1] M.T. Abu Osman, *On some product of fuzzy subgroups*, Fuzzy Sets and Systems, 24 (1987), 79 – 86.

[2] S.S. Ahn, *Neutrosophic BCC-ideals in BCC-algebras*, J. Computational Analysis and Appl., Vol. 28 No. 4 (2020), 605 – 614.

[3] S.S. Ahn, S-H. Kwon, *Topological properties in BCC-algebras*, Commun. Korean Math. Soc., 23 (2008), 169 – 178.

[4] M. Akram, K.P. Shum, *Intuitionistic Fuzzy Topological BCC-Algebras*, Advances in Fuzzy Math., 1 (2006), 1 – 13.

[5] N.O. Alshehri, S.M. Bawazeer, *On Derivations of BCC-algebras*, Int. J. of Algebra, 6 (2012), 1491 – 1498.

[6] N. Alp, A. Firat, *On right derivations of weak BCC-algebras*, Ars Combinatoria, 110 (2013).

[7] R. Ameri, R.A. Borzooei, R. Moradian, *Soft Ideals of BCC-algebras*, Int. Sci. Lett., 2 (2013), 63 – 68.

[8] Y. Arai, *On axiom systems of propositional calculi, XVII*, Proc. Japan Acad., 42 (1966), 351 – 354.

[9] K. Atanassov, *Intuitionistic fuzzy sets*, Fuzzy Sets and Systems, 20 (1986), 87 – 96.

[10] S.M. Bawazeer, N.O. Alshehri, R.S. Babusail, *Generalized Derivations of BCC-Algebras*, Int. J. of Math. and Math. Sciences, 2013, Article ID 451212.

[11] R.A. Borzooei, R. Ameri, M. Hamidi, *Fundamental BCC-algebras*, Annals of the Univ. of Craiova 40 (2013).

[12] R.A. Borzooei, W.A. Dudek, N. Koohestani, *On Hyper BCC-algebras*, Int. J. of Math. and Math. Sci., 2006, 1 – 18. Annals of the Univ. of Craiova 40 (2013).

[13] R.A. Borzooei, J. Shohani, *On BCC-subalgebras which are BCK-algebras*, East Asian Math. J., 20 (2004).

[14] W.M. Bunder, *BCK and related algebras and their corresponding logics*, J. Non-classical Logic, 7 (1983), 15 – 24.

[15] K.T. Atanassov, *Intuitionistic fuzzy sets*, Fuzzy Sets and Systems, 20 (1986), 87 – 96.

[16] M.A. Chaudhry, *Weakly positive implicative and weakly implicative BCI-algebras*, Math. Japonica, 35 (1990), 141 – 151.

[17] M.A. Chaudhry, *Branchwise commutative BCI-algebras*, Math. Japonica, 37 (1992), 163 – 170.

[18] M.A. Chaudhry, *On two classes of BCI-algebras*, Math. Japonica, 53 (2001), 269 – 278.

[19] M.A. Chaudhry, *On branchwise implicative BCI-algebras*, International J. Math. Math. Sci., 29 (2002), 417 – 425.

[20] W.H. Cornish, *A large variety of BCK-algebras*, Math. Japonica, 26 (1981), 339 – 342.

[21] K. Dar, M. Akram, *A BCC-algebra as a subclass of K-algebras*, Annals of Univ. of Craiova, 36 (2009), 12 – 16.

[22] Y.D. Du, X.H. Zhang, *Hyper BZ-algebras and semihypergroups*, J. of Alg. Hyperstr. and Log. Alg., 2 (2021), 13 – 28.

[23] I.M. Dudek, *On para-associative BCI-algebras*, Prace Naukowe WSP w Czestochowie ser. Matematyka, 1 (1987), 21 – 24.

[24] I.M. Dudek, W.A. Dudek, *Remarks on BCI-algebras*, Prace Naukowe WSP w Czestochowie ser. Matematyka, 2 (1996), 63 – 70.

[25] W.A. Dudek, *On some BCI-algebras with condition (S)*, Math. Japonica, 31 (1986), 25 – 29.

[26] W.A. Dudek, *On medial BCI-algebras*, Prace Naukowe WSP w Czestochowie, 1 (1987) 25 – 33.

[27] W.A. Dudek, *On group-like BCI-algebras*, Dem. Math., 21 (1988), 369 – 376.

[28] W.A. Dudek, *On BCC-algebras*, Logique et Analyse, 129-130 (1990), 103 – 111.

[29] W.A. Dudek, *On constructions of BCC-algebras*, Selected Papers on BCK- and BCI-algebras, 1 (1992), 93 – 96.

[30] W.A. Dudek, *On proper BCC-algebras*, Bull. Inst. Math. Acad. Sinica, 20 (1992), 137 – 150.

[31] W.A. Dudek, *The number of subalgebras of finite BCC-algebras*, Bull. Inst. Math. Acad. Sinica, 20 (1992), 129 – 135.

[32] W.A. Dudek, *Algebras inspired by logics*, Abstract of Talks of the International Conference on Mathematical Logic, Novosibirsk 1994, 45 – 46.

[33] W.A. Dudek, *On the axiom system for BCI-algebra*, Matematychni Studii, 3 (1994), 5 – 9.

[34] W.A. Dudek, *Algebras connected with the equivalential calculus*, Math Montisnigri, 4 (1995), 13 – 18.

[35] W.A. Dudek, *Algebras motivated by the equivalential calculus*, Rivista Mat. Pura et Appl., 17 (1996), 107 – 112.

[36] W.A. Dudek, *Remarks on the axioms system for BCI-algebras*, Prace Naukowe WSP w Czestochowie, ser. Matematyka, 2 (1996), 46 – 61.

[37] W.A. Dudek, *A computer method of computations of BCK- and BCI-algebras of small orders*, Proceedings of the XI Conference on Applied Mathematics, PRIM'96, Novi Sad 1997, 41 – 64.

[38] W.A. Dudek, *A new characterization of ideals in BCC-algebras*, Novi Sad J. Math., 29 (1999), 139 – 145.

[39] W.A. Dudek, *Algebras inspired by the equivalential calculus*, Italian J. Pure and Appl. Math., 8 (1999).

[40] W.A. Dudek, *On embedding Hilbert algebras in BCK-algebras*, Math. Moravica, 3 (1999), 25 – 28.

[41] W.A. Dudek, *Subalgebras in finite BCC-algebras*, Bull. of Ins. of Math. Acad. Sinica, 28 (2000), 201 – 206.

[42] W.A. Dudek, *Solid weak BCC-algebras*, Int. J. of Comp. Math., Vol. 88 No. 14 (2011), 2915 – 2925.

[43] W.A. Dudek, Y.B. Jun, *Fuzzy BCC-ideals in BCC-algebras*, Math. Montisnigri, 10 (1999), 21 – 30.

[44] W.A. Dudek, Y.B. Jun, *Normalizations of fuzzy BCC-ideals in BCC-algebras*, Math. Moravica, 3 (1999), 17 – 24.

[45] W.A. Dudek, Y.B. Jun, *On fuzzy BCC-ideals over a t-norm*, Mathematical Communications, 5 (2000), 149 – 155.

[46] W.A. Dudek, Y.B. Jun, *Fuzzifications of ideals in BCC-algebras*, Glasnik Matematički, 36(56) (2001), 127 – 138.

[47] W.A. Dudek, Y.B. Jun, *Intuitionistic Fuzzy Approach to BCC-algebras*, Buletinul Acad. Sci. Rep. Moldova, ser. Matematica 3(37) (2001), 7 – 12.

[48] W.A. Dudek, Y.B. Jun, *Radical theory in BCH-algebras*, Algebra and Discrete Math., 1 (2002), 69 – 78.

[49] W.A. Dudek, Y.B. Jun, *Zero invariant and idempotent fuzzy BCC-subalgebras*, Math. Slovava, 52 (2002), 145 – 156.

[50] W.A. Dudek, Y.B. Jun, *Intuitionistic fuzzy BCC-ideals of BCC-algebras*, Italian J. of Pure and Appl. Math., 15 (2004), 9 – 20.

[51] W.A. Dudek, Y.B. Jun, *On multiplicative fuzzy BCC-algebras*, J. Fuzzy Math., 13 (2005), 929 – 939.

[52] W.A. Dudek, Y.B. Jun, S.M. Hong, *On fuzzy topological BCC-algebras*, Disc. Math., General Algebra and Appl., 20 (2000).

[53] W.A. Dudek, Y.B. Jun, Z. Stojakovic, *On fuzzy ideals in BCC-algebras*, Fuzzy Sets and Systems, 123 (2001), 251 – 258.

[54] W.A. Dudek, B. Karamdin, S.A. Bhatti, *Branches and ideals of weak BCC-algebras*, Algebra Coll., 18 (Spec 1) (2011), 899 – 914.

[55] W.A. Dudek, K.H. Kim, Y.B. Jun, *Fuzzy BCC-subalgebras of BCC-algebras with respect to a t-norm*, Scientiae Mathematicae, 3 (2000), 99 – 106.

[56] W.A. Dudek, R. Rousseau, *Set-theoretic relations and BCH-algebras with trivial structure*, Zbornik Radova Prirod.-Mat. Fak. Univ. u Novom Sadu (now Novi Sad J. Math.), 25 (1995), 75 – 82.

[57] W.A. Dudek, J. Thomys, *On decompositions of BCH-algebras*, Math. Japonica, 35 (1990), 1131 – 1138.

[58] W.A. Dudek, J. Thomys, *On some generalizations of BCC-algebras*, Int. J. Computer Math., 89 (2012), 1596 – 1616.

[59] W.A. Dudek, X.H. Zhang, *On atoms in BCC-algebras*, Algebra and Stoch. Methods, 15 (1995), 81 – 85.

[60] W.A. Dudek, X.H. Zhang, *Initial segments in BCC-algebras*, International Algebraic Conference, Slovyans'k 1997, 29 – 30.

[61] W.A. Dudek, X.H. Zhang, *On ideals and congruences in BCC-algebras*, Czech. Math. J., 48 (123) (1998), 21 – 29.

[62] W.A. Dudek, X.H. Zhang, Y.Q. Wang, *Ideals and atoms of BZ-algebras*, Math. Slovaca, 59 (2009), 387 – 404.

[63] A. Firat, *On f-derivations of BCC-Algebras*, Ars Combinatoria, 97 (2010), 377 – 382.

[64] H.G. Forder, J.A. Kalman, *Implication in equational logic*, Math. Gaz., 46 (1962), 122 – 126.

[65] G. Grätzer, *Universal Algebra*, Springer-Verlag, 2008.

[66] R. Halaš, *BCC-algebras inherited from posets*, Multiple-valued Logic, 8(2) (2002).

[67] R. Halaš, L. Plojhar, *Annihilators on weakly standard BCC-algebras*, Int. J. of Math. and Math. Sci., 22 (2005), 3631 – 3643.

[68] R. Halaš, L. Plojhar, *Weakly standard BCC-algebras*, J. of Multiple-valued Logic and Soft Comp., 12(5-6) (2006), 521 – 537.

[69] J. Hao, *A theorem for estimating the number of subalgebras in a finite BCK-algebra*, Kobe J. Math., 3 (1986), 51 – 59.

[70] J. Hao, *Atlas of proper BCI-algebras of order $n \leq 5$*, Math. Japonica, 38 (1993), 589 – 591.

[71] J. Hao, *Ideal theory of BCC-algebras*, Scientiae Mathematicae, 1 (1998), 373 – 381.

[72] S.M. Hong, Y.B. Jun, E.H. Roh, *k-nil radical in BCI-algebras*, Far East J. Math. Sci., 5 (1997), 237 – 242.

[73] Y. Imai, K. Iséki, *On axiom system of propositional calculi*, Proc. Japan Acad., 42 (1966), 19 – 22.

[74] K. Iséki, *An algebra related with a propositional calculus*, Proc. Japan Acad., 42 (1966), 26 – 29.

[75] K. Iséki, *BCK-algebras with condition (S)*, Math. Sem. Notes, 5 (1977), 215 – 222.

[76] K. Iséki, *On BCI-algebras*, Math. Sem. Notes, 8 (1980), 125 – 130.

[77] K. Iséki, S. Tanaka, *Ideal theory of BCK-algebras*, Math. Japonica, 21 (1976), 351 – 366.

[78] K. Iséki, S. Tanaka, *An introduction to the theory of BCK-algebras*, Math. Japonica, 23 (1978), 1 – 26.

[79] K. Iséki, A.B. Thaheem, *Note on BCI-algebras*, Math. Japonica, 29 (1984), 255 – 258.

[80] Y.B. Jun, *Congruences on BCC-algebras*, Sci. Math. Japonicae, 58 (2003), 553 – 559.

[81] Y.B. Jun, *On nil radicals in BCI-algebras*, Sci. Math. Japonicae, 58 (2003), 125 – 130.

[82] Y.B. Jun, S.S. Ahn, *Applications of coupled N-structures in BCC-algebras*, J. of Comp. Analysis and Appl., 16(4) (2014).

[83] Y.B. Jun, S.S. Ahn, E.H. Roh, *BCC-algebras with pseudo-valuations*, Fac. of Sci. and Math. Uni. of Niš, www.pmf.ni.ac.rs/filomat, (2012), 243 – 252.

[84] Y.B. Jun, W.A. Dudek, *n-fold BCC-ideals of BCC-algebras*, Scientiae Mathematicae, 3 (2000), 173 – 178.

[85] Y.B. Jun, C.S. Kim, K.O. Yang, *Cubic sets*, Ann. Fuzzy Math. Inform., 4 (2012), 83 – 98.

[86] Y.B. Jun, K.H. Kim, *Rough set theory applied to BCC-algebras*, Int. Math. Forum, 2 (2007), 2023 – 2029.

[87] Y.B. Jun, K.H. Kim, *Imaginable fuzzy ideals of BCK- algebras with respect to a t-norm*, J. Fuzzy Math., 8 (2000), 737 – 744.

[88] Y.B. Jun, K.J. Lee, *Soft ideals in soft BCC-algebras*, Honam Math. J., 31 (2009), 335 – 349.

[89] Y.B. Jun, K.J. Lee, M.A. Öztürk, *Soft BCC-algebras*, J. Appl. Math. and Inform., 27 (2009), 1293 – 1305.

[90] Y.B. Jun, K.J. Lee, M.S. Kang, *Cubic structures applied to ideals of BCI-algebras*, Comp. and Math. with Appl., 62 (2011), 3334 – 3342.

[91] Y.B. Jun, K.J. Lee, C.H. Park, *Soft set theory applied to ideals in d-algebras*, Comp. Math. Appl., 57 (2009), 367 – 378.

[92] Y.B. Jun, E.H. Roh, H. Harizavi, *Hyper BCC-algebras*, Honam Math. J., 28 (2006), 57 – 67.

[93] Y.B. Jun, S.Z. Song, *Quasi BCC-algebras*, Kyungpook Math. J., 45 (2005), 115 – 121.

[94] Y.B. Jun, S.Z. Song, *Fuzzy ideals with operators in BCC-algebras*, Int. Review of Fuzzy Math., 1 (2006), 1 – 9.

[95] Y.B. Jun, X.L. Xin, *On derivators of BCI-algebras*, Information Sciences, 159 (2004), 167 – 176.

[96] Y.B. Jun, X.H. Zhang, *Fuzzy setting of ideals in BZ-algebras*, J. of fuzzy Math., 13(1) (2005), 149 – 158.

[97] Y.B. Jun, X.H. Zhang, *General forms of BZ-ideals and T-ideals in BZ-algebras*, Honam Math. J., 30 (2008), 379 – 390.

[98] B. Karamdin, *Anti fuzzy ideals in weak BCC-algebras*, East Asian Math. J., 26 (2010), 441 – 446.

[99] B. Karamdin, S.A. Bhatti, *Ideals and branches of BCC-algebras*, East Asian Math. J., 23 (2007), 247 – 255.

[100] B. Karamdin, S.A. Bhatti, *(m, n)-fold p-ideals in weak BCC-algebras*, East Asian Math. J., 26 (2010), 641 – 647.

[101] B. Karamdin, S.A. Bhatti, K.P. Shum, *Decompositions of Weak BCC-Algebras*, Algebra Colloquium, 18 (2011), 889 – 898.

[102] B. Karamdin, J. Thomys, *Quasi-commutative weak BCC-algebras*, Scientiae Mathematicae Japonicae, Online, e-2012, 97 – 116.

[103] K.H. Kim, H.J. Lim, *On multipliers of BCC-algebras*, Honam Math. J., 35 (2013), 201 – 210.

[104] K.H. Kim, E.H. Roh, *Quotient BCC-algebra induced by a fuzzy BCC-ideal*, (2003).

[105] Y. Komori, *The variety generated by BCC-algebras is finitely based*, Reports Fac. Sci. Shizuoka Univ., 17 (1983), 13 – 16.

[106] Y. Komori, *The class of BCC-algebras is not variety*, Math. Japonica, 29 (1984), 391 – 394.

[107] S. Kutukcu, S. Sharma, *On anti fuzzy structures in BCC-algebras*, (2008).

[108] J.G. Lee, K. Hur, S. M. Mostafa, *Cubic bipolar structures of BCC-ideal on BCC-algebras*, Annals of Fuzzy Math. and Inform., 20 (2020), 89 – 103.

[109] K.J. Lee, *Bipolar fuzzy subalgebras and bipolar fuzzy ideals of BCK/BCI-algebras*, Bull. Malays. Math. Soc., 32 (2009), 361 – 373.

[110] K.M. Lee, *Bipolar-valued fuzzy sets and their operations*, Proc. Int. Conf. on Intelligent Techn., Bangkok, Thailand (2000), 307– 312.

[111] S.M. Lee, K.H. Kim, *A note on f-Derivations of BCC-algebras*, Pure Math. Sci., 1 (2012), 87 – 93.

[112] A. Łyczkowska-Hanćkowiak, *Fuzzy QA-ideals in weak BCC-algebras*, Dem. Math. 34 (2001), 513 – 524.

[113] P.K. Maji, A.R. Roy, R. Biswas, *An application of soft sets in a decision making problem*, Comput. Math. Appl., 44 (2002), 1077– 1083.

[114] F. Marty, *Sur une généralization de la notion de groupe*, 8th Congress Math. Scand., Stockholm 1934, 45 – 49.

[115] S. Mehrshad, *Completion of BCC-algebras*, Open J. Math. Sci. 4 (2020), 337 – 342.

[116] S. Mehrshad, N. Kouhestani, *A quasi-uniformity On BCC-algebras*, Annals of the Univ. of Craiova, Math. and Comp. Sci. Series, 44 (2017), 64 – 77.

[117] J. Meng, Y.B. Jun, *BCK-algebras*, Kyung Moon SA, Seoul, 1994.

[118] J. Meng, Y.B. Jun, E.H. Roh, *The role of $B(X)$ and $L(X)$ in the ideal theory of BCI-algebras*, Indian J. pure Appl. Math., 28 (1997), 741 – 752.

[119] J. Meng, X.L. Xin, *Commutative BCI-algebras*, Math. Japonica, 37 (1992), 569 – 572.

[120] J. Meng, X. L. Xin, *Positive implicative BCI-algebras*, Pure Appl. Math., 9 (1993), 19 – 23.

[121] D. Molodtsov, *Soft set theory - First results*, Comput. Math. Appl., 37 (1999), 19 – 31.

[122] S. Mostafa, M. Hassan, *Fuzzy Derivations BCC-Ideals on BCC-Algebras*, Pure and Appl. Math. J., 4 (2015), 225 – 232.

[123] X. Ougen, *Fuzzy BCK-algebra*, Math. Japon., 36 (1991), 935–942.

[124] Ch. Prabpayak, U. Leerawat, *On derivations of BCC-algebras*, Kasetsart J. (Nat. Sci.), 43 (2009), 398 – 401.

[125] H. Rasiowa, *An algebraic approach to non-classical logics*, North-Holland, Amsterdam 1974.

[126] K.S. So, S.S. Ahn, *Complicated BCC-algebras and its derivation*, Honam Math. J., 34 (2012), 263 – 271.

[127] S. Tanaka, *On axiom systems of propositional calculi*, Proc. Japan Acad., 42 (1966), 355 – 357.

[128] J. Thomys, *On some generalization of BCC-algebras*, Prace Naukowe WSP w Czestochowie, ser. Matematyka, 2 (1996), 89 – 97.

[129] J. Thomys, *Derivations of weak BCC-algebras*, Sci. Math. Japonicae, 73 (2011), 175-181.

[130] J. Thomys, *f-derivations of weak BCC-algebras*, Int. J. Algebra, 5 (2011), 325-334.

[131] J. Thomys, X.H. Zhang, *On weak-BCC-algebras*, Hindawi Publ. Corp., The Sci. World J., 2013, Art.ID 935097.

[132] L. Tiande, Z. Changcheng, *p-radicals in BCI-algebras*, Math. Japonica, 30 (1985), 511 – 517.

[133] H.S. Wall, *Hypergroups*, Amer. J. of Math., 59 (1937), 77 – 98.

[134] Y.Q. Wang, L. Han, B. Shen, *On zero-symmetric BZ-algebras*, J. East China Uni. of Sci. and Techn., 23 (1997), 755 – 761 (in Chinese).

[135] Y.Q. Wang, Z.Q. Shao, X.H. Zhang, *BCC-algebras and BCK-algebras*, J. East China Uni. of Sci. and Techn., 32(4), 471 – 474.

[136] Y.Q. Wang, X.H. Zhang, *Properties of periods of elements in zero-symmetric BZ-algebras*, J. Northwest Normal Uni. (Nat. Sci.), 39 (2003), 27 – 31 (in Chinese).

[137] A. Wroński, *BCK-algebras do not form a variety*, Math. Japonica, 28 (1983), 211 – 213.

[138] A. Wroński, *An algebraic motivation for BCK-algebras*, Math. Japonica, 30 (1985), 187 – 193.

[139] A. Wroński, J.K. Kabziński, *There is no largest variety of BCK-algebras*, Math. Japonica, 29 (1984), 545 – 549.

[140] C. Xi, *On a class of BCI-algebras*, Math. Japonica 35 (1990), 13 – 17.

[141] O.G. Xi, *Fuzzy BCK-algebras*, Math. Japonica 36 (1991), 935 – 942.

[142] Y. Xiao, T. Zou, *On fuzzy BCC-ideals and quotient BCC-algebra induced by a fuzzy BCC-ideal*, Southeast Asian Bull. of Math., 30(1) (2006), 165 – 175.

[143] J.M. Xu, X.H. Zhang, *On anti-grouped ideals in BZ-algebras*, Pure Appl. Math., 14 (1998), 101 – 102.

[144] R.F. Ye, *On BZ-algebras*, Selected Papers on *BCI, BCK*-algebras and Comp. Log., Shanghai Jiaotong Univ. Press, (1991), 21 – 24.

[145] R.F. Ye, X.H. Zhang, *On ideals in BZ-algebras and its homomorphism theorems*, J. East China Univ. Sci. Techn., 19 (1993), 775 – 778 (in Chinese).

[146] H. Yisheng, *BCI-algebra*, Science Press, Beijing 2006.

[147] Y. Yu, J.N. Mordeson and S.C. Cheng, *Elements of L-algebra*, Creighton Uni., Center for Res. in Fuzzy Math. and Comp. Sci., 1994.

[148] L.A. Zadeh, *Fuzzy sets*, Inform. control., 8 (1965), 338 – 353.

[149] L.A. Zadeh, *The concept of a linguistic variable and its application to approximate reasoning*, Inform. Sci., 8 (1975), 199 – 249.

[150] D. Zelent, *BZ-algebras of small orders*, preprint.

[151] D. Zelent, *On Traczyk's BCK-sequences*, Ann. Univ. Paedagog. Crac. Stud. Math., 21 (2022), 43-49.

[152] J. Zhan, T. Zhisong, *T-fuzzy multiply positive implicative BCC-ideals of BCC-algebras*, Int. J. of Math. and Math. Sci., 42 (2003).

[153] X.H. Zhang, *BCC-algebra and integral pomonoid*, J. of Math. Research and Exposition Vol. 19 (1999).

[154] X.H. Zhang, *Fuzzy BZ-ideals and fuzzy anti-grouped ideals*, J. of Fuzzy Math., 11 (2003), 915 – 932.

[155] X.H. Zhang, *BCC-algebras and residuated partially-ordered groupoid*, Math. Slovaca, 63 (2013), 397 – 410.

[156] X.H. Zhang, W.A. Dudek, *Fuzzy BIK^+-logic and non-commutative fuzzy logics*, Fuzzy Systems and Math., 23(4) (2009), 8 – 20.

[157] X.H. Zhang, H. Jiang, S.A. Bhatti, *On p-ideals of a BCI-algebra*, Punjab Univ. J. Math., 27 (1994), 121 – 128.

[158] X.H. Zhang, Y.B. Jun, *The role of $T(X)$ in the ideal theory of BCI-algebras*, J. Bull. Korean Math. Soc. 34 (1997), 199 – 204.

[159] X.H. Zhang, W.H. Li, *On pseudo-BL algebras and BCC-algebras*, J. Soft. Comput.,10 (2006), 941 – 952.

[160] X.H. Zhang, W.H. Liu, *Two notes on ideals of BZ-algebras*, J. Northwest Univ. Suppl., 24 (1994), 103 – 104 (in Chinese).

[161] X.H. Zhang, Y.Q. Wang, *On necessary and sufficient conditions of zero-symmetric BZ-algebras*, Pure and Applied Math., 19(3) (2003), 253 – 256 (in Chinese).

[162] X.H. Zhang, Y.Q. Wang, B.Y. Shen, *BZ-algebras of order 5*, J. East China Univ. of Sci. and Techn., 29 (2003), 1559 – 1570 (in Chinese).

[163] X.H. Zhang, Y.Q. Wang, W.A. Dudek, *T-ideals in BZ-algebras and T-type BZ-algebras*, Indian J. pure Appl. Math., 34 (2003), 1559 – 1570.

[164] X.H. Zhang, Z. Wei, *BCC-algebras and BCK-algebras*, (1995).

[165] X.H. Zhang, R.F. Ye, *BZ-algebras and groups*, J. Math. Phys. Sci., 29 (1995), 223 – 233.

[166] X.H. Zhang, Z.C. Yue, W.B. Hu, *On fuzzy BCC-ideals*, Sci. Math. Jap., 54 (2001), 349 – 353.

[167] H. Zhou, X. Mao, X.H. Zhang, Y. Zhao, *On an open problem of hyper BCC-algebras*, J. of Comp. Inf. Syst., 9(5) (2013), 1827 – 1834.

Index